SOLAR HEATING AND COOLING OF BUILDINGS

By Joe J. Harrell, Jr.

Accelerating developments in solar energy design are placing tough demands on building designers. Increasingly complex solar energy systems make it essential for you to have a complete understanding of components as well as the system as a whole. This encyclopedic volume clearly describes how solar energy system components function and which type of component is best for a specific application. The author skillfully explores design techniques used for sizing system components and evaluates the methods used to determine the economics of a specific solar design.

All the techniques, data, and money-saving tips that can help you create an optimum solar design are here —

- a life cycle costing strategy for determining the economics of a solar energy system
- data processing methods and actual computer programs that streamline design analysis
- vital meteorological data for numerous U.S. cities
- available government grants and lucrative business opportunities in the solar field
- state-by-state listing of tax and other legislative incentives in the solar energy field

... and much more. By using these proven guidelines and design procedures you'll exploit all the design opportunities and avoid all the pitfalls when selecting a solar energy system.

Systems discussed include solar energy collectors, water heaters, and space heating systems. Surpassing all other books in the field, this volume gives you extensive coverage of how programmable hand-held calculators and computers can be used to solve solar design problems. You'll also find complete program listings of actual computer programs that can be run on your own calculator or small computer.

(Continued on back flap)

SOLAR HEATING AND COOLING OF BUILDINGS

SOLAR HEATING AND COOLING OF BUILDINGS

Joe J. Harrell, Jr.

VNR VAN NOSTRAND REINHOLD COMPANY
NEW YORK   CINCINNATI   TORONTO   LONDON   MELBOURNE

Van Nostrand Reinhold Company Regional Offices:
New York Cincinnati

Van Nostrand Reinhold Company International Offices:
London Toronto Melbourne

Library of Congress Catalog Card Number: 81-7534
ISBN: 0-442-21658-0

Manufactured in the United States of America

Published by Van Nostrand Reinhold Company Inc.
135 West 50th Street, New York, N.Y. 10020

Published simultaneously in Canada by Van Nostrand Reinhold Ltd.

15 14 13 12 11 10 9 8 7 6 5 4 3 2 1

Library of Congress Cataloging in Publication Data
Harrell, Joe J. Jr
    Solar heating and cooling of buildings.

    Includes bibliographies and index.
    1. Solar heating.   2. Solar air conditioning.
I.   Title.
TH7413.H37   697′.78   81-7534
ISBN 0-442-21658-0   AACR2

PREFACE

This book was written to assist both the professional (i.e., consulting, design, and contracting firms) and the homeowner in analyzing aspects of solar energy use in both commercial and residential buildings. I have tried to provide some depth to the various topics discussed rather than a cursory overview. This is one reason why this book attempts to only cover the aspects of solar energy as it relates to being incorporated in a building.

A number of sections on construction are presented which illustrate the building of solar control devices, solar insolation measurement devices, solar collectors, etc. My main purpose for presenting these is to show how such devices work. For instance, one section describes the construction of a flat plate solar collector. By observing the schematic diagram illustrating the various parts and how they are put together, the reader can quickly see that there is no real mystery to the workings of a solar collector. Likewise, the sections describing the construction of various types of control devices used in actuating pumps which circulate fluids help to illustrate how solar system control devices work. If so inclined, the reader can use the circuit diagrams and parts list to inexpensively build any of the various devices illustrated in this book.

Chapter 1 serves as an introductory chapter which describes the historical background of solar energy usage and how solar energy can be used today to meet some of our growing energy needs.

Chapter 2 deals with solar energy collection methods. In this chapter various types of solar collectors are described. In discussing the various types, attention is directed to collector design, types of controls, and the required maintenance. An important aspect of Chapter 2 concerns the topic of collector efficiency described in Section 2.5. This section lays the groundwork for Chapter 6 dealing with solar system design. In Section 2.5 the reader is introduced to procedures used in determining solar collector efficiency. Methods of comparing the operating characteristics of various collectors are also described which can be used by an individual to select the correct type of solar collector based on the building's solar system design requirements.

Chapter 3 describes the types of solar water heating systems available and the advantages and disadvantages of each.

Chapter 4 concerns solar space heating and describes and contrasts solar air type and liquid type space heating systems.

Chapter 5 deals with combined solar heating and cooling systems. The main type of solar cooling system described is absorption refrigeration. Less attention is directed to the solar assisted heat pump and the solar Rankine-cycle engine.

Chapter 6 describes the design of solar systems. Procedures are given for determining such things as the sun's specific altitude and azimuth location and estimates of solar radiation falling upon an inclined surface such as a solar collector. The design steps and procedures listed in the various sections of Chapter 6 build upon one another so that ultimately an individual can determine how much of the building's thermal load can be carried by a specific solar collector size based upon such things as solar collector efficiency, monthly solar insolation, etc.

Chapter 7, in dealing with economics, describes procedures for determining the most economically feasible solar collector to use in a solar system. Items such as solar system financing and tax incentives are also discussed.

Chapter 8 on computer applications describes various computer programs which are available for use in solar system design. Although all of the design procedures used in Chapter 6 and solar system economic analysis procedures described in Chapter 7 require nothing more than a simple hand-held calculator, there are times when computer programs are helpful. For instance, the solar system designer who does not have the time to go through hand calculation procedures for every system he designs will benefit from computer programs. In this instance, computer programs can reduce both the time element and mental fatigue caused by lengthy hand calculations. The computer programs described in Section 8.3 are for all types of data processing equipment ranging in size from programmable hand-held calculators to large computer systems.

Chapter 9 deals with data processing systems. Basically this chapter covers the data processing (e.g., computer) systems used to run the programs discussed in Chapter 8.

Chapter 10, on purchasing solar equipment, lists a number of buying tips to heed when purchasing a solar system.

Chapter 11 concerns solar energy business opportunities for the professional. In this chapter, estimates are given for projected growth of the solar energy market. Furthermore, government contracts, grants, and financial assistance are also discussed.

Chapter 12 lists sources of solar energy information.

Appendix I presents conversion factors used in converting to and from Système International (SI) units and English units. Heat equivalents for various substances are also shown.

Appendix II lists meteorological data necessary for computations in Chapter 6.

Appendix III illustrates solar radiation maps of the United States for all twelve months. The insolation data presented in the charts can be used to get a general feel for relative solar insolation levels across the United States for the various months. (Appendix II should be consulted to obtain numerical data for design calculations rather than trying to interpolate between the isopleths on the maps for a particular geographical location.)

Appendix IV lists a heat-gain computer program which can be run on a programmable calculator.

Appendix V lists a printout of the U.S. Department of Housing and Urban Development's RSVP computer program used in estimating solar economics.

Appendix VI is a solar design computer program that I wrote based on the F-Chart method. The program as listed was written to run on a Hewlett-Packard 2000 ACCESS minicomputer using Hewlett-Packard timeshare BASIC.

Appendix VII is a listing of a life cycle costing computer program written in BASIC.

Appendix VIII graphically presents properties of aqueous glycol solutions.

Appendix IX shows construction details for a wooden, rock storage bin.

Appendix X shows sample drawings of concrete storage tanks.

Appendix XI discusses the fusion processes in the sun and shows a calculation procedure for determining the amount of energy liberated by the sun.

Appendix XII gives a state-by-state listing of solar legislation.

Appendix XIII illustrates some sun charts which can be used in determining the location of the sun at a given latitude and hour of the day.

Appendix XIV lists the design worksheets used in Chapters 6 and 7 of this book.

Appreciation is given to a number of people who gave me support and assistance while writing this book:

To my father who encouraged me to excel and strive for excellence regardless of the endeavor.

To Larry S. Hager, Senior Editor at Van Nostrand Reinhold Company for his patience and support while writing this book.

To Robert W. Roose, Senior Editor for Heating/Piping/Air Conditioning Magazine for his constructive suggestions for the original manuscript.

To David Ziller, Manuscript Editor at Van Nostrand Reinhold Company for his work in preparing the manuscript for publication.

To Joyce Gibler for her unfailing assistance in typing the original manuscript.

To Nell Smith for her assistance in alphabetizing and organizing the numerous references at the end of each chapter in this book and also for her positive encouragement.

To Lucy Van Swearingen for her assistance in taking some of the pictures illustrated in this book.

And last, but not least, to my mother for her continued support and encouragement.

CONTENTS

SOLAR HEATING AND COOLING OF BUILDINGS

CHAPTER 1
INTRODUCTION

1.1 THE ENERGY SITUATION

Since the onset of fuel shortages in the early 1970s, people have had to face the grim reality that there is a finite amount of oil in the ground and when that is gone there will be no more for another million years. This has prompted many to begin re-evaluating their lifestyles and cutting down on energy consumption either by their own choice or government mandate. The current energy problem lies in the supply and demand of oil. As the demand for oil increases and the supply decreases, economic theory dictates that prices will rise to reach an equilibrium representing the market value of the commodity (oil).

Thus, the price of gasoline and fuel oil will continue to increase—as most people are painfully aware every time they stop at a gas station. The Central Intelligence Agency (CIA) supports this unpleasant prediction in a press release (October, 1979), saying that the price for a barrel of oil could increase to $52 per barrel by 1995—based on the 1979 value of the dollar.

The oil crisis has prompted the U. S. Government to enact legislation which will ease the energy squeeze. One such law was the National Energy Act passed by Congress on October 15, 1978, which is composed of five bills:

1. The National Energy Conservation Policy Act of 1978.
2. The Powerplant and Industrial Fuel Use Act of 1978.
3. The Public Utilities Regulatory Policy Act.
4. The Natural Gas Policy Act of 1978.
5. The Energy Tax Act of 1978.

The passage of the National Energy Act represents a positive commitment of the U. S. Government to reduce oil import needs by 1985 and to promote the use of alternate sources of energy such as solar power.

After the passage of this Act, former President Carter said, "We have declared to ourselves and the world our intent to control our use of energy, and thereby to control our own destiny as a nation."

Former Energy Secretary James R. Schlesinger (in DOE publication OPA-003) said:

The NEA represents an historic turning point. The era of cheap and abundant energy is recognized to be over. For the first time, energy conservation is recognized as an indispensable ingredient in national energy policy. With the NEA, we will save 2.5 to 3 million barrels a day by 1985, compared to what we would otherwise have required for an estimated balance of payments savings of approximately $14 billion in current dollars (as much as $20 billion in 1985 dollars).

The purpose of the National Energy Act is to put into place a policy framework for decreasing oil imports by:

— replacing oil and gas with abundant domestic fuels in industry and electric utilities,

— reducing energy demand through improved efficiency,

— increasing production of conventional sources of domestic energy through more rational pricing policies, and

— building a base for the development of solar and renewable energy sources.

Another government mandate which became effective July 16, 1979 was the Emergency Building Temperature Restrictions Regulations. This regulation places temperature restrictions on nonresidential buildings. Exempt from this regulation are residential buildings, hotels and other lodging facilities, hospitals and other health-care facilities, elementary schools, nursery schools, and day-care centers. These restrictions include maximum and minimum temperature settings for space heating and cooling, and maximum temperature settings for domestic hot water. [Figure 1.1 shows the letter issued by the Secretary of the Department of Energy (DOE) explaining the Emergency Building Temperature Restrictions Regulations.]

Additional energy legislation includes the Military Construction Act of 1979, sponsored by Senator Gary Hart (D., Colo.), which requires that solar heating systems be installed in all new military housing units and that solar systems must also be installed in 25% of the new, nonresidential, military construction projects.

As the price of fossil fuels (e.g., oil) increase and nuclear problems surface (e.g., Metropolitan Edison Company's Three Mile Island Nuclear-Power Plant mishap on March 28, 1979), the idea of using solar energy as a viable energy resource will seem better and better. Realistically, from an economic, political, and present technology standpoint, solar power *alone* cannot supply all of our energy needs but it can make a substantial contribution to the present energy sources being used such as oil, nuclear, and hydroelectric production. President Carter announced on June 20, 1979 a goal of meeting 20% of our energy needs, with the use of solar *and* renewable resources, by the year 2000. Although this goal does not mean that solar alone will supply 20% of our energy needs by the year 2000, it does show a positive commitment that solar energy will be used to help supply our energy requirements. (The Department of Energy (DOE) has predicted that solar *alone* could supply 13% of the predicted U. S. energy demand in the year 2000.)

The Department of Energy was established on October 1, 1977 as a means of unifying the numerous energy programs and policies of the various government agencies. (The Energy Research and Development Administration

Department of Energy
Washington, D.C. 20585

July 1979

MY FELLOW CITIZENS:

There is abundant evidence that serious problems lie ahead for our country unless we act now to avoid them. Unless we act now, there may not be enough fuel oil to adequately heat homes, schools and businesses during the coming winter. Unless we act now, chronic fuel shortages, together with rapidly increasing fuel prices, will threaten our Nation with recession, higher infla-tion and unemployment.

Nearly half of the oil we use is imported and roughly two-thirds of that is supplied by the member nations of OPEC, the Organization of Petroleum Exporting Countries. OPEC has proven its ability to reduce production and raise prices virtually at will. The price increases announced in June were the steepest since the embargo of 1973.

The Nation must respond to this threat. Our growing dependence on uncertain and costly supplies of imported oil must be reversed. Accordingly, the President has committed the United States to hold oil imports below 8.5 million barrels per day through 1980. He announced that decision after his June meeting in Tokyo with the leaders of six oil-dependent industrial nations, whose concerns reflect our own.

Secondly, the President has determined that a severe energy supply interruption exists and has ordered the immediate imposition of the Standby Emergency Building Temperature Restrictions Plan.

This Plan sets strict limits on heating and cooling temperatures in millions of non-residential buildings throughout the country. I recognize that every citizen may in some way be inconve-nienced by the Plan. Each of us may be a little warmer this summer and a little cooler next winter. But it also will help provide needed supplies of heating oil for the winter. It is estimated that, with full compliance nationwide, oil use can be reduced by as much as *400,000* barrels daily. At current OPEC prices, this represents a potential savings to the Nation of more than $2 billion during the projected 9-month life of this program.

The measures called for in this effort are essential, and will prove both effective and equitable. They will help us develop new approaches in our use of energy and move us as a Nation toward wiser energy management.

The Secretary
Department of Energy

Figure 1.1. Emergency Building Temperature Restrictions Regulations Letter.

(ERDA) is now part of DOE.) By reviewing the DOE energy budget break-downs one can get an idea of where the emphasis is being placed. Table 1.1 gives a breakdown of the major program groups.

Table 1.1. DOE Energy Budget. (Source: U. S. Department of Energy[39]).

	Budget Authority (In Millions)	
	FY 1979	FY 1980
Energy Technology	$ 3,625	$ 3,583
Basic Sciences	431	474
Conservation	671	555
Regulation and Information	276	323
Defense Activities	2,685	3,022
Government-Owned Operations	248	149
Policy and Management	258	308
Less: Supplementals and Other Adjustments	-403	-
Subtotal	$ 7,791	$ 8,415
Strategic Petroleum Reserve	3,008	8
TOTAL DOE FUNDING	$10,799	$ 8,423

Table 1.2 gives a further breakdown of the Energy Technology program category. As can be seen, solar energy funding for fiscal year (FY) 1980 is $597 million while the main budget outlay for nuclear fission is over one billion dollars.

Table 1.3 provides a further breakdown of the solar energy figures presented in Table 1.2.

The 1979 and 1980 totals in Table 1.3 represent *direct* solar funding. There are other program groups where there are some solar-related activities, such as in the Environment and Basic Energy Research program categories presented in Table 1.2. The solar-related funding of these categories amounts to $31 million in FY 1979 and $49 million in FY 1980. This additional funding means that the total DOE solar funding is $559 million in FY 1979 and $646 million in FY 1980. Table 1.4 gives a more in-depth breakdown of DOE funding and also other federal solar funding not connected with the DOE budget.

With the government taking a step in the right direction, companies have begun to follow in the government's footsteps. Large and reputable companies such as General Electric, Grumman Aerospace, ITT, and Owens-Illinois, to name a few, are producing solar devices such as collectors, storage tanks, and even complete heating and cooling systems. Industry and government involvement are positive signs that solar energy has grown from the neglected stepchild to a cherished offspring.

Table 1.2. DOE Energy Technology Program Budget. (Source: U. S. Department of Energy[39]).

	Budget Authority (In Millions)	
	FY 1979	FY 1980
Fossil Energy	$ 791	$ 796
Solar Energy	528	597
Geothermal	130	111
Hydroelectric	29	18
Magnetic Fusion	356	364
Nuclear Fission	1,204	1,037
Environment	245	278
Basic Energy Research	220	276
Other Technology Programs	122	106
TOTAL	$ 3,625	$ 3,583

1.2 HISTORY OF SOLAR ENERGY USAGE

Constructive use of solar energy dates back to the earliest periods in history. Early cave dwellers probably used solar energy to dry and thus preserve foods. Xenophon in his Memorabilia (III, viii, 8–14) recorded some of the disciplines professed by Socrates (470–399 B.C.) with regard to homes:

Again his dictum about houses . . . was a lesson in the art of building houses as they ought to be. He approached the problem thus: 'When one means to have the right sort of house, must he contrive to make it as pleasant to live in and as useful as can be?' And this being admitted, 'Is it pleasant,' he asked, 'to have it cool in summer and warm in winter?' And when they agreed with this also, 'Now in houses with a south

Table 1.3. DOE Solar Energy Budget. (Source: U. S. Department of Energy[39]).

	Budget Authority (In Millions)	
	FY 1979	FY 1980
Solar Technology	$ 315	$ 383
Biomass	44	58
Solar Applications	169	156
TOTAL DOE DIRECT SOLAR FUNDING	$ 528	$ 597

Table 1.4. DOE Budget for Solar Technology and Biomass (in tenths of millions). (Source: U. S. Department of Energy[39]).

	FY 1979 Budget Authority	FY 1980 Budget Authority
Solar Thermal Electric..................	$102.1	$121.0
Photovoltaic Energy Development........	105.8	130.0
Wind Energy Conversion.................	61.9	67.0
Ocean Systems.........................	38.9	35.0
Solar Energy Research Institute Building.............................	3.0	27.0
Biomass...............................	43.2	57.0
Program Direction & Support............	3.5	4.0
TOTAL, SOLAR TECHNOLOGY AND BIOMASS	358.4	441.0

Solar Applications

Systems Development....................	41.0	47.0
Demonstrations: Agricultural and Industrial Process Heat..............	11.0	14.0
Demonstrations: Residential and Commercial Buildings.................	55.0	35.5
Demonstrations: Federal Buildings.....	25.7	23.5
Demonstrations: Federal Photovoltaic Utilization Programs.................	15.0	-
Commercialization.....................	2.7	5.0
Market Development & Training..........	16.6	27.0
Program Directions & Support..........	2.5	3.6
TOTAL, SOLAR APPLICATIONS	169.5	155.6
Other DOE Solar Related Funding........	31.5	49.4
TOTAL DOE SOLAR FUNDING	559.4	646.0

Other Federal Solar Funding

Agency for International Development...	16.0	42.0
Department of Agriculture.............	22.0	27.0
Small Business Administration.........	5.0	14.0
Federal Buildings Investments.........	20.0	25.0
Tennessee Valley Authority............	8.0	16.0
Tax Credits..........................	88.0	74.0
TOTAL, OTHER FEDERAL SOLAR FUNDING	159.0	198.0
TOTAL FEDERAL SOLAR PROGRAMS	$718.4	$844.0

aspect, the sun's rays penetrate into the porticoes in winter, but in summer the path of the sun is right over our heads and above the roof, so that there is shade. If, then this is the best arrangement, we should build the south side loftier to get the winter sun and the north side lower to keep out the cold winds. To put it shortly, the house in which the owner can find a pleasant retreat at all seasons and can store his belongings safely in presumably at once the pleasantest and the most beautiful.[1]

Supposedly the Greek physicist Archimedes was the first person to use solar energy on a large scale to set fire to the attacking Roman Fleet at Syracuse in 212 B.C. using a combination of small mirrors to focus the sun's energy on the advancing Roman Fleet. About 1800 years later in 1600 A.D. Solomon de Caux of France built a solar engine used to pump water. In the 1880s Charles Albert Tellier, also of France, built the first flat-plate collector and some fifty years later, the first building to be heated with solar energy was at the Massachusetts Institute of Technology (MIT) in 1938. These are just a few of the many instances of the use of solar energy over the years, long before the present energy crisis threw solar energy into the limelight.

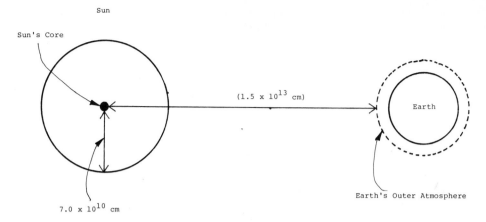

Figure 1.2. Sun–Earth Distance.

1.3 THE SUN—PROVIDER OF SOLAR ENERGY

The sun is basically a giant fusion machine supplying the earth with "free" heat and light. The amount of energy being produced every second by the sun is staggering; on the order of 5.2×10^{23} horsepower which is equal to 3.8×10^{23} Btu's per second or 9.3×10^{25} (gram) calories per second.* This huge amount of energy is produced in the center of the sun (core) where the fusion process is sustained by great temperatures (15 million degrees Kelvin) and pressures (3.7 trillion pounds per square inch).

The solar energy reaching the earth's outer atmosphere can be calculated using the inverse square law which states that the intensity (I) of radiation (P) (e.g., heat or light energy) decreases with the square of the distance (R) stated in the form:

$$I = P/(4\pi R^2) \tag{1.1}$$

The $4\pi R^2$ of the equation represents the surface area of the sphere of energy produced by the sun's interior at a distance of 1.5×10^{13} centimeters. See Fig. 1.2. To find the rate of energy reaching the outer layer of the sphere which touches the outer layer of the earth's atmosphere, the rate of energy produced by the sun's interior is divided by the surface area of the sphere. This will give an energy figure measured (3.3×10^{-2} cal/cm²sec) *per* unit area of the sphere. Using the distance values found in Fig. 1.2, the solar intensity reaching the earth's upper atmosphere is approximately 2.0 cal/cm²min.

$$
\begin{aligned}
I &= (9.3 \times 10^{25} \text{ cal/sec})/[4\ (1.5 \times 10^{13}\text{cm})^2] \\
&= 3.3 \times 10^{-2} \text{ cal/cm}^2\text{sec} \\
&= 2.0 \text{ cal/cm}^2\text{min}
\end{aligned}
\tag{1.2}
$$

*See Appendix XI for a discussion of the solar fusion process and how the numerical figures showing the sun's energy production were derived.

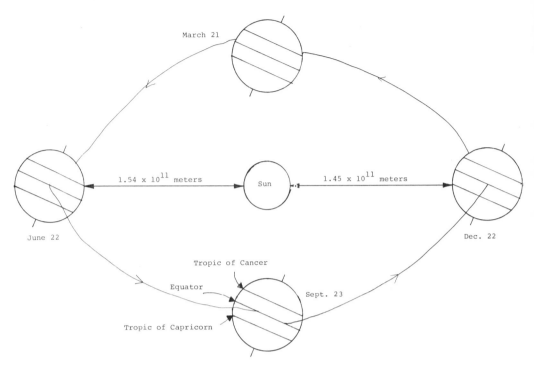

Figure 1.3. Relative Tilt of Earth During Four Seasons.

This calculated value is close to the actual value recorded by high-flying aircraft which measured the solar intensity as being approximately 1.94 cal/cm²min (which is 1353 W/m², 1.94 L/min, 428 Btu/ft²hr, or 4871 kJ/m²hr). This intensity is referred to as the solar constant (*Isc*), which is defined as the energy received on a surface normal to the sun's rays, outside of the earth's atmosphere at the earth's average distance from the sun.

The solar energy received at the earth's surface, called insolation is less than the solar constant. Factors affecting the amount of solar radiation reaching a particular location on the earth's surface are: amount of solar radiation absorbed and reflected in the earth's atmosphere, time of year, hour of day, and latitude.

Before proceeding further, the spacial relationship of the sun to the earth needs to be discussed. Figure 1.3 illustrates this relationship. In the illustration the earth is tilted 23.45° on its axis during each complete revolution around the sun (i.e., one year). In Fig. 1.4, as the earth rotates around the sun, the direct (normal) rays of the sun hitting the earth travel from 23.45° north latitude (on June 22), crossing the equator (on September 23), then move to 23.45° south latitude (on December 22), then through the equator again (on March 21), and finally arrive back at 23.45° north latitude (on June 22).

Figure 1.5 can help give a better idea of what is meant by direct (normal) sun rays. This illustration shows the angle that the sun's rays hit the earth on June 22. If someone was standing in Havana, Cuba about noon (suntime), the

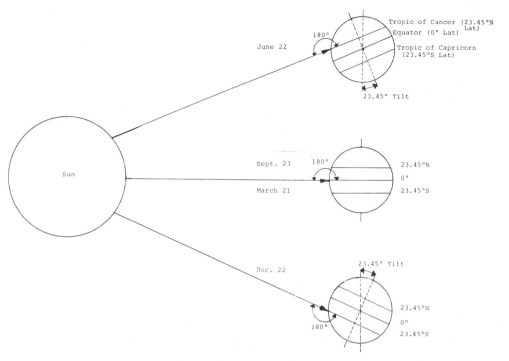

Figure 1.4. Angles of Solar Incident Radiation Upon Earth.

sun would appear directly overhead. This is because Havana is almost at the Tropic of Cancer (23.45° north latitude)*—the same degree the earth's tilt is toward the sun on this date. Since the sun's rays are falling directly on 23.45° north latitude on June 22, someone standing in Tampa, Florida (located at 27.95° north latitude) at noon suntime would receive the rays traveling at an *angle* through the earth's atmosphere rather than perpendicular (normal) to the earth's atmosphere as when falling on Cuba. The observer in Tampa would say that the sun's rays do not appear *directly* overhead but appear 4° south of the vertical. If the observer were to be in Detroit, Michigan (42.33° north latitude) on June 22 at noon (suntime), he would say that the sun's rays appear to strike the earth at a 71° angle—as measured from the earth's surface rather than from directly overhead.

Since the earth's tilt causes the sun's rays to only fall perpendicular to the earth's atmosphere between the latitude bands of 23.45° north latitude (Tropic of Cancer) and 23.45° south latitude (Tropic of Capricorn), this banded area will receive more solar radiation than other parts of the globe. The question now may be how does the angle that the sun hits the earth's atmosphere (and the earth's surface) affect the amount of solar radiation received by the area. Figure 1.6 can help to answer this question. In the first diagram of Fig. 1.6,

*Havana is actually 23.13° north latitude, but for the discussion it is assumed the latitude of Havana is close to being 23.45° latitude to simplify the point to be made.

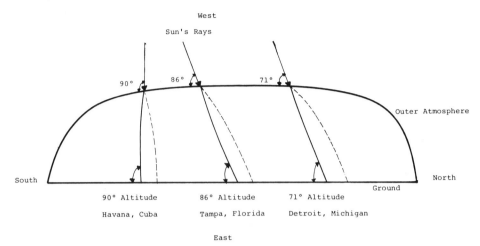

Figure 1.5. Sun's Angle of Incidence on June 22.

the sun's rays are hitting the earth directly, such as they would at the Tropic of Cancer (23.45° north latitude) on June 22 or the equator (0° latitude) on September 23. Assuming three rays of sunlight fall within distance (d) at an angle of 90°, then only two would fall within distance (d) at an angle of, say 45°. This means that per unit area, the first diagram of Fig. 1.6 would receive more sunlight. Therefore latitude and time of year play an important part in determining how much solar radiation strikes a particular area or location at the earth's upper atmosphere.

Besides latitude and the angle the sun hits the upper atmosphere, the atmosphere plays a large part in determining how much solar energy reaches the earth's *surface*. Figure 1.7 illustrates the atmospheric factors that affect the amount of sunlight reaching the earth. On a clear day 80% of the sunlight reaching the upper atmosphere is transmitted to the ground, while with clouds this can vary from 45% to 0% depending upon the extent of cloud cover. The atmosphere also affects the amount of solar radiation reaching the earth's surface during different hours of the day. As one may recall, the sun's rays appear weaker when the sun is rising or setting than when it is overhead. This is due to the attenuating effects of the atmosphere. Figure 1.8 illustrates this

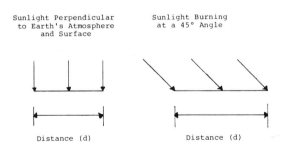

Figure 1.6. Intensity of Solar Radiation as a Function of Incident Angle.

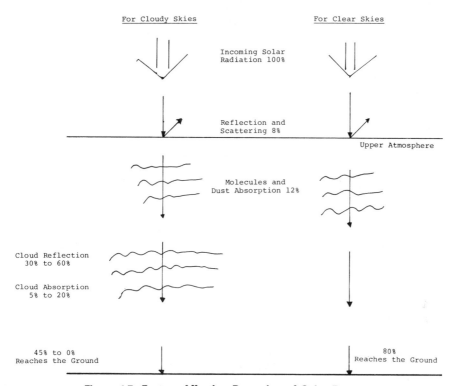

For Cloudy Skies

For Clear Skies

Incoming Solar
Radiation 100%

Reflection and
Scattering 8%

Upper Atmosphere

Molecules and
Dust Absorption 12%

Cloud Reflection
30% to 60%

Cloud Absorption
5% to 20%

45% to 0%
Reaches the Ground

80%
Reaches the Ground

Figure 1.7. Factors Affecting Reception of Solar Energy.

point. The noon rays of the sun travel through the least amount of atmosphere and thus appear the brightest, whereas the morning and afternoon sun rays appear weaker because they travel the greatest distance through the atmosphere to reach the earth's surface. Using the formula for finding the hypotenuse of a triangle (e.g., $C^2 = A^2 + B^2$), approximate distances may be found. When the sun has risen in the morning to 45° altitude, sunlight has to travel about 1.4 times as far to reach the ground as when the sun has risen directly overhead at noon.

1.4 MEASUREMENT OF SOLAR ENERGY

Solar energy arriving at the earth's surface is composed of both direct (also called beam) and diffuse (also called sky) solar radiation. The direct radiation reaches the earth as parallel beams of sunlight having directional order whereas the diffuse radiation appears to come from all parts in the sky. Diffuse radiation is generally the result of clouds present which change the direction of the incoming parallel rays of sunlight and give them a random directionality. (See Fig. 1.7.) Global radiation is a term sometimes used to describe the total solar radiation obtained by both direct and diffuse radiation. A solar collector on a building will be exposed to direct, diffuse, and reflected solar radiation. The reflected radiation from the ground is a function of the albedo (or reflectiveness ratio) of the earth's surface. If snow covers the ground, the

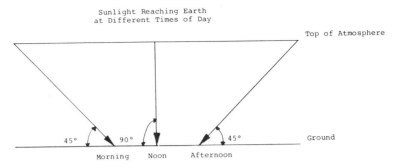

Figure 1.8. Sunlight Reaching Earth at Different Times of Day.

albedo will be high and if the ground cover is dark, less light will be reflected and the albedo will be less.

Two commonly used solar radiation measurement devices are the pyrheliometer and the pyranometer. The pyrheliometer measures direct solar radiation, while the pyranometer measures global (direct and diffuse) radiation. Figure 1.9 illustrates an Eppley Angstrom pyrheliometer and Fig. 1.10 illustrates two versions of the Eppley pyranometer. If a shadow band stand is used, as shown in Fig. 1.11, the pyranometer becomes an instrument for measuring only diffuse sky radiation.

These instruments receive solar radiation and then typically produce an analog direct current (DC) signal which is transmitted to a device such as an electronic integrator that converts the analog DC signal into an output signal which is then fed to a chart recorder. While the quality and accuracy of these devices is high, with a price tag reflecting these attributes, the device in Fig. 1.12 can be constructed to give a general indication of the relative solar insolation received at a given location. This device has a claimed accuracy or

Figure 1.9. Eppley Amstrom Pyrheliometer. (Source: The Eppley Laboratory, Inc.).

Figure 1.10. Eppley Pyranometer. (Source: The Eppley Laboratory, Inc.).

Figure 1.11. Eppley Shadow Band Stand. (Source: The Eppley Laboratory, Inc.).

Figure 1.12. Homebuilt Solar Radiation Detector. (Reprinted by permission of *Popular Electronics*, December 1976. See Ref. 18).

±5% and can be built for less than $35. Figure 1.13 illustrates the circuit diagram and parts list. Figure 1.14 shows the paper face plate which is placed over the original face plate of the 0 to 50 milliammeter used in the solar radiometer. As shown in Fig. 1.14 the meter face plate reads in Langleys per minute. The Langley is a measurement used in determining the amount of solar radiation (insolation) received. It was named in honor of Samuel Pierpont Langley, an American physicist, astronomer and pioneer designer of airplanes.

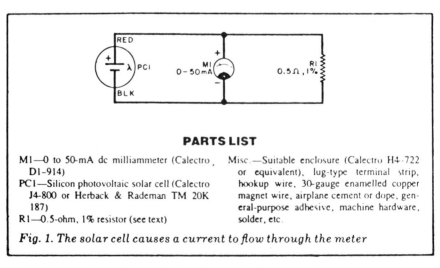

PARTS LIST

M1—0 to 50-mA dc milliammeter (Calectro D1–914)
PC1—Silicon photovoltaic solar cell (Calectro J4-800 or Herback & Rademan TM 20K 187)
R1—0.5-ohm, 1% resistor (see text)

Misc.—Suitable enclosure (Calectro H4-722 or equivalent), lug-type terminal strip, hookup wire, 30-gauge enamelled copper magnet wire, airplane cement or dope, general-purpose adhesive, machine hardware, solder, etc.

Fig. 1. The solar cell causes a current to flow through the meter

Figure 1.13. Schematic and Parts List for Homebuilt Detector. (Reprinted by permission of *Popular Electronics*, December 1976. See Ref. 18).

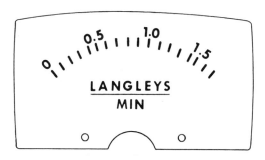

Figure 1.14. Faceplate for Homebuilt Solar Radiation Detector. (Reprinted by permission of *Popular Electronics,* December 1976. See Ref. 18).

A Langley is equal to 1 (gram) cal/cm², 4.184 J/cm² or about 3.69 Btu/ft². A Langley per minute is about 1 (gram) cal/cm² min, 2510 kJ/m²hr, 697 W/m² or 221 Btu/ft²hr.

1.5 WAYS OF HARNESSING SOLAR ENERGY

Some of the methods envisioned to harness the sun's energy seem to border on the realm of science fiction. One such method devised would incorporate huge solar collectors in space to obtain solar energy which would then be beamed by microwave to ground stations that would then convert the microwave energy into a usable form such as electricity. Other more down-to-earth schemes require vast expanses of (silicon) solar cells to produce huge amounts of electrical power which could be used to supply adjacent population centers. At present, converting solar energy directly into electricity is expensive. Hopefully through advances in technology and with mass production the practicality of such a system could be realized. Another method of collecting solar energy involves using many mirrors, located over a large area of land, that focus the sun on a pinpoint area which would receive enough concentrated solar energy to melt steel. This concentrated energy could be used to superheat water to produce steam and run a steam engine (turbine) or the energy could be used in foundry applications. This list could go on but the intent is to present some of the methods for harnessing solar energy which do not directly concern the heating and cooling of buildings.

1.6 REFERENCES AND SUGGESTED READING

1. Anderson, Bruce, *Solar Energy Fundamentals in Building Design,* New York: McGraw-Hill Book Company, 1977, pp. 3–4.
2. Baker, Robert H., and Fredrick, Lawrence W., *Astronomy,* New York: Van Nostrand Reinhold, 1971, pp. 52–53, 64–66, 70, 489–491.
3. Beiser, Arthur, *Physics,* Menlo Park, California: Cummings Publishing Company, Inc., 1973, p. 294.
4. Benson, James, "Nuclear, Solar and Jobs," *Solar Age,* **3,** 8:15–17, (1978).
5. Campbell, Stu, *Build your Own Solar Water Heater,* Charlotte, Vermont: Garden Way Publishing Company, 1978, pp. 14–22.

6. Commoner, Barry, *The Politics of Energy,* New York: Alfred A. Knopf, 1979.
7. Critchfield, Howard J., *General Climatology,* Englewood Cliffs, New Jersey: Prentice-Hall, Inc., 1974, pp. 13–20.
8. Delinger, W. G., "The Definition of the Langley," *Solar Energy,* **18,** 369–370 (1976).
9. Dolmatch, Theodore B. (ed.), *Information Please Almanac,* New York: Information Please Publishing, Inc., 1979, pp. 513–514.
10. Duffie, John A., and Beckman, William A., *Solar Energy Thermal Processes,* New York: John Wiley & Sons, Inc., 1974, pp. 1, 3–4, 22–25.
11. Fisher, Arthur, "What Are We Going to Do About Nuclear Waste?," *Popular Science,* **213,** 6:90–97, (1978).
12. Gamon, George, *A Star Called the Sun,* New York: The Viking Press, Inc., 1964, p. 21.
13. Gibson, Edward G., *The Quiet Sun,* Washington, D.C.: U. S. Government Printing Office, 1972.
14. Hafemeister, David W., "Nonproliferation and Alternative Nuclear Technologies," *Technology Review,* **81,** 3:58–62, (1979).
15. Henry, John P., Jr., Harless, V. Eugene, and Kopelman, Jay B., "World Energy: A Manageable Dilemma," *Harvard Business Review,* **57,** 3:150–161, (1979).
16. Henderson, Hazel, "Economics: A Paradigm Shift is in Progress," *Solar Age,* **3,** 8:18–21, (1978).
17. Hoyle, Fred, *Astronomy and Cosmology–A Modern Course,* San Francisco, Cal.: W. H. Freeman and Company, 1975, pp. 6, 11, 41–43.
18. Jochem, Warren, "Measure the Sun's Energy with a Solar Radiometer," *Popular Electronics,* **10,** 6:45–57, (1976).
19. Jordon, Richard C., and Liu, Benjamin, Y. H., eds., *Applications of Solar Energy for Heating and Cooling of Buildings,* New York: American Society of Heating, Refrigerating and Air-Conditioning Engineers, Inc., 1977, pp. 1.1–1.3, 3.7.
20. Kondratyev, K. Ya, *Radiation in the Atmosphere,* New York: Academic Press, Inc., 1969, pp. 1–2, 252–253, 304–305, 342–344, 474.
21. Kreider, Jan F., and Kreith, Frank, *Solar Heating and Cooling: Engineering, Practical Design and Economics,* Washington, D.C.: Hemisphere Publishing, Inc., 1977, p. 7.
22. "Langley, Samuel Pierpont," *The Encyclopedia Americana,* (1953), XVI, 721.
23. Marcus, Steven J., and Phillips, Leonard A., "Nuclear Power: Can We Live With It?," *Technology Review,* **81,** 7, (1979).
24. Menzel, Donald H., *Our Sun,* Cambridge, Mass.: Harvard University Press, 1959, pp. 72–73.
25. Mutz, Lloyd, and Duveen, Aneta, *Essentials of Astronomy,* New York: Columbia University Press, 1977.
26. Neiburger, Morris, Edinger, James G., and Bonner, William D., *Understanding Our Atmospheric Environment,* San Francisco, Cal.: W. H. Freeman and Company, 1973.
27. Norris, D. J., "Calibration of Pyranometers," *Solar Energy,* **14,** 99–108, (1973).
28. Norris, D. J., "Calibration of Pyranometers in Inclined and Inverted Positions," *Solar Energy,* **16,** 53–58, (1974).
29. Oddo, Sandra, "In Conversation with Ted Taylor," *Solar Age,* **3,** 8:22–25 1978.
30. *The Pensacola Journal,* "CIA Predicts 'Real' Gas Crunch by 1982," October 18, 1979, p. 5A.
31. Reuyl, John S., Harman, Willis W., Carlson, Richard C., Levine, Mark D., and Witwer, Jeffrey G., *Solar Energy in America's Future* (2nd ed.) Washington, D.C.: U. S. Government Printing Office, 1977.
32. Sheet Metal and Air Conditioning Contractors National Association, *Fundamentals of Solar Heating,* Washington, D.C.: U. S. Government Printing Office, 1978, pp. 2.2–2.4.
33. Smith, Elske V. P., and Jacobs, Kenneth C., *Introductory Astronomy and Astrophysics,* Philadelphia, Penn.: W. B. Sanders, 1973, pp. 58–63.
34. Stiefel, Michael, "Soft and Hard Energy Paths: The Roads Not Taken?," *Technology Review,* **82,** 1:56–66, (1979).
35. Strahler, Arthur N., *The Earth Sciences,* New York: Harper and Row, Inc., 1963, pp. 58–60, 70, 202–204.

36. Tetra Tech, Inc., *Energy Fact Book,* Washington, D.C.: U. S. Government Printing Office, 1976.

37. Thekaekara, Matthew P. (ed.), *The Solar Constant and the Solar Spectrum Measured from a Research Aircraft,* NASA Technical Report R-351, Goddard Space Flight Center, Greenbelt, Maryland.

38. U. S. Central Intelligence Agency, *The International Energy Situation: Outlook to 1985,* Washington, D.C.: Government Printing Office, 1977.

39. U. S. Department of Energy, *Budget Highlights,* FY 1980 Budget to Congress, DOE/CR0004.

40. U. S. Department of Energy, *1980 Congressional Budget Request,* Vol. 6, Part II, "Energy Conservation."

41. U. S. Department of Energy, Office of Consumer Affairs, *The Energy Consumer.*

42. U. S. Department of Energy, *How to Comply with the Emergency Building Temperature Restrictions.*

43. U. S. Department of Energy, *Introduction to Solar Heating and Cooling, Design and Sizing,* Washington, D.C.: U. S. Government Printing Office, 1978, pp. 9–19.

44. U. S. Department of Energy, Office of Public Affairs, *The National Energy Act–Information Kit,* (DOE/OPA-0003).

45. U. S. Department of Housing and Urban Development, *Solar Dwelling Design Concepts,* Washington, D.C.: U. S. Government Printing Office, 1976, pp. 9–15, 42–45.

46. *U. S. News and World Report,* "Fuels for America's Future," LXXXVII, (7), 33–34, 36–38, 1979.

47. Watkins, Joel S., Bottino, Michael L., and Morisawa, Marie, *Our Geological Environment,* Philadelphia, Penn.: W. B. Saunders Company, 1975, pp. 135–139.

48. Watt Engineering Ltd., *On the Nature and Distribution of Solar Radiation,* Washington, D.C.: U. S. Government Printing Office, 1978, pp. 19–26, 38–41, 43, 46, 84.

49. Wiesner, C. J., *Hydrometeorology,* London: Chapman and Hall Ltd., 1970, pp. 5–15.

50. Willrich, Mason, and Taylor, Theodore B., *Nuclear Theft: Risks and Safeguards,* Cambridge, Mass.: Ballinger Publishing Company, 1974.

51. Zirin, Harold, *The Solar Atmosphere,* Waltham, Mass.: Blaisdell Publishing Co., 1966.

CHAPTER 2
SOLAR ENERGY COLLECTION METHODS

2.1 LIQUID FLAT PLATE COLLECTORS

2.1.1. Collector Design

The basic flat plate collector consists of:

1. Cover plate
2. Air space
3. Absorber plate
4. Insulation
5. Frame

Each of these items represent an important part of the flat plate collector. Careful design consideration needs to be afforded each of these items because they will affect both collector efficiency and lifespan. Figure 2.1 illustrates the basic details of a typical flat plate collector and Fig. 2.2 shows a closeup of a collector. In Fig. 2.1 the five basic parts of the collector can be seen: the cover plate consisting of tempered glass having a low iron content, the air space, the absorber plate which in this diagram is a copper roll bond absorber plate having fluid channels integrated within the absorber plate itself, and the aluminum frame.

1. Cover plate. The cover plate's basic function is to allow short wave radiation from the sun to easily pass through, be absorbed by the absorber

Figure 2.1. Section View of Flat Plate Solar Collector. (Courtesy Grumman Energy Systems, Inc., Bohemia, New York[23]).

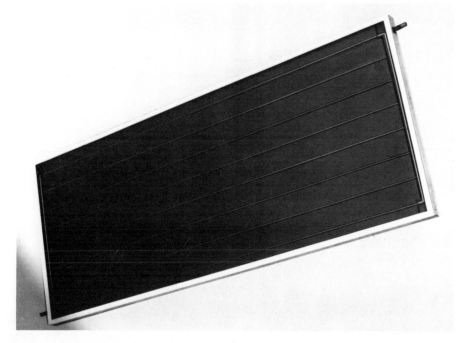

Figure 2.2. Closeup of a Flat Plate Solar Collector. (Courtesy Grumman Energy Systems, Inc., Bohemia, New York[23]).

plate, and then to keep the reradiated long wave radiation emitted by the absorber plate from escaping back through the cover plate. This principle is known as the greenhouse effect. Everyone has probably had some experience with this effect when the sun is shining brightly, and although it may be cold outside, it is warm in a car with the windows rolled up. Many materials may be used for the cover plate but glass still seems to be the favorite. The main virtues of glass are that it does not chemically break down in the presence of ultraviolet radiation from the sun and does not allow much absorbed solar energy to be reradiated from the collector. In other words, the glass is rather opaque to long wave radiation. Some drawbacks of glass are that it is usually heavier and more breakable than the plastics which are generally less expensive and easier to work with. Table 2.1 illustrates some of the properties of materials used for collector cover plates. As can be seen in the table the solar transmission is generally higher, the strength is greater, and the weight is less for the plastics than for glass. The drawbacks to the plastics can outweigh their benefits in many solar application cases. Although the solar light transmission is initially higher for the plastics than for the glass plates, their solar transmission (in percent) drops as they age. This is due to the detrimental effects of ultraviolet solar radiation breaking down the plastics. The plastics generally have lower operating temperatures than for glass. Although a collector having a plastic cover may operate at a "safe" temperature while fluid is flowing in the collector, if a breakdown were to occur in the pumping mechanism to halt fluid flow for an extended period of time during a period of high solar insolation, stagnation temperatures could exceed the maximum operating tem-

Table 2.1. Properties of Materials Used for Cover Plates.[a] (Source: U. S. Department of Energy[56]).

Test	Polyvinyl Fluoride	Polyethylene Terephthalate or Polyester	Polycarbonate	Fiberglass Reinforced Plastics	Methyl Methacrylate	Fluorinated Ethylene-Propylene	Ordinary Clear Lime Glass (Float) (0.10-0.13% Iron)	Sheet Lime Glass (0.05-0.06% iron)	Water White Glass (0.01% Iron)
Solar Transmission (%)	92-94	85	82-89	77-90	89	97	85	87	85-91
Maximum Operating Temperature (°F)	227	220	250-270	200° produces 10% transmission loss	180-190	248	400	400	400
Tensile Strength (psi)	13000	24000	9600	15000-17000	10600	2700-3100	1600 annealed 6400 tempered	1600 annealed 6400 tempered	1600 annealed 6400 tempered
Thermal Expansion Coefficient (in./in./°F $\times 10^4$)	24	15	37.5	18-22	41.0	8.3-10.5	4.8	5.0	4.7-8.6
Elastic Modulus (psi $\times 10^4$)	0.26	0.55	0.345	1.1	0.45	0.5	10.5	10.5	10.5
Thickness (in.)	0.004	0.001	0.125	0.040	0.125	0.002	0.125	0.125	0.125
Weight (lb/ft²) For above thickness	0.028	0.007	0.77	0.30	0.75	0.002	1.63	1.63	1.65
Greatest Load Area (psf/ft²)	...	...	...	...	...	...	...	...	30:30 annealed 100:28 tempered
Length of Life (yr)	In 5 yr retains 95% of total transmission	4	...	7-20	...	...	...	...	...

[a]These values were obtained from the following references.

D. P. Grimmer and S. W. Moore, "Practical Aspects of Solar Heating: A Review of Materials Use in Solar Heating Applications," paper presented at Society for Advancement in Materials Process Engineering Meeting. October 14-16, 1975, Albuquerque, New Mexico.

T. Kobayashi and L. Sargent, "A Survey of Breakage-Resistant Materials for Flat-Plate Solar Collector Covers," paper presented at U.S. Section of ISES Meeting, Ft. Collins, Colorado, August 20-23, 1974.

A. F. Scoville, "An Alternate Cover Material for Solar Collectors," paper presented at ISES Congress and Exposition, Los Angeles, California, July 1975.

C. W. Clarkson and J. S. Herbert, "Transparent Glazing Media for Solar Energy Collectors," paper presented at U. S. Section of ISES Meeting, Ft. Collins, Colorado, August 20-23, 1974.

Modern Plastics Encyclopedia (McGraw-Hill 1975-1976).

R. B. Toenjes, "Integrated Solar Energy Collector Final Summary Report," Los Alamos Scientific Laboratory report LA-6143-MS (November 1975).

perature of the plastics. Glass, having a higher maximum operating temperature, is less susceptible to damages due to elevated stagnant temperatures. Another problem of plastics is that they are more transparent to long wave radiation emitted by the absorber plate than glass and hence allow more collected solar radiation to escape back through the cover plate and into the atmosphere. Alternatives to using glass for collector covers, as compiled by Kobaysaki and Sargent, include:

Plastics:
  Polymethyl methacrylate or acrylic (Acrylite®, Lucite®, Plexiglas®)
  Polycarbonate (Lexan®, Merlon®)
  Polyethylene terephthalate or polyester (Mylar)
  Polyvinyl fluoride (Tedlar)
  Fluorinated ethylene–propylene, or fluorocarbon (Teflon, FEP)
  Polymide (Kapton®)
  Polyethylene

Other Materials:
  Fiberglass—reinforced polyester (Sun-lite®)
  Plastic—plastic laminates
  Plastic—glass laminates

Most attention in this book will be directed to glass because it appears to have the least amount of problems. The plastics and other materials generally do not have the service life of glass. This is due mainly to the deterioration of materials by the sun's ultraviolet radiation. Many of these plastic products are also easily scratched or marred.

Of the total solar radiation striking a solar collector, a percentage will be absorbed or reflected by the cover plate and the rest being transmitted. Table 2.2 shows some properties of various types of glass when the sun strikes the glass at a normal incidence angle. Note that the sum of the reflectance loss, absorptance loss, and light transmittance add up to 100%.

In choosing glass for the collector plate, attention should be directed to the percentage of iron oxide (Fe_2O_3) present. As shown in Table 2.2, the higher

Table 2.2. Properties of Glass. (Source: Kreider[29]).

Property	Type of glass		
	Ordinary float	Sheet lime	Crystal white
Iron oxide content, %	0.10-0.13	0.05	0.01
Refractive index	1.52	1.51	1.50
Light transmittance (normal), %	85-81	87-85	90.5
Glass thickness, in.	0.125-0.1875	0.125-0.1875	0.1875
Reflectance, loss, %	8.2-8.0	8.1-8.0	8.0
Absorptance loss, %	6.8-11.0	4.9-7.0	1.5

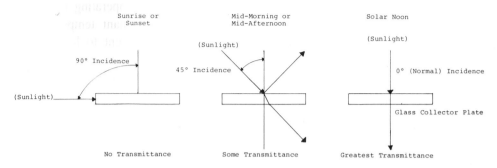

Figure 2.3. Collector Plate Transmittance as a Function of Solar Incidence Angle.

the iron oxide content, the greater the absorptance losses, and the lower the percentage of light transmitted. The greater amount of iron oxide particles in the glass, the less "transparent" the glass is. By looking at the edge of a piece of glass, one can judge the relative percentage of iron oxide present. The greener the glass edge appears, the higher the iron oxide content.

As shown in Table 2.2, reflectance losses at normal incident generally run about 8% regardless of the type of glass used. At varying angles of incidence the percentages of reflection will vary and thus the percentage of transmitted solar radiation will change as shown in Fig. 2.3. Note that zero degrees incidence represents the normal incidence angle.

By using a few simple equations the percentage of light reflected by a glass collector cover plate can be determined as shown in Fig. 2.4. When light passes at an angle from an air medium, through a different medium such as a glass solar collector cover plate, and into the air again it is bent (refracted) toward the normal. Snell's law can be used to predict the angle of refraction:

$$\text{sine } i/\text{sine } r = n_2/n_1 \tag{2.1}$$

or

$$\text{sine } r = \text{sine } i \ (n_1/n_2) \tag{2.2}$$

where

i = angle of incidence

r = angle of refraction

n_1 = index of refraction of first medium (e.g., air)

n_2 = index of refraction of second medium (e.g., glass)

The first medium (n_1) is air having an index of refraction of 1.00 and the second medium (n_2) is glass having indexes of refraction of 1.50, 1.51, and 1.52 depending upon the type of glass used (see Table 2.1). When the angle of refraction (r) is determined, Fresnel's formula can then be used.

$$R = 1/2(R1 + R2) \tag{2.3}$$

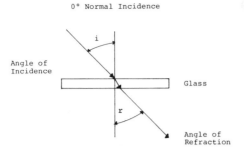

Figure 2.4. Angle of Refraction.

where

$$R = \text{ratio of reflected light to incident light}$$

$$R1 = \text{sine}^2(i - 4)/\text{sine}^2(i + r) \tag{2.4}$$

$$R2 = \text{tangent}^2(i - 4)/\text{tangent}^2(i + r) \tag{2.5}$$

R1 and R2 represent the reflections, respectively, for the two components of polarization.* The average of these two component values represents the ratio of the reflected light in the incident light. The formula will work for all angles of incidence except when $i = $ o (normal incidence).

When $i = $ o and $n_1 = 1$, the following formula must be used:

$$R1 = (n_2 - 1/n_2 + 1)^2 \tag{2.6}$$

When $i = $ o there will be only one reflective component of polarization. Therefore only R1 will be used.

The following formula can be used to determine the degree of reflection for N number of glass cover plates used in a flat-plate solar collector.

$$R_N = 1 - \{[1 - R1/1 + (2N - 1) \, R1] + [1 - R2/1 + (2N - 1)]\}/2 \tag{2.7}$$

where $R_N = $ total reflection for N cover plates.

If $i = $ o (normal incidence) then only one reflective polarization component is available, therefore:

$$R_N = 1 - \{(1 - R1)/[1 + (2N - 1) \, R1]\} \tag{2.8}$$

By determining R_N for various angles of incidence, reflectance curves, as shown in Fig. 2.5, can be constructed. Notice that for one cover, at zero degrees (normal) incidence, the reflection is approximately 8% (as illustrated in Table 2.2).

---

*More specifically, $R1$ represents the reflective components of light polarized in the plane of incidence and $R2$ represents the reflective component of light polarized perpendicular to the plane of incidence.

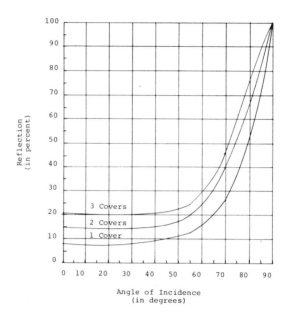

Figure 2.5. Collector Plate Reflectances.

As mentioned earlier, the sum of the percentages of reflected, absorbed, and transmitted solar radiation for a glass cover plate add up to 100%. The transmittance (τ), neglecting absorptance losses, may be determined by subtracting individual valves obtained in equations 2.7 and 2.8 from 1.00. Equations 2.9 and 2.10, representing minor variations of equations 2.7 and 2.8, can also be used to determine the transmittance (τ) by reflectance, neglecting absorptance losses.

$$\tau_N - \{(i - R1)/[1 + (2N - 1)\ R1)] + [(1 - R2)/(1 + 2N - 1)\ R2]\}/2 \tag{2.9}$$

where τ_N = light transmittance for N glass collector plates.

If i = o then:

$$\tau_N + (1 - R1)/[1 + (2N - 1)\ R1] \tag{2.10}$$

Figure 2.6 illustrates the percentage of light transmitted (neglecting absorptance losses) for varying angles of incidence. As the angle of incidence increases, the reflectance losses increase, decreasing light transmittance.

The number of cover plates used for a collector is a consideration when designing a flat-plate solar collector. Table 2.3 is a guide for determining the number of cover plates to use. The logic behind Table 2.3 is that the greater the difference between the outside temperature and the temperature of the metal absorber inside the flat collector (ΔT), the greater the outward flow of heat from the collector's absorber plate. The extra cover plates provide addi-

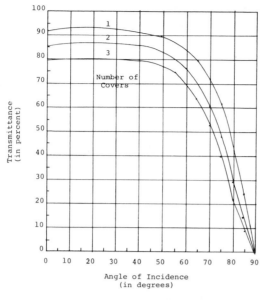

Transmittance of solar energy, neglecting
absorption, through glass plates having
an index of refraction of 1.50.

Figure 2.6. Collector Plate Transmittances.

tional barriers to stop the outflow of collected heat from the collector to the surrounding outside air.

Using Table 2.3, about the only application where no cover plate would be needed would be in the heating of swimming pools. With a swimming pool the ΔT would be low and therefore the thermal losses would be low. If a cover plate is used, the attenuating effects of a glass cover (from absorption and reflection losses) may outweigh the benefits gained in reducing thermal losses by using the cover plate. In applications such as solar water heating, space heating, and air conditioning where the ΔT is larger, the use of at least

Table 2.3.   Guide for Determining Number of Collector Cover Plates.
(Source: Anderson[4]).

Optimum Number of Cover Plates	
Collector Temperature minus Outdoor Temperature (°F)	Number of Covers
-10 to 10	0
10 to 60	1
60 to 100	2
100 to 150	3

Table 2.4. Properties of Typical Absorber Substrate Materials (Typical values: Standard specifications should be consulted for specific types or alloys). (Source: U. S. Department of Housing and Urban Development[60]).

Material Property	Aluminum	Copper	Mild Carbon Steel	Stainless Steel
Elastic Modulus, Tension psi x 10^6	10	19	29	28
Density lbs./cu.in.	0.098	0.323	0.283	0.280
Expansion Coefficient (68–212°F) in/in/°F x 10^{-6}	13.1	9.83	8.4	5.5
Thermal Conductivity (77–212°F) Btu/hr·ft^2·°F·ft	128	218	27	12
Specific Heat (212°F) Btu/lb·°F	0.22	0.09	0.11	0.11

one collector cover plate is mandatory because the glass absorption and reflection losses are low compared to the thermal losses incurred without a cover plate.

2. Airspace. The second part of the collector consists of the airspace. The airspace serves to reduce the internal convective losses. Cover plates are generally spaced 1/2 to 1 inch apart and 3/4 inch is a typical spacing.

3. Absorber Plate. The absorber plate represent the third part of the flat plate collector. Generally it is composed of a metal such as aluminum or copper and coated with either a selective or nonselective absorber coating. Table 2.4 shows some properties of typical absorber substrate materials. Note that copper has a higher thermal conductivity than aluminum or steel.

Figure 2.7 shows some types of absorber plate designs. Integral tube and absorber designs, such as diagram A in Fig. 2.7, give excellent heat transfer from absorber to fluid in the tubes because the tubes are actually part of the absorber plate. This type of absorber design is well suited for "closed" systems where a heat exchanger is used and a noncorrosive heat transport fluid is employed. For "open" systems where there is no heat exchanger and the fluid used is potable water, these "integral" designs, with small fluid transport tubes, can clog up due to scaling. Figure 2.8 illustrates another common absorber plate design. This type is similar to Diagram D in Fig. 2.7 where the tubes are bonded to the absorber plate. Collectors with these bonded tubes (as opposed to integral type absorber designs) need to be checked for proper adhesion to the absorber plate because a thermally poor bond will not properly and efficiently transfer the collected heat to the liquid flowing in the tubes.

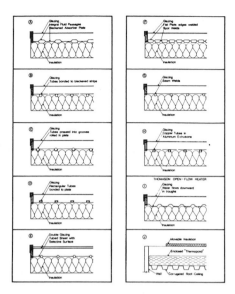

Figure 2.7. Absorber Plate Designs for Liquid-Type Systems. Reprinted with permission from the 1978 Solar Energy Utilization for Heating and Cooling 58, ASHRAE HANDBOOK and Product Directory.

Note that in Fig. 2.8 the tubes are large enough to overcome many of the clogging problems that beset smaller tubes when used in open type systems. Figure 2.9 illustrates a collector absorber plate design whereby the water flows in troughs rather than in tubes. This design is the Thomason Solaris type collector (shown in Diagram J of Fig. 2.7) invented and patented by Dr. Harry E. Thomason and his son, Jack Thomason.

The flow pattern of the collector is a necessary design consideration. Figure

Figure 2.8. Flat Plate Collector.

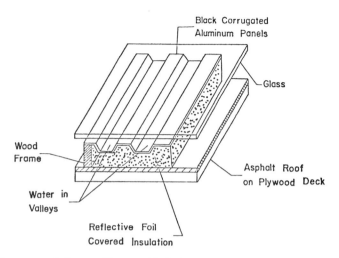

Figure 2.9. Thomason Collector. (Source: U. S. Department of Housing and Urban Development[63]).

2.10 illustrates some common absorber tube patterns. Item A of Fig. 2.10 is an easily designed pattern suitable from the do-it-yourselfer since there are no joint connections in the collector itself. The serpentine pattern gives the collector uniform fluid flow but the pressure drop is greater than for the other designs shown. Although the other patterns do not have the degree of pressure drop that type A does, they do suffer from varying degrees of flow distribution problems. This is due to nonuniform flow rates between the various absorber plate tubes. Of the tubes in a collector that do have lower flow rates, the absorber plate contacting these tubes will be hotter, creating "hot spots" which lower the efficiency of the collector because of a higher ΔT.

In most solar installations, more than one solar collector is used. This brings up the question of how to connect the collectors. The two basic types of

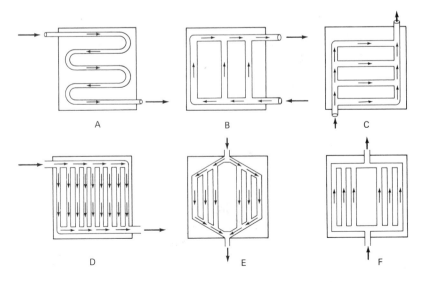

Figure 2.10. Collector Liquid Flow Patterns.

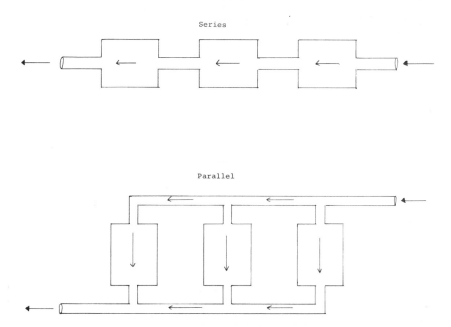

Figure 2.11. Parallel and Series Collector Connections.

connections are the parallel and series as shown in Fig. 2.11. When collectors are joined together, pressure drops must be considered. For the collectors connected in parallel, the pressure drop for the whole array of collectors is equal to the pressure drop across any one collector in the array. For collectors connected in series, the pressure drop is equal to the sum of the pressure drops for each collector.

The final discussion on absorbers centers around the types of absorber coatings to use. The main function of the absorber coating is to retain the absorbed heat and serve to reduce longwave emittance (ϵ). The emittance (ϵ) of a material is a numerical figure expressing the material's capacity for giving off long-wave radiation. An ideal absorber coating would have an emittance of zero representing no long-wave (thermal) radiation being given off. The absorptance (α) of a material represents the amount of energy absorbed. It is also expressed in numerical form. The perfect absorber would have an absorptance (α) of 1.0, indicating complete absorption. In practice this figure would be less because the material would reflect and radiate some of the energy rather than absorb it all. Table 2.5 gives emittances and absorptances of various materials.

Naturally, a material with a high absorptance and a low emittance will have a high efficiency. Some selective surfaces come close to representing the ideal absorber. As a rule, flat plate collector efficiency decreases as the temperature of the collector increases. This is because collector heat losses become greater as a result of large temperature differentials (ΔT) between the collector and the surrounding air. Selective absorber surfaces help to keep losses down as the temperature differential increases by retaining more of the absorbed heat due to their lower emittances than would nonselective absorber coatings having higher emittances.

Table 2.5. Emittances and Absorptances of Various Materials. (Source: Anderson[3])

Class I Substances: Absorptance to Emittance Ratios (α/ε) Less than 0.5

Substance	Short-wave Absorptance	Long-wave Emittance	α/ε
Magnesium carbonate, $MgCO_3$	0.025-.04	0.79	0.03-.05
White plaster	.07	0.91	.08
Snow, fine particles, fresh	.13	0.82	.16
White paint, .017 in. on aluminum	.20	0.91	.22
Whitewash on galvanized iron	.22	0.90	.24
White paper	.25-.28	0.95	.26-.29
White enamel on iron	.25-.45	0.9	.28-.5
Ice, with sparse snow center	.31	0.96-0.97	.32
Snow, ice granules	.33	0.89	.37
Aluminum oil base paint	.45	0.90	.50
White powdered sand	.45	0.84	.54

Class II Substances: Absorptance to Emittance Ratios (α/ε) Between 0.5 and 0.9

Substance	Short-wave Absorptance	Long-wave Emittance	α/ε
Asbestos felt	.25	0.50	.50
Green oil base paint	.5	0.9	.56
Bricks, red	.55	0.92	.60
Asbestos cement board, white	.59	0.96	.61
Marble, polished	.5-.6	0.9	.61
Wood, planed oak	--	0.9	--
Rough concrete	.60	0.97	.62
Concrete	.60	0.88	.68
Grass, green, after rain	.67	0.98	.68
Grass, high and dry	.67-.69	0.9	.76
Vegetable fields and shrubs, wilted	.70	0.9	.78
Oak leaves	.71-.78	0.91-.95	.78-.82
Frozen soil	--	0.93-.94	--
Desert surface	.75	0.9	.83
Common vegetable fields and shrubs	.72-.76	0.9	.82
Ground, dry plowed	.75-.80	0.9	.83-.89
Oak woodland	.82	0.9	.91
Pine forest	.86	0.9	.96
Earth surface as a whole (land and sea, no clouds)	.83	--	--

Class III Substances: Absorptance to Emittance Ratios (α/ε) Between 0.8 and 1.0

Substance	Short-wave Absorptance	Long-wave Emittance	α/ε
Grey paint	.75	.95	.79
Red oil base paint	.74	.90	.82
Asbestos, slate	.81	.96	.84
Asbestos, paper		.93-.96	--
Linoleum, red-brown	.84	.92	.91
Dry sand	.82	.90	.91
Green roll roofing	.88	.91-.97	.93
Slate, dark grey	.89	--	--
Old grey rubber	--	.86	--
Hard black rubber	--	.90-.95	--
Asphalt pavement	.93	--	--
Black cupric oxide on copper	.91	.96	.95
Bare moist ground	.9	.95	.95
Wet sand	.91	.95	.96
Water	.94	.95-.96	.98
Black tar paper	.93	.93	1.0
Black gloss paint	.90	.90	1.0
Small hole in large box, furnace or enclosure	.99	.99	1.0
"Hohlraum," theoretically perfect black body	1.0	1.0	1.0

Table 2.5. (Continued)

Class IV Substances: Absorptance to Emittance Ratios (α/ε)
Greater than 1.0

Substance	Short-wave Absorptance	Long-wave Emittance	α/ε
Black silk velvet	.99	.97	1.02
Alfalfa, dark green	.97	.95	1.02
Lamp black	.98	.95	1.03
Black paint, 0.017 in. on aluminum	.94-.98	.88	1.07-1.11
Granite	.55	.44	1.25
Graphite	.78	.41	1.90

High Ratios, but Absorptances Less Than .80

Dull brass, copper, lead	.2-.4	.4-.65	1.63-2.0
Galvanized sheet iron, oxidized	.8	.28	2.86
Galvanized iron, clean, new	.65	.13	5.00
Aluminum foil	.15	.05	3.00
Magnesium	.3	.07	4.3
Chromium	.49	.08	6.13
Polished zinc	.46	.02	23.0
Deposited silver (optical reflector) untarnished	.07	.01	

Class V Substances: Selective Surfaces[a]

Substance	Short-wave Absorptance	Long-wave Emmittance	α/ε
Plated metals:[b]			
Black sulfide on metal	.92	.10	9.2
Black cupric oxide on sheet aluminum	.08-.93	.09-.21	
Copper (5×10^{-5} cm thick) on nickel or silver-plated metal			
Cobalt oxide on platinum			
Cobalt oxide on polished nickel	.93-.94	.24-.40	3.9
Black nickel oxide on aluminum	.85-.93	.06-.1	14.5-15.5
Black chrome	.87	.09	9.8
Particulate coatings:			
Lampblack on metal			
Black iron oxide, 47 micron grain size, on aluminum			
Geometrically enhanced surfaces:[c]			
Optimally corrugated greys	.89	.77	1.2
Optimally corrugated selectives	.95	.16	5.9
Stainless steel wire mesh	.63-.86	.23-.28	2.7-3.0
Copper, treated with $NaClO_2$ and NaOH	.87	.13	6.69

Figure 2.12 helps to illustrate the merits of using selective absorber coatings and/or additional cover plates at higher collector operating temperatures. As shown in Fig. 2.12, by using the standard lamp black paint with one glass cover plate, collector efficiency rapidly goes down as collection temperatures increase, whereas the selective surface with one cover plate performs considerably better as temperatures increase.

At *lower* operating temperatures, the use of special selective coatings and additional cover plates does not increase collector efficiency to a great extent. At higher operating temperatures though, the use of additional cover plates and/or selective surfaces is necessary. Otherwise, flat plate collector operating efficiencies drop to a very low level. Unless high operating temperatures *are*

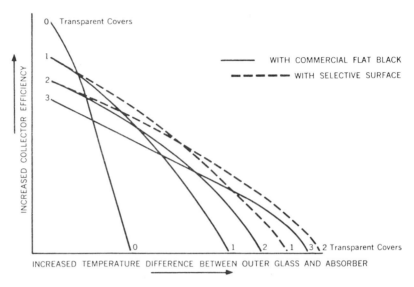

Figure 2.12. Solar Energy Collection as a Function of Absorber Coating and Additional Cover Plates. (Source: Anderson[3]).

necessary, such as for solar absorption chiller air conditioning systems, the flat plate collectors should be operated at reduced temperatures where their efficiencies are higher.

There are numerous selective and nonselective absorber coatings to choose from. In choosing an absorber coating, the design consultant or individual buyer should determine from manufacturers' literature the operating characteristics of the product such as absorptance and emittance values, maximum operating temperature of the coating, and estimated life of the product. This information can help in deciding whether one product justifies a higher cost.

In choosing an absorber coating, the compatibility between the absorber coating material and the metallic absorber plate should be known. Incompatibility between materials may cause the absorber coating to peel off. Absorber coatings compatible with copper may not be compatible with, say, aluminum and *vice versa*. For the individuals applying absorber coating themselves, manufacturers' suggested methods of surface preparation, primer application, and final coating applications should be followed.

Table 2.6 shows several recipes for selective black surfaces and is included to demonstrate what is involved in formulating and applying a selective coating.

4. Insulation. Insulation plays a large part in curbing heat loss due to conduction in a solar collector. It is generally placed on the back side of the absorber plate and around the edges of the collector frame. By leaving air space between the absorber plate and the insulation, thermal efficiency is increased and the insulation is not in direct contact with the hot absorber plate.

The type of insulation used in a solar collector is a major consideration. If

Table 2.6. Formulas for Selective Black Surfaces. Reprinted by Permission, American Society of Heating, Refrigerating and Air-Conditioning Engineers, Inc.

(1) Nickel-Black

The metal base must be perfectly clean: this is effected using standard chemical cleaning techniques as used in the electro-plating industry.

The black coating is then obtained by immersion as a cathode in an aqueous electrolytic bath containing, per liter:

    75 gms nickel sulfate ($NiSO_4 \cdot 6H_2O$)'
    28 gms zinc sulfate;
    24 gms ammonium sulfate;
    17 gms ammonium thiocynate;
     2 gms citric acid.

The pH of the solution should be about 4 and a pure nickel anode is used. The bath is operated at 30°C. Electrolysis is carried on for 2-4 minutes at 2 ma per cm^2 the exact time depending upon the nature of the base metal and the temperature.

In cold weather there is danger of precipitation from the solution given above so that in recent work the same solution but half as concentrated has been employed instead--with the same results. (The citric acid may be omitted.)

Better results are obtained by the two-layer technique. Thus on galvanized iron the electrolysis is carried on at 1 ma per cm^2 for 1 minute followed by 1-2 minutes at 2 ma per cm^2.

(2) Copper Oxide on Aluminum

An aluminum base is first covered with an oxide layer by anodizing. For this purpose the aluminum body is immersed as cathode in an aqueous solution containing 3% by volume of sulfuric acid and 3% by volume of phosphoric acid, with carbon as an anode. An electric current of 6 millamperes/cm is passed during 20 to 30 seconds through the solution, then the current is reversed for a few seconds to give partial anodizing. After rinsing, the aluminum body is immersed for 15 minutes at 85-90°C in an aqueous solution containing per liter:

    25 gms of copper nitrate $Cu(NO_3)_2 \cdot 6H\,O$;
     3 gms of concentrated nitric acid;
    15 gms of potassium permanganate.

After this treatment, the aluminum body is withdrawn, dried and heated to about 450°C for some hours, until the surface color has become almost black.

This treatment is rather sensitive to the type, composition and grain structure of the aluminum so that the results are not equally good on all grades of commercial aluminum.

(3) Copper Oxide on Copper

The formulation is quoted from Ref. 5 and is very similar to that of Salem and Daniels.

Before blackening, the copper is buffed to remove dirt and oxide layers to yield a clean bright surface. After being degreased in a boiling bath of metal cleaner it is washed in clean water and rubbed with a soft wire brush to remove gritting particles.

It is then treated, for various times--between 3 and 13 minutes[*]--in the blackening bath at a solution temperature of 140-145°C.

The bath comprises:

    16 oz sodium hydroxide (NaOH);
     8 oz sodium chlorite ($NaClO_2$)

per gallon (imperial) of water.

-----

[*]At 3 minutes the measured α = 0.79; E - 0.05. At 8 minutes α = 0.89, E = 0.17. Longer times increase E with little increase of α.

Table 2.7. Upper Temperature Limits for Insulation Materials. (Source: U. S. Department of Energy[56]).

Material	Density (lb/CF)	Thermal Conductivity at 200°F (Btu/hr-SF-°F/in.)	Temperature Limits (°F)
Fiberglass with Organic Binder	0.6	0.41	350
	1.0	0.35	350
	1.5	0.31	350
	3.0	0.30	350
Fiberglass with Low Binder	1.5	0.31	850
Ceramic Fiber Blanket	3.0	0.4 at 400°F	2300
Mineral Fiber Blanket	10.0	0.31	1200
Calcium Silicate	13.0	0.38	1200
Urea-Formaldehyde Foam	0.7	0.20 at 75°F	210
Urethane Foam	2-4	0.20	250-400

the collector temperatures exceed the upper temperature limits of the insulation it could ignite or release toxic fumes depending upon the composition of the insulation. Outgassing can also take place at temperatures above the insulation's maximum operating temperature. This happens when the high temperatures break down the insulation and cause fumes which deposit residue on the collector cover plate resulting in a reduction of transmitted light. Table 2.7 illustrates some properties of various insulation materials. Although the urea–formaldehyde foam and the urethane foam have low thermal conductivity, their maximum operating temperatures could be exceeded in a high temperature collector or during collector stagnation due to a pump or temperature sensor failure during periods of high solar insolation. The mineral fiber blanket and the fiberglass with low binder are better choices. Figure 2.13 illustrates the performance difference between a moderately insulated collector and a well insulated collector. As can be seen, the most noticeable difference between the two graphs occurs at higher collector temperatures where thermal losses are increased.

5. Frame. The collector frame serves to support the components of the solar collector. It may be made of metal or wood. The main criteria is that the frame will not rust or rot thereby causing maintenance problems. Aluminum is generally used for metal frames and redwood or cypress for wooden frames.

Sealants can be used to caulk cracks in the collector and prevent glass from resting directly on the metal collector frame which can become hot. The sealants also allow the glass room to expand and contract in the frame without breaking due to changes in the temperature of the glass. Table 2.8 illustrates some properties of sealing compounds. Note that the silicones have the highest service temperature range (-85 to $500°$ F) of the sealing compounds shown in Table 2.8. Figure 2.14 shows a simple flat plate that can easily be constructed.

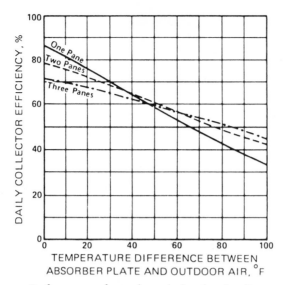

Performance of a moderately-insulated collector.

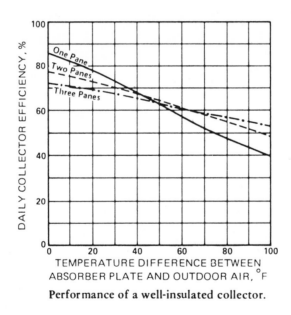

Performance of a well-insulated collector.

Figure 2.13. Performance of Insulated and Uninsulated Collectors. (Source: Anderson[4]).

For do-it-yourselfers wanting to build their own collectors, the Environmental Information Center of the Florida Conservation Foundation, Inc., has an excellent publication giving a step by step method for building a flat plate solar collector.

2.1.2. Collector Controls

Collector controls are an important part of the solar flat plate collector. Since the flat plate collector is nonfocusing, elaborate tracking controls are not

Table 2.8. Properties of Sealing Compounds. (Source: U. S. Department of Housing and Urban Development[60]).

	Butyl			Acrylic		Polysulfides		Polyurethanes		Silicones
	Oil Base	Skinning Type	Non-Skinning Type	Solvent-Release Type	Water-Release Type	One Component	Two Component	One Component	Two Component	One and Two Component
Chief ingredients	Selected oils, fillers, plasticizers, binders, pigment	Butyl polymers, inert reinforcing pigments, non-volatile plasticizers and polymerizable dryers	Butyl polymers, inert reinforcing pigments, non-volatilizing and non-drying plasticizer	Acrylic polymers with limited amounts of fillers & plasticizers	Acrylic polymers with fillers and plasticizers	Polysulfide polymers, activators, inert fillers, curing agents, and non-volatilizing plasticizers	Base polysulfide polymers, activators, pigments, plasticizers, fillers. Activator: accelerators, extenders, activators	Polyurethane prepolymer & filler pigments & plasticizers	Base: polyurethane prepolymer, filler, pigment, plasticizers. Activator: accelerators, extenders, activators	Silicone polymer, pigment & fillers
Primer required	In certain applications	none	none	none	none	usually	usually	usually	always	usually
Curing process	solvent release, oxidation	solvent release, oxidation	no curing: remains permanently tacky	solvent release	water evaporation	chemical reaction with moisture in air & oxidation	chemical reaction with curing agent	chemical reaction with moisture in the air	chemical reaction with curing agent	chemical reaction and/or moisture
Tack-free time (hrs)	6	24	remains indefinitely tacky	36	36	24	36-48	36	24	0.3 to 1
[1/] Cure time days	continuing	continuing	N/A	14	5	14-21	7	14	3-5	1 to 7
Max. cured elongation	15%	40%	N/A	60%	not available	300%	600%	300%	400%	100 to 450%
Recommended max. joint width movement %	25% decreasing with age	±7 1/2%	N/A	±10%	±5%	±25%	±15%	±15%	±25%	±25%
Max. joint width	1"	3/4"	N/A	3/4"	5/8"	1"	1"	3/4"	1"	5/8"
Resiliency	low	low	low	low	low	high	high	high	high	high
Resistance to compression	very low	moderate	low	very low	low	moderate	moderate	high	high	high
[2/] Resistance to extension	very low	low	low	very low	low	moderate	high	moderate	high	high
Service temp. range °F	-20° to 150°	-20° to 180°	-20° to 180°	-20° to 180°	-20° to 180°	-40° to 200°	-60° to 200°	-25° to 250°	-40° to 250°	-85 to 500°
Normal application temp. range (surface) (°F)	+40° to +120°	+40° to 120°	+40° to 120°	+40° to 120°	+40° to 120°	+40° to 120°	+40° to 120°	+40° to 120°	+40° to 120°	0° to +160°
Weather resistance	poor	fair	fair	very good	very good	very good	very good	very good	very good	excellent
Ultra-violet resistance, direct	poor	good	good	very good	not available	good	good	poor to good	poor to good	excellent
Cut, tear, abrasion resistance	N/A	N/A	N/A	N/A	N/A	good	good	excellent	excellent	fair
[3/] Life expectancy	5 to 10 years	10 years +	10 years +	20 years +	20 years +	20 years +	20 years +	20 years +	20 years +	20 years +
Hardness Shore A	20-80	20-40	N/A	20-40	30-35	25-35	24-45	25-45	25-45	20 to 60
Applicable specifications	TT-C-598C	TT-S-001657	NAAMM SS-1a-68	TT-S-00230C	TT-S-00230C	TT-S-00230C	TT-S-00227e	TT-S-00230C	TT-S-00227e	TT-S-00230C TT-S-001543 TT-S-00227E

1/ Cure time as well as pot life are greatly affected by temperature and humidity. Low temperatures and low humidity create longer pot life and longer cure time; conversely, high temperatures and high humidity create shorter pot life and shorter cure time.

2/ Resistance to extension is better known in technical terms as modulus. Modulus is defined as the unit stress required to produce a given strain. It is not constant but, rather, changes in values as the amount of elongation changes.

3/ Life expectancy is directly related to joint design, workmanship and conditions imposed on any sealant. The length of time illustrated is based on joint design within the limitations outlined by the manufacturer, and good workmanship based on accepted field practices and average job conditions. A violation of any one of the above would shorten the life expectancy to a degree. A total disregard for all would render any sealant useless within a very short period of time.

36

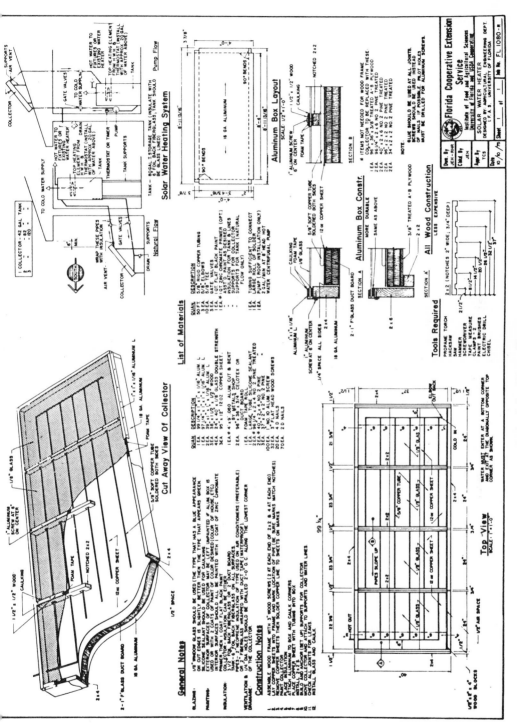

Figure 2.14. Basic Solar Water Heater.

37

Table 2.9. Description of Solar Collectors Plotted in Fig. 2.34. (Source: U. S. Department of Commerce[55]).

Absorber Material	Collector Number from Figure 2.34	Manufacturer and Remarks	Absorber Surface Coating	Transparent Covers Number	Transparent Covers Material	Stagnation Temperature °F*
Aluminum	1	NASA/Honeywell	Black nickel	2	glass	466
Aluminum	2	MSFC	Black nickel	2	Tedlar	313
Aluminum	3	NASA/Honeywell	Black paint	1	glass	274
Aluminum	4	NASA/Honeywell (mylar honeycomb)	Black paint	2	glass	475
Aluminum	5	NASA/Honeywell	Black paint	2	glass	355
Aluminum	6	PPG	Black paint	2	glass	268
Glass	**7	Owens (evacuated tube)	Selective surface	1	glass	1,150
Steel	8	Solaron (data furnished by manufacturer) Heat transfer fluid is air	Black paint	2	glass	355

*Values are calculated assuming that incident solar radiation, S, is 300 Btu/(ft^2)(hr) and that ambient temperature, T_a, is 70°F.

**With the exception of solar collectors number 7 and 8, the absorber plates are tube-in-plate.

necessary to follow the movement of the sun across the sky. The main controller used on the flat plate collector is a differential temperature controller that regulates the flow of the liquid through the collector. One major use of the controller with respect to the solar flat plate collector is to keep from allowing the collector to become overheated. When overheating does occur due to no fluid flow (stagnation) for extended periods during times of high solar insolation, collector parts can become warped and absorber plate coatings can peel off from the intense heat buildup. Table 2.9 illustrates some stagnation temperatures of various types of collectors including an air-type flat plate collector and an evacuated tube type collector discussed later in this chapter.

2.1.3. Maintenance Factors

The amount of maintenance is dependent upon the materials used to construct the collector and the type of operating environment to which the collector is exposed. Copper seems to be the most reliable absorber material, while aluminum and steel will give more maintenance problems. Corrosion can be reduced in aluminum and steel absorbers by adding chromatebased inhibitors to the water or antifreeze but their use is no absolute guarantee that corrosion will cease to be a problem. In areas where freezing is probable, the use of antifreeze will be necessary to prevent broken pipes when the sun goes down and freezing temperatures can cause untreated water to freeze and expand. Corrosion problems probably represent the biggest headaches to be dealt with in liquid-type collectors.

Table 2.10.   Corrosion Inhibitor Solution. (Source: U. S. Department of Commerce[55]).

Concentration	Optimum Percent by Weight
Mercaptobenzothiazole (technical grade, 92% min)	15.1
Sodium borate decahydrate $Na_2B_4O_7 \cdot 10\ H_2O$ (borax)	75.7
Anhydrous disodium phosphate $Na_2H\ PO_4$	9.2
	100.0

Table 2.10 illustrates a corrosion inhibitor solution that can be added to water at a concentration of 1.5% by weight. This solution will give the water a pH of between 7.5 and 8.0. Table 2.11 shows various types of fluids that may be used for collectors. Note the different corrosion characteristics.

Table 2.11.   Collector Coolant Characteristics. (Source: U. S. Department of Energy[56]).

Characteristic	Air	Water	85% Ethylene Glycol & Water	50% Propylene Glycol % Water	Thermia 15 Paraffinic Oil	UCON (Polyglycol) 50-HB-280-X	Dowtherm J	Therminol 60
Freezing Point	---	32°F	-37°F	-26°F	---	---	-100°F	-90°F
Pour Point	---	---	---	---	10°F	-35°F	---	---
Boiling Point (at atm pressure)	---	212°F	265°F	369°F	700°F	600°F	358°F	---
Corrosion	Noncorrosive	Corrosive to Fe or AL Requires inhibitors.			Noncorrosive	Noncorrosive	Noncorrosive	Noncorrosive
Fluid Stability	---	Requires pH or inhibitor monitoring.			Good[a]	Good[b]	---	---
Flash Point	---	None	None	215°F	455°F	500°F	145°F	310°F
Bulk Cost ($/gal) (December 1975)	---	---	2.35	2.50	1.00	4.40	4.00	4.00
Thermal Conductivity (Btu/hr-ft°F at 100°F)	0.0154	0.359	0.17	0.22	0.76	0.119	---	---
Heat Capacity (Btu/lb-°F at 100°F)	0.24	1.0	0.66	0.87	0.46	0.45	0.47	0.42
Viscosity (lb/ft-hr at 100°F)	0.04626	1.66	15.7	7.5	28.5	143.1	0.7	0.59

[a]Requires an isolated cold expansion tank or nitrogen-containing hot expansion tank to prevent sludge formation.
[b]Contains a sludge formation inhibitor.

2.2 AIR FLAT PLATE COLLECTORS

2.2.1. Collector Design

In the 1940s and 1950s solar air-type flat plate collectors were being developed and installed in buildings by engineer Dr. George Löf, architect-engineering associates Dr. Maria Telkes and Eleanor Raymond, and husband and wife architects Raymond Bliss and Mary Donovan.

The air-type collector can be used for both space heating and water heating applications. (For example, the air-type system marketed by Solaron Corporation.) The main advantage of air-type collectors as opposed to liquid-type collectors is that they are free from the freezing and corrosion problems inherent in liquid-type systems. When using the air system solely for heating, the heated air can be pumped directly into the rooms, thus avoiding heat transfer inefficiencies when liquid collectors are used. (With the liquid collectors, a heat exchanger must be used to transfer heat from the liquid to the air which is then blown into the rooms of the building to be heated.) The disadvantage of the air-type collector is that the air cannot hold as much heat energy as water. Therefore, to transfer the same amount of heat as the liquid system, the air system must use large air ducts to transfer the heat energy collected in the solar collector back into the house. Air-type systems also use larger horsepower motors, than those used for liquid-type systems, to transport the heated air to the storage bin.

Basically, the air collector design is similar to the liquid flat plate collector design. The air-type collector consists of:

1. Cover plate
2. Air space
3. Absorber plate
4. Insulation
5. Frame

1. Cover Plate. In the air-type flat plate collector the main purpose of the cover plate is to act as a seal to reduce air leakage.

2. Air Space. The air spaces found between the cover plate and absorber and/ or between the absorber plate and the back of the collector serve as air ducts to carry the unheated air through the collector to be heated. Figure 2.15 illustrates four kinds of air-type collectors. Note the airflow for each diagram. In diagrams K and N the airflow is between the absorber and the insulation, while in diagrams L and M it is through the absorber material.

3. Absorber Plate. The absorber plate in the air-type collector is used to collect solar energy which is then transferred to the air flowing through the collector. Since air is blown over the entire absorber to pick up collected heat, there is no need for the absorber material to be metal. Items which can be used for absorber material include metal scraps attached to plywood, aluminum beer cans, crushed rocks or glass—the list is endless. The main criteria are

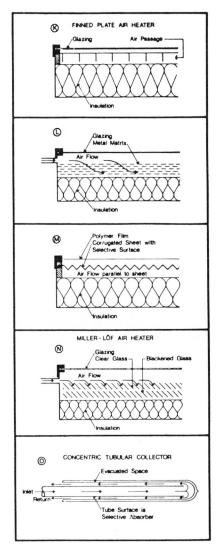

Figure 2.15. Air-Type Solar Collectors. Reprinted with permission from the 1978 Solar Energy Utilization for Heating and Cooling 58, ASHRAE HANDBOOK and Product Directory.

that the whole absorber be black, be heated directly by the sun, and come in contact with the air circulated through the collector. Although any material may be used for the absorber, metal is preferred for collectors in which the sun cannot shine on all of the absorber surface reached by the circulating air. This is because with a completely metal absorber the areas not heated by the sun will still receive heat through conductance by the rest of the metal that *does* directly receive the sun's rays. If the absorber were not metal, the areas not directly receiving the sun's rays would not make any contribution to heating up the air. If a selective surface is warranted, metal will be the choice of the absorber material since the application of a selective coating to non-metallic materials is extremely difficult.

In designing the absorber plate, one should strive to create turbulence in the

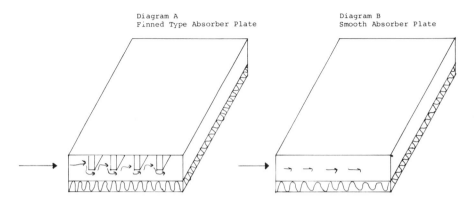

Figure 2.16. Effect of Absorber Design on Air Flow Turbulence.

air flowing through the collector. Turbulent air flow will result in better heat transfer from the absorber to the air flowing through the collector. The degree of turbulence created is affected by the surface composition of the absorber material and the absorber design. Making the texture of the absorber coarse and rough increases turbulence. Figure 2.16 illustrates the effect absorber shape has on turbulence. Like an automobile, the sleeker the shape, the less air drag and turbulence. For the finned absorber plate design shown in Figs. 2.15 and 2.16, the air currents are forced to take nonlinear paths in reaching the outlet passage of the collector. When the air is broken up in this fashion, air velocities and direction become erratic, causing turbulent air flow.

By restricting the air flow to create turbulence, a drop in pressure occurs across the collector. To compensate for the drop in pressure, a larger fan must be used. The additional electricity used to power a larger fan may negate any thermal efficiency gains realized by *increased* disruption of air flow through the collector.

4. Insulation. Insulation is as necessary in the air-type collector as it is in the liquid-type collector. Insulation is generally applied to the bottom and sides of the collector to prevent heat loss.

5. Frame. As in the liquid-type collector, the frame also serves the same purpose in the air-type collector; it holds all of the workings of the collector in place (e.g., absorber plate, cover plate, etc.). The frame should be fairly airtight to prevent collector air losses as air is forced through the collector under pressure.

Figures 2.17 and 2.18 illustrate some air-type collector designs. Figure 2.17 represents a simple design which can easily be constructed and placed in a window. Instead of using a fan, convection currents created by the heated air cause circulation to occur in the collector. Figure 2.18 represents a general design used for large roof-top collectors. Cold air is forced through the collector and as it travels around the absorber plate it is heated and exits the collector to be routed inside the building.

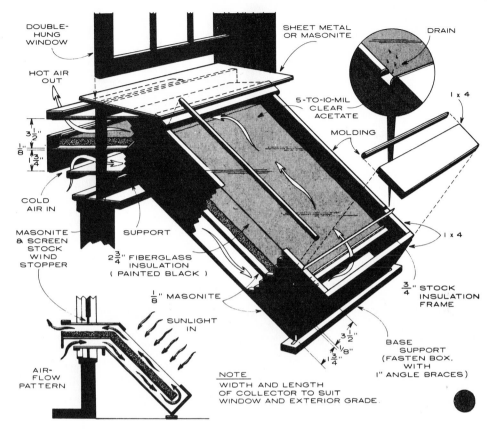

**Figure 2.17. Window Type Air Collector. Reprinted from Popular Science with permission ©
1980, Times Mirror Magazines, Inc.**

2.2.2. Collector Controls

The basic collector control is a thermostat that controls a fan which regulates
the flow of air through the collector. As in the liquid-type collector, the air
collector generally remains in a fixed position and therefore does not use a
mechanical device to make the collectors follow the sun as the sun moves
across the sky. If the air flow is interrupted for a period of time due to a
malfunction, the possible harmful results due to high stagnant temperatures
created will not generally be as severe as in the case of the liquid type flat
plate collector.

2.2.3. Maintenance Factors

Since air is used as the transfer medium rather than water in the case of the
liquid-type collector, the problems of corrosion and freezing are eliminated as
potential maintenance problems. Dr. George Löf built an air-type system that
has been in continuous operation since 1957. Solaron Corporation, a manufacturer of air-type collectors, says that their solar air heating system is expected

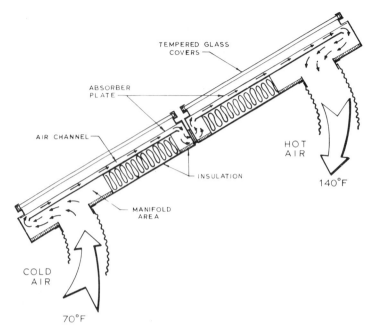

Figure 2.18. Air Type Flat Plate Collector. (Source: Solaron Corporation[50]).

to last as long as the building itself. They stress that very little maintenance is required other than changing filters, lubricating the motor, and checking the fan belt.

2.3 EVACUATED TUBE COLLECTORS

2.3.1. Collector Design

The evacuated tube collector offers a means of providing high temperature operation with high efficiency. There are numerous vendors on the market with evacuated tube collectors. Among them are General Electric with their Solartron collector and Owens-Illinois which manufactures the Sunpak Collector. Figures 2.19 and 2.20 illustrate the designs of the Solartron and Sunpak Collectors, respectively. Figure 2.21 shows the General Electric Solartron

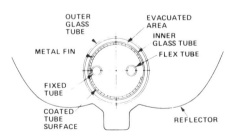

vacuum tube and collector

Figure 2.19. GE Solartron Evacuated Tube Collector. (Source: General Electric Co.[21]).

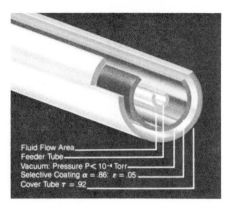

Fluid Flow Area
Feeder Tube
Vacuum: Pressure $P < 10^{-4}$ Torr
Selective Coating $\alpha = .86$; $\varepsilon = .05$
Cover Tube $\tau = .92$

Figure 2.20. Owens-Illinois Sunpak Evacuated Tube Collector. (Source: Owens-Illinois, Inc.[43]).

Collector. Note the two fluid connection taps. Table 2.12 illustrates the characteristics of the General Electric Solartron Collector. Figure 2.22 illustrates the Owens-Illinois Sunpak Collector. These collectors can use either air or liquid as the transfer fluid. General characteristics of the Owens-Illinois Sunpak Collector are shown in Table 2.13.

The evacuated tube collectors are nonfocusing and will function using both beam and diffuse solar radiation. Figure 2.23 illustrates the collection method used by the evacuated tube collectors.

2.3.2. Collector Controls

The only control device affecting the collector is one which regulates fluid flow through the collector tubes. Since the collectors are nonfocusing, no

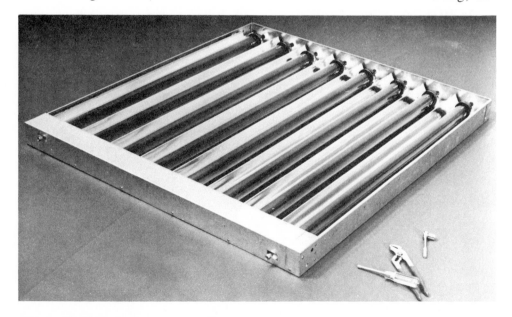

Figure 2.21. GE Solartron Evacuated Tube Collector. (Solartron® TC-100 vacuum tube collector. Photo courtesy GE Space Systems Division. ® registered Trademark of the General Electric Company).

Table 2.12. Characteristics of the General Electric Solartron Collector. (Source: General Electric Co.[21,22]).

Weight (dry)	57 lb
Module Installation Area	17.4 ft^2
Module Collecting Aperture	14.8 ft^2
Glass Tubes	008 sodalime
Absorber Coating Absorptance (α)	>0.85
Emittance (ε)	<0.05 at 212°F
Fluid Operating Temperature[A,B]	100 to 300°F
Minimum Pressure Drop	5 PSI
Recommended Heat Transfer Fluid	Water with 35 to 50% (by volume) Prestone II ^{TM}
Insulation	Fiberglass
Vacuum Pressure	P < 10^{-4} torr
Design Flowrate	0.22 gpm at 180°F
Wind Velocity (max)	100 MPH
Ice Load (max)	13 PSF
Snow Load (max)	20 PSF
Combined Load (max)	33 PSF
Frame Composition	18 Ga. Aluminized Steel

[A]Operating temperature should be limited to 300°F to prevent decomposition of the ethylene glycol.

[B]Claimed collector will survive inadvertant stagnation conditions.

Figure 2.22. Owens-Illinois Sunpak Evacuated Tube Collector. (Source: Owens-Illinois, Inc.[43]).

**Table 2.13. Characteristics of the Owens-Illinois Sunpak Collector.
(Source: Owens-Illinois, Inc.[43]).**

			Standard Collector	Drainable Collector
Weight (dry)			100 lb	50 lb
Module Installation Area			32 ft^2	16 ft^2
Module Collecting Aperture			27.4 ft^2	14 ft^2
Glass Tubes			Kimble KG-33	Borosilicate Glass
Absorber Coating	Absorptance	(α)	0.86	
	Emmittance	(ϵ)	0.05	
	Transmittance	(τ)	0.92	
Operating Conditions Temperature Range ^{a}			-40°F to 240°F	
Maximum Outlet Pressure			30 PSIG	
Minimum Pressure Drop			5 PSI	
Recommended Heat Transfer Fluid			Water	
Manifold Insulation^b			Polyurethane Foam	
Vacuum Pressure			$P < 10^{-4}$ Torr	
Module Hardware			Anodized Aluminum	

^{a}Claimed coatings and tubes can stand stagnation temperatures
of 650°F indefinitely.

^{b}2 lbs/ft^3 density, k = 0.012 Btu/hr ft°F.

controller is needed to reorient the collectors with changes in the sun's position
in the sky.

2.3.3. Maintenance Factors

Owens-Illinois recommends that for most locations in the United States, water
without antifreeze protectants be used in their Sunpak system. The fluid
passageways for the Sunpak Collector are glass, while the material used to
connect each collector tube together in the manifold is hard drawn type M
copper tubing.

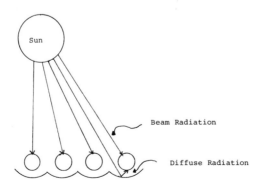

Figure 2.23. Solar Energy Collection of Evacuated Tube Collector.

For their Solartron system, General Electric advises using a mixture of 35 to 50% (by volume) Prestone II™ ethylene glycol with distilled, deionized, or demineralized water. They also recommend that the water used should contain no more than the following:

Chloride	100 PPM
Sulfates	100 PPM
Bi-carbonates	100 PPM
Total hardness	250 PPM

Since an antifreeze solution is recommended for the Solartron collector, General Electric further advises that operating temperatures be limited to 300°F to prevent decomposition of the ethylene glycol. In the Solartron collector, copper tubing is used both in the collector itself and in the header piping (used to connect each evacuated tube) for circulating the ethylene glycol solution. General Electric recommends using type L copper tubing and wrought copper fittings throughout the entire collector loop of their Solartron collector system to minimize galvanic reactions.

For areas of high vandalism, collector covers can be used, such as the collector window kits sold by General Electric. These collector window kits can be purchased either in ultraviolet-stabilized Lexan® or acrylic. General Electric recommends their kit *only* when collector protection is absolutely necessary. (When a protective cover is used, the solar radiation received by the evacuated tube collectors will be less.)

2.4 CONCENTRATING COLLECTORS

2.4.1. Collector Design

A concentrating collector functions by collecting incoming solar radiation and focusing (reflecting) it onto an area smaller than the parabolic reflector area of the collector. By focusing the incoming solar radiation on the absorber tube of a concentrating collector, such as the type shown in Fig. 2.24, high temperatures are achieved.

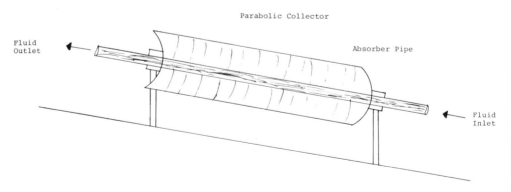

Figure 2.24. Concentrating Solar Collector.

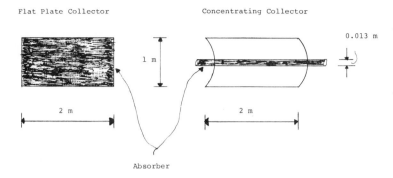

Figure 2.25. Flat Plate Collector versus Concentrating Collector.

The following example helps to illustrate the operating characteristics of the concentrating collector compared to a flat plate collector. Although the magnitude of solar radiation reaching the collector absorber surfaces will be less due to optical losses for the cover plate and parabolic reflector, for the sake of simplicity, these losses and variations will not be considered.

The two collectors in the example are illustrated in Fig. 2.25. The collection area is assumed to be the same for both collectors, 2.0 m², and the solar intensity is 945 W/m². (The collection area is the product of length times width for the flat plate collector and for the concentrating collector it is length times the aperture width opening.) Therefore the total energy falling on both collectors is 1890 W.

$$(945 \text{ W/m}^2) \ (2.0 \text{ m}^2) = 1890 \text{ W} \tag{2.11}$$

Since the flat plate collector uses an absorber having a dimension of 2.0 × 1.0 m, the solar intensity on the absorber plate will be 945 W/m².

$$1890 \text{ W/2.0 m}^2 = 945 \text{ W/m}^2 \tag{2.12}$$

For the concentrating collector having absorber dimensions of 2.0 × 0.013 m, the absorber area is 0.026 m². The solar intensity on the concentrating collector absorber tube is 72,692 W/m².

$$1890 \text{ W/0.26 m}^2 = 72,692 \text{ W/m}^2 \tag{2.13}$$

Therefore, the intensity of solar energy falling on the concentrating collector absorber tube is about 77 times greater than the intensity falling on the flat plate collector absorber plate shown in Fig. 2.26.

Notice that in the example both collector areas are identical (2.0 m²) and each receives the same amount of energy (1890 W). The difference between the two collectors is how they utilize the collected energy. For the flat plate collector, the collected energy is spread over the whole absorber area (2.0 m²), whereas for the concentrating collector the collected energy is focused on an absorber surface of only 0.026 m². The concentrating collector does not

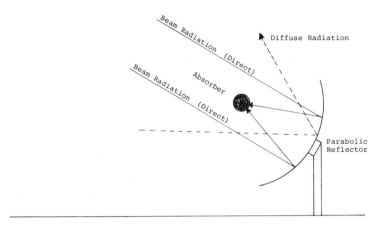

Figure 2.26. Direct and Diffuse Component of Solar Energy on Concentrating Collectors.

receive any more solar radiation than the flat plate collector of comparable size.

The higher energy concentration of the concentrating collector raises the temperature of the absorber which results in greater collector fluid temperatures. Although thermal losses increase with absorber temperatures, the effects of the smaller absorber area of the concentrating collector in decreasing thermal losses more than offsets the increased losses due to higher operating temperatures. Therefore, the efficiency will be higher for the concentrating collector as opposed to the flat plate collector when both are operated at high temperatures.

Unlike the flat plate collector which utilizes both beam and diffuse solar radiation, focusing (concentrating) collectors utilize only direct normal (beam) radiation due to optical constraints of the parabolic reflector. Therefore, the concentrating collector must utilize some type of tracking mechanism to follow the sun as it moves across the sky. This adds to the costs of the collector system. Figure 2.26 illustrates the effects of direct and diffuse radiation on a solar concentrating collector.

A variation of the parabolic concentrating collector is shown in Fig. 2.27.

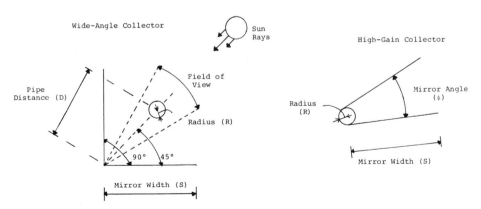

Figure 2.27. Concentrating Collectors Using Flat Mirrors.

Table 2.14. Wide Angle Collector Design Parameters.

Field of View	D, Pipe Distance	S, Mirror Width	Theta, Central Shadow
50.00	4.13 R	10.96 R	27.19
51.00	4.24 R	11.46 R	26.56
52.00	4.34 R	11.99 R	25.93
53.00	4.46 R	12.57 R	25.29
54.00	4.58 R	13.20 R	24.65
55.00	4.70 R	13.87 R	24.01
56.00	4.84 R	14.61 R	23.36
57.00	4.98 R	15.41 R	22.71
58.00	5.13 R	16.28 R	22.06
59.00	5.29 R	17.24 R	21.40
60.00	5.46 R	18.28 R	20.74
61.00	5.65 R	19.44 R	20.08
62.00	5.85 R	20.71 R	19.41
63.00	6.06 R	22.13 R	18.75
64.00	6.29 R	23.70 R	18.08
65.00	6.53 R	25.46 R	17.40
66.00	6.80 R	27.44 R	16.73
67.00	7.09 R	29.67 R	16.05
68.00	7.41 R	32.20 R	15.37
69.00	7.76 R	35.09 R	14.69
70.00	8.14 R	38.42 R	14.00
71.00	8.57 R	42.27 R	13.31
72.00	9.04 R	46.75 R	12.62
73.00	9.57 R	52.03 R	11.93
74.00	10.16 R	58.31 R	11.24
75.00	10.83 R	65.85 R	10.55
76.00	11.60 R	75.03 R	9.85
77.00	12.49 R	86.37 R	9.15
78.00	13.53 R	100.59 R	8.45
79.00	14.76 R	118.79 R	7.75
80.00	16.23 R	142.62 R	7.05

By using flat mirrors, two types of concentrating collectors can be built. The variables used are self-explanatory. Tables 2.14 and 2.15 are used for building the two concentrating collectors in Fig. 2.27. (These tables were generated by this writer using equations from Miller, Reference 34.)

In Table 2.14 the field of view represents the angle that the direct rays of sunlight must fall within to be effectively reflected onto the absorber tube. The Theta Central Shadow refers to the shadow angle caused by the absorber pipe. In Table 2.15 optical gains are shown for two mirrors having reflectivities (R)

Table 2.15. High-Gain Collector Design Parameters.

Phi, Mirror Angle	S, Mirror Angle	G, Optical Gain: R=0.98	G, Optical Gain: R=0.90
30.00	11.06 R	3.58	3.04
29.00	11.96 R	3.71	3.13
28.00	12.95 R	3.84	3.23
27.00	14.07 R	3.99	3.34
26.00	15.32 R	4.14	3.45
25.00	16.73 R	4.31	3.57
24.00	18.32 R	4.49	3.70
23.00	20.14 R	4.68	3.84
22.00	22.23 R	4.89	3.99
21.00	24.62 R	5.12	4.15
20.00	27.40 R	5.37	4.32
19.00	30.65 R	5.64	4.50
18.00	34.47 R	5.95	4.71
17.00	39.01 R	6.28	4.93
16.00	44.44 R	6.66	5.17
15.00	51.03 R	7.09	5.44
14.00	59.12 R	7.57	5.73
13.00	69.20 R	8.12	6.07
12.00	81.96 R	8.75	6.44
11.00	98.42 R	9.49	6.87
10.00	120.17 R	10.37	7.36
9.00	149.70 R	11.42	7.93
8.00	191.17 R	12.70	8.62
7.00	251.94 R	14.32	9.47
6.00	345.98 R	16.40	10.55
5.00	502.66 R	19.18	12.03

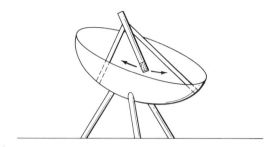

Figure 2.28. SRTA Solar Collector.

of 0.98 and 0.90. For further information on the two collector designs shown in Fig. 2.27 the reader should refer to Miller.

The two most promising concentrating collector systems in solar heating and cooling are the SRTA (stationary reflector/tracking absorber) and the CPC (compound parabolic concentrator). Basically the SRTA uses a large parabolic mirror which remains stationary but has a small cylindrical absorber which moves instead. By moving the absorber rather than the whole collector as in most concentrating devices, a relatively simple and inexpensive tracking system may be used. Figure 2.28 illustrates the SRTA collector and Fig. 2.29 shows the principles of operation of the system. By moving only the absorber to follow the sun's movement, the SRTA collector can still effectively collect solar energy even though the absorber may be at the extreme edge of the parabolic mirror. The compound parabolic concentrator (CPC) is another type of nontracking solar collector. It consists of a parabolic reflecting surface which channels the incoming solar radiation onto the absorber tube. Figure 2.30 illustrates the CPC. The CPC can operate even on cloudy days by collecting both direct and diffuse solar rays. Note that the CPC has a glass

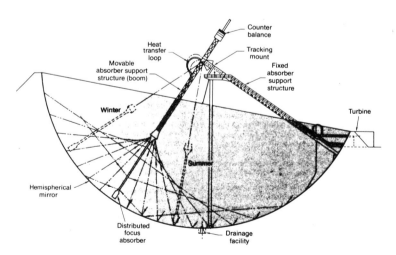

Figure 2.29. Operation of the SRTA Solar Collector. (Source: Office of Technology Assessment[42]).

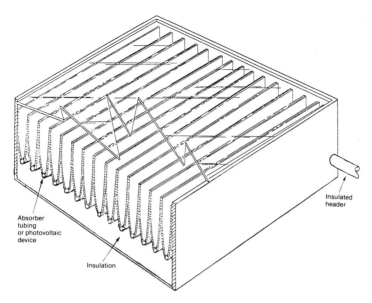

Figure 2.30. One-Axis Compound Parabolic Collector. (Source: Office of Technology Assessment[42]).

cover plate like a flat plate collector. The same types of absorption and reflection losses from glass cover plates hold true for the CPC.

2.4.2. Collector Controls

The main collector control adjusts the tracking of the collector (or just the absorber in the case of the SRTA). The other collector control is the thermal sensing device which regulates fluid flow through the absorber.

2.4.3. Maintenance Factors

With a concentrating collector which employs a metallic reflector open to the atmosphere, deterioration of the reflecting material is a problem which may be caused by one or more of the following: dust in the atmosphere settling on the reflector, smog which may chemically react with the metallic surface, and the general oxidation effects of air and water. Table 2.16 gives some selected data on various reflector materials.

2.5 COLLECTOR EFFICIENCY

Collector efficiency is an important aspect of solar system design. By knowing the efficiency for various collectors, the designer can determine which type will function best at the temperature required.

The efficiency (n) of a solar collector is the ratio of solar energy removed by the collector (Qu) to the solar energy falling on the collector (i). The useful heat (Qu) delivered by the solar collector can be expressed in equation form as modeled by Hottel and Whillier.[54,65]

Table 2.16. Concentrating Collector Reflector Materials. (Source: Kreider[29]).

Type of Material	Original reflectance	Exposure time, weeks	Reflectance		
			Uncleaned surface	Cleaned surface	Degradation %
Aluminized fiber glass (General Dynamics)	0.92	55	0.79	0.82	11
Aluminized acrylic (3M Company)	0.86	50	0.83	0.85	1
Aluminum plexi-glass (Ram Products)	0.80	39	0.71	0.78	3
Anodized aluminum (Alcoa Aluminum)	0.82	34[a]	--	0.79	4

[a] Accelerated radiation test was used.

$$Qu = \dot{m}Cp\,(T_o - T_i) \qquad (2.14)$$

where

Qu = useful energy removed (delivered) by collector (W, Btu/hr)

$\dot{m}$ = fluid flow rate (kg/s, lb m/hr)

CP = Specific heat of the fluid (J/kg°C, Btu/lb m°F)

T_o = Outlet fluid temperature (°C, °F)

T_i = Inlet fluid temperature (°C, °F)

Under steady conditions, the useful heat (Qu) delivered by the solar collector is basically equal to the solar energy absorbed by the collector absorber plate minus the heat loss from the absorber plate. This can also be expressed in equation form.

$$Qu = A[I\,\tau\alpha - UL\,(\bar{T}p - Ta)] \qquad (2.15)$$

where

A = collector area (m², ft²)

I = incident solar energy reaching the collector (W/m², Btu/hr ft²)

τ = Cover plate transmittance or fraction of solar radiation reaching the absorber plate (no dimensions)

α = Fraction of solar energy absorbed by the absorber plate, absorptivity (no dimensions)

UL = Overall collector heat loss coefficient (W/°C, Btu/hr°F)

$\bar{T}p$ = Average temperature of the upper absorber plate (°C, °F)

Ta = Atmospheric (ambient) temperature (°C, °F)

In many cases it is necessary to relate Qu to the inlet (T_i) and ambient (T_a) temperatures of the collector. This involves replacing $\bar{T}_p$ in Eq. 2.15 with T_i. For the typical liquid solar collector, the average plate temperatures ($\bar{T}p$) are 10 to 20° higher than the inlet liquid temperature (Ti). For the air type collector this temperature difference is 30 to 50°. Since $\bar{T}_p$ does not directly equal T_i, a correction factor (F_R) must be used.

$$Qu = F_R A[I\tau\alpha - UL (T_i - T_a)] \qquad (2.16)$$

where

F_R = Heat removal (recovery) factor having
a value between 0 and 1.0

The heat removal factor, F_R, is essentially a ratio of the heat actually removed at a given collector plate temperature and the heat that would be removed if the collector plate were at the same temperature as the inlet fluid (T_i). Therefore, if the collector absorber plate temperature was equal to the inlet fluid temperature, then F_R would be 1.0.

Since collector efficiency (n) is the ratio of solar energy removed by the collector (Qu) to the solar energy falling on the collector (I), Eq. 2.16 can be divided by I and A to obtain Eq. 2.17.

$$\text{Collector Efficiency } (n) = Qu/IA = F_R \tau\alpha - F_R U_L (T_i - T_a)/I \quad (2.17)$$

For a given collector that is working at a constant fluid flow rate ($\dot{m}$), the values of A, F_R, τ, α, and U_L will remain nearly constant regardless of temperature and solar conditions. Therefore, a graph can be constructed to show instantaneous collector efficiencies at varying operating conditions as shown in Fig. 2.31. When a collector's instantaneous efficiency is plotted as a graph, the values of $F_R\tau\alpha$ and $F_R U_L$ can be determined. These two values are necessary for the solar system design methodology presented in Chapter 7. The y intercept (b) corresponds to the value of $F_R\tau\alpha$ and the slope (m) is equal to $-F_R U_L$. In determining collector efficiency numerous data points are collected and then plotted as shown in Fig. 2.31. By using a simple least squares linear regression, the data can be fitted to a straight line (or first order equation).

Generally efficiency graphs can be obtained from the collector manufacturer. One good source of liquid flat plate collector efficiency data is the Florida Solar Energy Center. For a fee, they test solar collectors submitted by collector manufacturers. If the collector passes, the manufacturer receives collector rating labels which can be attached to other collectors that are the same model as was tested. Table 2.17 shows a typical test results summary sheet. Notice the first and second order efficiency equations. The first order equation corresponds to Fig. 2.31 in that it yields a straight line plot. The second order

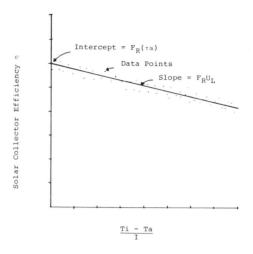

Figure 2.31. Graph of Instantaneous Collector Efficiencies.

equation is a refinement over the first order efficiency equation. The collector efficiency equations in Table 2.17 are based on solar radiation falling at normal incidence to the collector cover plate. The incident angle modifier is used to adjust the efficiency curves for changes in the sun's incident angle. For example, in Table 2.17 the first order collector efficiency at normal incidence is:

$$n = 69.81 - 690.93 \ (T_i - T_a)/I \qquad (2.18)$$

If the collector efficiency is needed for a 45° incident angle, then the incident angle modifier is used to adjust the efficiency equations in Table 2.17. The incident angle modifier $(K\tau\alpha)$ for 45° is 0.96. This number is then multiplied by the first term in each of the two efficiency equations to adjust them to the 45° incident angle. The new first order equation would then become:

$$n = 67.02 - 690.93 \ (T_i - T_a)/I \qquad (2.19)$$

Figure 2.32 illustrates the graph of the solar collector described in Table 2.17. Graph A of Fig. 2.32 is the plot of the first order equation at normal incidence while Graph B is for the modified first order equation at a 45° incidence. Note that in Fig. 2.32 at a given value for $(T_i - T_a)/I$ the efficiency for the collector receiving sun at a normal incidence will be higher than the one receiving solar energy at a 45° incidence angle.

To gain a feel for collector efficiency graphs, an example is presented. The solar intensity (I) reaching the collector in the example is assumed to be 950 W/m². Using Fig. 2.34, the x intercept occurring when efficiency (n) equals zero means that:

$$(T_i - T_a)/I = 0.10104 \qquad (2.20)$$

Table 2.17.   Solar Collector Test Data Sheet. (Source: Florida Solar Energy Center[15,16]).

SUMMARY INFORMATION SHEET

FLORIDA SOLAR ENERGY CENTER
300 STATE ROAD 401, CAPE CANAVERAL, FLORIDA 32920, (305) 783-0300

FSEC #78048A

MANUFACTURER	
Grumman Energy Systems, Inc. 4175 Veteran's Memorial Highway Ronkonkoma, New York 11779	Collector Model 100 F

This solar collector was evaluated by the Florida Solar Energy Center (FSEC) in accordance with prescribed methods and was found to meet the minimum standards established by
FSEC. This evaluation was based on solar collector tests performed at Approved Engineer
ing Test Laboratories, Sauqus, California. The purpose of the tests is to verify initia
performance conditions and quality of construction only. The resulting certification is
not a guarantee of long term performance or durability.

DESCRIPTION

Gross Length	2.962 meters	9.06 feet
Gross Width	0.933 meters	3.06 feet
Gross Depth	0.140 meters	0.46 feet
Gross Area	2.578 square meters	27.75 square feet
Transparent Frontal Area	2.190 square meters	23.57 square feet
Volumetric Capacity	1.59 liters	0.42 gallons
Weight (empty)	78.9 kilograms	174.0 pounds
Number of Cover Plates	One	
Flow Pattern	Parallel	

Incident Angle Modifier $K\tau a = 1.0 - 0.09 \ (1/\cos\theta - 1)$

Efficiency Equations   First Order $\eta = 69.81 - 690.93 \ (Ti-Ta)/I$

Second Order $\eta = 69.40 - 609.86 \ (Ti-Ta)/I - 1422.69 \left[(Ti-Ta)/I\right]^2$

Tested per ASHRAE 93-77   Units of Ti-Ta/I are $^{\circ}C/Watt/m^2$

MATERIALS

Enclosure	Aluminum extrusions alloy 6063- T5 with aluminum sheet alloy 3003-N14 bottom
Glazing	Water white tempered glass
Absorber	Aluminum extrusions with copper tube, water passages, PPG Duracron Super 600 L/G black acrylic paint
Insulation	Fiberglass 7.62 cm thick

RATING

The collector has been rated for energy output on measured performance and an assumed standard day. Total solar energy
available for the standard day is 5045 watt-hour/m^2 (1600 BTU/ft^2) distributed over a 10 hour period.

Output energy ratings for this collector based on the second-order efficiency curve are:

Collector Temperature	Energy Output			
Low Temperature, 35°C (95°F)	28,800	Kilojoules/day	27,300	BTU/day
Intermediate Temperature, 50°C (122°F)	20,600	Kilojoules/day	19,600	BTU/day
High Temperature, 100°C (212°F)	2,400	Kilojoules/day	2,300	BTU/day

Therefore:

$$T_i - T_a = (950 \text{ W/m}^2) \ (0.10104°C \text{ m}^2/\text{W}) \tag{2.21}$$

$$T_i = 96°C + T_A \tag{2.22}$$

Using Eq. 2.22, T_i can be calculated when collector efficiency equals zero
such as occurs under stagnation conditions when collector fluid flow ceases for
a period of time.

If the ambient temperature (T_a) in a winter month is 2°C (35°F), then from
Eq. 2.21, T_i equals 98°C (208°F). During a summer month when the outside

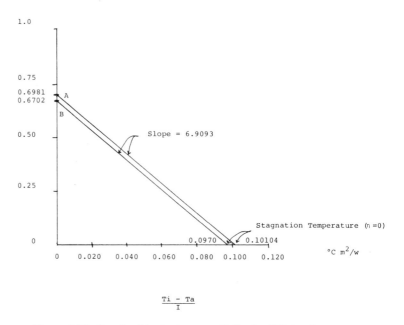

$$\frac{Ti - Ta}{I}$$

Figure 2.32. Graph of Instantaneous Collector Efficiencies.

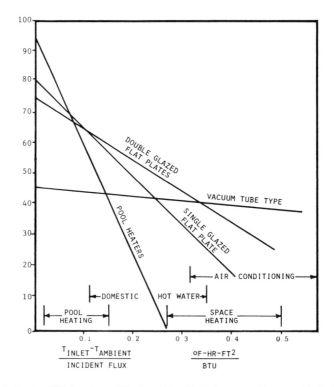

Figure 2.33. Collector Efficiencies of Various Liquid Collectors. (Source: U. S. Department of Energy[58]).

temperature is 32°C (90°F), T_i would then equal 128°C (262°F). By estimating what the maximum solar intensity and ambient temperatures could be at a particular time of the year, one can get a general idea of the temperature the solar collector can reach if stagnation were to occur at this time. Realize that the collector in the previous example is the one described in Table 2.17 and that it only has one collector cover and uses black acrylic paint. By using additional collector covers and a selective absorber coating such as black chrome, the stagnation temperatures would be much higher.

The abscissa (x axis) of the graph of the collector efficiency can be used to give an indication of how a collector will perform at a given operating range. Figure 2.33 illustrates this concept:

When comparing graphs of various collectors, the intended use and operating temperature range must be a prime consideration. Therefore, when comparing various types and brands of collectors, the efficiencies of the collectors should be determined based on operating temperature requirements. For example, in Fig. 2.33 the "highest" collector efficiency is found in the pool heater collectors which have no cover plates. If the collector is to be used for solar (absorption chiller) air conditioning, it would be a mistake to pick the pool heater over the vacuum tube type collector just because the pool heater collector had the highest efficiency. Further investigation would show that in Fig. 2.33 the efficiency of the pool heater collector is zero at the high temperatures required for solar air conditioning whereas the vacuum tube type collector has a relatively high efficiency. Notice that the vacuum tube type collector has the lowest slope of the other collectors shown in Fig. 2.33. This means that efficiency decreases at a lower rate as operating temperatures increase. Figure 2.34 illustrates some collector efficiencies for specific collectors. The eight efficiency curves shown are for the collectors listed in Table 2.9.

When comparing collector efficiency curves obtained from collector manufacturers or other sources such as the Florida Solar Energy Center, care must be taken to ensure that the observer is comparing oranges to oranges and apples to apples. In other words, collector efficiency curves obtained from various collector manufacturers may be based on different collector efficiency test conditions. There has been an attempt to standardize collector testing by the introduction of a collector testing procedure devised by the American Society of Heating, Refrigerating, and Air Conditioning Engineers (ASHRAE 93-77). This testing procedure provides specific test methods for determining collector efficiency. It should be noted that the National Bureau of Standards also has a collector testing procedure (NBSIR 74-635). The NBSIR 74-635 test sometimes yields different results from the ASHRAE 93-77 procedure.

Figures 2.35 and 2.36 show operating conditions for various types of collectors. Note that both graphs are for the same type collectors. The discrepancies in efficiencies for specific collector types is due to differences in how the collector efficiencies were determined. In Fig. 2.35 the collector efficiency is based on collective aperture area (or exposed collector plate area), while in

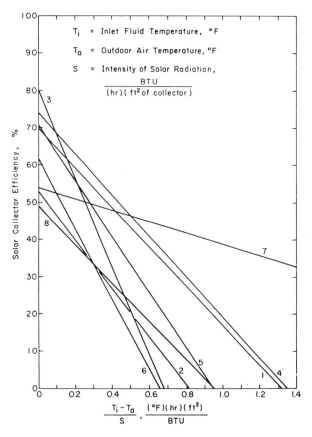

Figure 2.34. Eight Collector Efficiency Curves. (Source: U. S. Department of Commerce[55]).

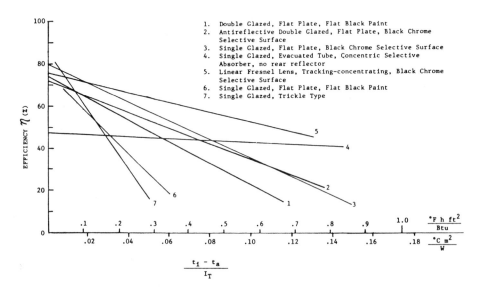

Figure 2.35. Typical Thermal Efficiency Curves for Liquid Collectors Based on Collector Aperture Area. (Source: U. S. Department of Housing and Urban Development[60]).

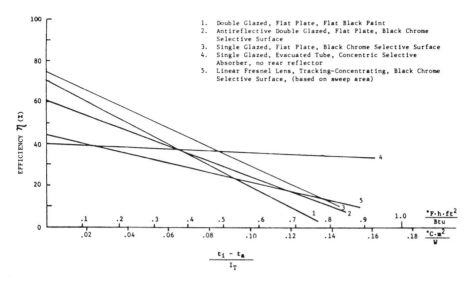

Figure 2.36. Typical Thermal Efficiency Curves for Liquid Collectors Based on Collector Gross Frontal Area. (Source: U. S. Department of Housing and Urban Development[60]).

Fig. 2.36 it is based on gross frontal area. Therefore, if one collector manufacturer based collector efficiency on aperture area and another manufacturer having essentially the same type collector based collector efficiency on gross frontal area, comparing the two collector efficiencies could be difficult and misleading. The abscissa and ordinates of separate manufacturer collector efficiency graphs should also be compared. Naturally, if there are differences, some modifications are necessary to make direct comparisons. For example, if two collector manufacturers had collector efficiency graphs having the same y axis corresponding to efficiency (n) but one manufacturer had an x axis showing $(T_i - T_a)/I$ and another had an x axis based on average all day efficiency using average incident solar radiation $(T_i - T_a)/I$ Avg., no *direct* comparison could be made. The angle of incident solar radiation to the collector during the tests should also be determined. In most cases collector testing is done with solar collectors directly oriented normal to the sun's rays. This will give higher collector efficiencies than if the incident solar radiation was received at other incidence angles (see Fig. 2.32). For additional factors to consider in relating different manufacturers collector efficiency ratings, the reader should consult the ASHRAE 93-77 test procedures publication.

There are times that collector efficiency curves will be based on graph abscissas other than $(T_i - T_a)/I$. Air type collectors in particular may base collector efficiencies on $\{[((T_i - T_o)/2] - T_a\}/I$ or $(T_o - T_a)/I$. For these cases, determining the values for $F_R \tau \alpha$ and $F_R U_L$ involve more than finding the y intercept (b) and slope ($-m$) as can be done when interpreting collector efficiency curves plotted against $(T_i - T_a)/I$.

When a collector efficiency curve utilizes the average of inlet and outlet temperatures, the modifications necessary to determine $F_R \tau \alpha$ and $F_R U_L$ are:

$$F_R\tau\alpha = b[1/(1 + \dot{m}A/2Cc)] \qquad (2.23)$$

$$F_RU_L = m[1/(1 + \dot{m}A/2Cc)] \qquad (2.24)$$

where

Cc = (volumetric flow rate) (density) (time conversion) (specific heat) = collector fluid capacitance rate = $\dot{m}$ Cp, (W/°C, Btu/hr°F)

For liquids, density = (8.33 lb/gal) (specific gravity). For air, density = 0.75 lb/ft^3 at 70°F and 1 atmosphere specific heat = 0.24 Btu/lb°F.

If a collector efficiency curve has the abscissa based on $(T_o - T_a)/I$, then:

$$F_R\tau\alpha = b[1/(1 + \dot{m}A/Cc)] \qquad (2.25)$$

$$F_RU_L = m[1/(1 + \dot{m}A/Cc)] \qquad (2.26)$$

2.6 REFERENCES AND SUGGESTED READING

1. American Society of Heating, Refrigerating and Air Conditioning Engineers, *Methods of Testing to Determine the Thermal Performance of Solar Collectors,* 1977.
2. American Society of Heating, Refrigerating and Air Conditioning Engineers, *Solar Energy Utilization for Heating and Cooling,* Washington, D.C.: U. S. Government Printing Office, 1974, pp. 59.10–59.13, 59.15–59.116.
3. Anderson, Bruce, *Solar Energy-Fundamentals in Building Design,* New York: McGraw-Hill Book Company, 1977, pp. 146–214, 352–356.
4. Anderson, Bruce, and Riordan, Michael, *The Solar Home Book,* Harrisville, New Hampshire: Cheshire Books, 1976, pp. 152–173, 180–183.
5. Beckman, William A., Klein, Sanford A., and Duffie, John A., *Solar Heating Design By the F Chart Method,* New York: John Wiley & Sons, 1977, pp. 10–21, 25, 38–39.
6. Better Heating and Cooling Bureau and Sheet Metal and Air Conditioning Contractors' National Association, Inc., *Solar Installation Standards for Heating and Air Conditioning Systems,* 1977.
7. Clarkson, C. W., and Herbert, J. S., "Transparent Glazing Media for Solar Energy Collectors", paper presented at U. S. Section of International Solar Energy Society Meeting, Ft. Collins, Colorado, August 20–23, 1974.
8. Conrat, Richard, "Building a Solar Collector", *Organic Gardening and Farming,* February, 132–147, 1978.
9. Copper Development Association, Inc., *Solar Energy Systems Design Handbook,* 1977.
10. Duffie, John A., and Beckman, William A., *Solar Energy Thermal Processes,* New York: John Wiley & Sons, Inc., 1974, p. 103.
11. Edwards, D. K., *Solar Collector Design,* Philadelphia, Pennsylvania: Franklin Institute Press, 1977.
12. Energy Research and Development, Solar Program Assessment, *Environmental Factors— Solar Heating and Cooling of Buildings,* Washington, D.C.: U. S. Government Printing Office, March, 1977, pp. 31–33.
13. Environmental Information Center of the Florida Conservation Foundation, Inc., *Build Your Own Solar Water Heater,* Winter Park, Florida: Superior Minute Print, 1976, pp. 7–17, 20–21.
14. Florida Solar Energy Center, *Operation of the Collector Certification Program,* Cape Canaveral, Florida: State University System of Florida, 1979, pp. 1–7.
15. Florida Solar Energy Center, *Summary Test Package* Vol. II, No. 1, Cape Canaveral, Florida: State University System of Florida, June, 1979.

16. Florida Solar Energy Center, *Summary Test Results FSEC Certified Solar Collectors,* Cape Canaveral, Florida: State University System of Florida, February, 1979.

17. Florida Solar Energy Center, *Test Methods and Minimum Standards for Solar Collectors,* Cape Canaveral, Florida: State University System of Florida, February, 1979.

18. Federal Energy Administration, *Buying Solar,* Washington, D.C.: U. S. Government Printing Office, June, 1976, pp. 1–3.

19. Graham, B. J., "Evacuated Tube Collectors," *Solar Age,* **4,** 11:13–17, (1979).

20. Grimmer, D. P., and Moore, S. W., "Practical Aspects of Solar Heating: A Review of Materials Used in Solar Heating Applications," paper presented at Society for Advancement in Materials Process Engineering Meeting, Albuquerque, New Mexico, October 14–16, 1975.

21. General Electric, Philadelphia, Pennsylvania, *Solartron Vacuum Tube Collectors,* December, 1978.

22. General Electric, Philadelphia, Pennsylvania, *Solartron TC-100 Vacuum Tube Solar Collector-Commercial and Industrial Application Guide,* August, 1979, pp. 1–2, 5.

23. Grumman Energy Systems, Inc., Bohemia, New York, *Grumman Sunstream Solar Collectors,* 1978.

24. Hill, James E., and Kusuda, Tamami, *Method of Testing for Rating Solar Collectors Based on Thermal Performance NBSTR-74-635,* Interim Report prepared for the National Science Foundation, December 1974,.

25. International Telephone and Telegraph Corporation, Fluid Handling Division, Training and Education Department, *Solar Heating Systems Design Manual,* 1976.

26. Jordan, Richard C., and Liu, Benjamin Y. H., editors, *Applications of Solar Energy for Heating and Cooling of Buildings,* New York: American Society of Heating, Refrigerating and Air Conditioning Engineers, Inc., 1977, pp. 6.1–6.9, 8.1–8.13, 10.1–10.18.

27. Kobayashi, T., and Sargent, L., "A Survey of Breakage-Resistant Materials for Flat-Plate Solar Collector Covers," a paper presented at U. S. Section of ISES Meeting, Ft. Collins, Colorado, August 20–23, 1974.

28. Kolbe, Harry, "A Practical Solar Collector You Can Build", *Mechanix Illustrated,* **74,** 601:45, (1978).

29. Kreider, Jon F., and Kreith, Frank, *Solar Heating and Cooling: Engineering, Practical Design and Economics,* Washington, D.C.: Hemisphere Publishing Co., 1977, pp. 57–104.

30. Libby-Owens-Ford Co., Toledo, Ohio, *Designing With the LOF Sun Angle Calculator,* 1974.

31. Libby-Owens-Ford Co., Toledo, Ohio, *Glass for Construction,* January, 1979.

32. Massena, Roy P., Root, Douglass E., Starr, Stanley, O., and Walker, Robert L., *Solar Water and Pool Heating Installation and Operation,* Cape Canaveral, Florida: Florida Solar Energy Center, 1979, pp. 3.6–3.15, 4.27–4.28, 5.40, 5.58–5.60.

33. Merrill, Richard, and Gage, Thomas, *Energy Primer–Solar, Water, Wind and Biofuels,* New York: Dell Publishing Co., Inc., 1978, pp. 23–35.

34. Miller, Walter E. Sr., "Concentrating Solar Collectors You Can Build," *Popular Science,* **211,** 3:95, (1977).

35. Minnesota Mining and Manufacturing Company, St. Paul, Minnesota, *Application and Maintenance of Nextel Solar Absorber and Scotchcal Solar Reflector for Solar Collector Applications,* Instruction Bulletin No. 55, September, 1978.

36. Minnesota Mining and Manufacturing Company, St. Paul, Minnesota, *Nextel Coatings/NEX-8,* June, 1979.

37. Minnesota Mining and Manufacturing Company, St. Paul, Minnesota, *Products for Solar Collectors,* September, 1978.

38. *Modern Plastics Encyclopedia,* New York: McGraw-Hill, 1975–76.

39. Moran, Edward, "Bill Rankins and David Wilson's Solar Window", *Solar Energy Handbook,* New York: Popular Science, 1978, pp. 12–14.

40. Napholtz, Stephen G., "Solar Collector Paints—How Do They Compare?," *Popular Science,* **214,** 5:134, (1979).

41. Newton, Alwin B., "Getting the Most Out of ASHRAE 93-77", *Solar Age,* **3,** 11:26–29, (1978).

42. Office of Technology Assessment, *Application of Solar Technology to Today's Energy Needs,* Washington, D.C.: U. S. Government Printing Office, June, 1978, pp. 285–290, 295–301, 309–326.

43. Owens-Illinois, Inc., Toledo, Ohio, *A Pipeline to the Sun From Owens, Illinois, Sun Pak,* 1979.

44. Payne, Jack, and McConaghe, Tom, "Solar Water Heater for Your Vacation Home," *Solar Energy Handbook,* New York: Popular Science, 1978, pp. 72–73.

45. PPG Industries, Inc., Pittsburg, Pennsylvania, *Plans and Specs,* **9,** 2, 1978.

46. Roberts, G. T., "Heat Loss Characteristics of an Evacuation Plate-in-Tube Collector," *Solar Energy,* Vol. 22, New York: Pergamon Press, 1979, pp. 137–140.

47. Root, Douglass E., *Medium-Temperature Flat-Plate Solar Energy Collectors for Use in Florida,* Cape Canaveral, Florida: Florida Solar Energy Center, 1979.

48. Scoville, A. E., "An Alternate Cover Material for Solar Collectors," paper presented at International Solar Energy Society Congress and Exposition, Los Angeles, California, July, 1975.

49. Shuttleworth, John, ed., "Solar on a Shoestring: A Build-It-Yourself Solar Collector," *The Mother Earth News,* 60:96, 1979.

50. Solaron Corporation, Denver, Colorado, *Application Engineering Manual,* 1977, pp. 1, 8.

51. Tentarelli, K. D., "Focusing Collectors Heat a Northern Home," *Popular Science,* **209,** 6:60, (1976).

52. Toenjes, R. B., "Integrated Solar Energy Collector Final Summary Report", Los Alamos Scientific Laboratory Report LA-6143-MS, November, 1975.

53. U. S. Department of Commerce, NBS Technical Note 899: *Development of Proposed Standards for Testing Solar Collectors and Thermal Storage Devices,* Washington, D.C.: U. S. Government Printing Office, February, 1976.

54. U. S. Department of Commerce, *Solar Heating and Cooling of Residential Buildings-Design of Systems,* Washington, D.C.: U. S. Government Printing Office, October, 1977, pp. 4.8, 7.2–7.4, 7.7, 11.1–11.28.

55. U. S. Department of Commerce, *Solar Heating and Cooling of Residential Buildings-Sizing, Installation and Operation of Systems,* Washington, D.C.: U. S. Government Printing Office, October, 1977, pp. 3.5–3.10, 5.1–5.24.

56. U. S. Department of Energy, *DOE Facilities Solar Design Handbook,* Washington, D.C.: U. S. Government Printing Office, January, 1978, pp. 16–21, 70–77, 79.

57. U. S. Department of Energy, *Fundamentals of Solar Heating,* Washington, D.C.: U. S. Government Printing Office, January, 1978, pp. 3.1–3.9.

58. U. S. Department of Energy, *Introduction to Solar Heating and Cooling Design and Sizing,* Washington, D.C.: U. S. Government Printing Office, August, 1978, pp. 1.43–1.44, 1.51–1.52, 5.36–5.56, 6.5, 6.8, 6.37, 6.57–6.58, 7.6–7.7.

59. U. S. Department of Energy, Solar Technology Transfer Program, *Space Heating Handbook with Service Hot Water and Heat Load Calculations: Solcost,* July, 1978.

60. U. S. Department of Housing and Urban Development, *HUD Minimum Property Standards Supplement,* Washington, D.C.: U. S. Government Printing Office, 1977, pp. A6, A24–A26, A50, A65, A78, A88, A89, A96, A120–122, B2–3, B9, C10, C20–24.

61. U. S. Department of Housing and Urban Development, *Solar Dwelling Design Concepts,* Washington, D.C.: U. S. Government Printing Office, May, 1976, pp. 20–23.

62. U. S. Department of Housing and Urban Development, Solar Factsheet, "Solar Collector Types and Components," Washington, D.C.: U. S. Government Printing Office, 1980.

63. U. S. Department of Housing and Urban Development, "Special Solar Announcement", George Washington University and DOE Published Results of Tests on Low Cost Solar Heating System.

64. Watson, Donald, *Designing and Building a Solar House,* Charlotte, Vermont: Garden Way Publishing Company, 1977, pp. 50–75.

65. Winn, C. B., and Johnson, G. R., "Solar Energy Analysis Programs for Programable Handheld Calculators, Texas Instruments SR-52 Edition," Report TR-99, SEEC, Inc., Fort Collins, Colorado, 80522.

66. Wright, William A., "The Collector Efficiency Curve," *Solar Age,* **4,** 9:44–49, (1979).

CHAPTER 3
SOLAR WATER HEATERS

3.1 DESCRIPTION OF THE WATER HEATER SYSTEM

A solar water heater represents the cheapest way to use solar energy (as opposed to using solar energy to heat or cool a home). Using solar energy to heat household water dates back to the 1930s where solar water heaters were predominately used in California and Florida. This was before low priced electricity and natural gas became available. The era of "cheap" energy has passed and once again people are beginning to look for alternate energy sources to petroleum products.

As dissatisfaction with higher electric bills grows, the demand for solar water heaters is experiencing a rebirth. Respected companies such as A. O. Smith, Revere Copper & Brass, Inc., Rheem Water Heater Division and State Industries, Inc., to name a few, are offering complete solar water heating systems.

The main elements of a solar water heater consist of:

1. Collector panels to heat the water (or air) flowing through them.
2. Circulation pump to transfer cold water through the collector and then transfer the heated water back into the storage tank.
3. Storage tank to contain the heated water until it is used.
4. Valves to control the flow of liquid circulating through the pipes.
5. Thermostatic control to regulate the pump which maintains a flow of liquid through the system.

3.1.1. Open and Closed Systems

There are two types of solar water heater systems: open (Fig. 3.1) and closed (Fig. 3.2). In the open system, cold water from the storage tank is pumped through the collectors on the roof and the heated water is then returned to the storage tank. With the closed system, oil or antifreeze is pumped from the heat exchanger through the solar collector and back into the heat exchanger. The heat exchanger, located in the storage tank, transfers the heat from the circulated (antifreeze) solution to the stored water.

The advantage of the open system is the simplicity; in terms of lower operating costs due to fewer moving parts than a closed system and better thermal efficiency because no heat exchanger is used. The main disadvantage with the open system is the *general* inability to cope with freezing tempera-

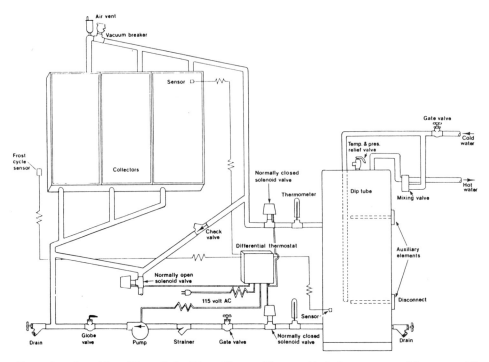

Figure 3.1. Open Type Direct Solar Water Heater. (Source: U. S. Department of Commerce[20]).

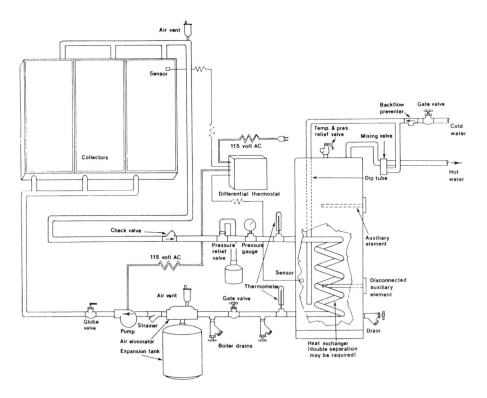

Figure 3.2. Closed Loop, Antifreeze Solar Water Heater. (Source: U. S. Department of Commerce[20]).

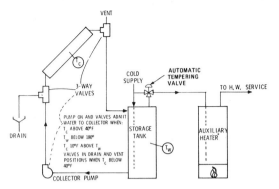

Figure 3.3. Solar Water Heater with Freeze Protection by Automatic Collector Drainage. (Source: U. S. Department of Energy[24]).

tures (see Fig. 3.1). Open systems can be used in freezing weather *if* appropriate drain-down systems are employed such as shown in Fig. 3.3. In Fig. 3.3 valves are used to drain the collector and exposed piping in freezing weather. The valves should be mechanically rather than electrically operated to prevent breakdowns in the event of a power outage.

The main advantages of the closed system are that no drain down is necessary and corrosion problems are less frequent than for the open system. Since a heat exchanger is used in the closed system, thermal efficiency will be reduced and the system may require 10% to 15% larger solar collectors than the ones used in the open system. The only thing to watch for in the closed system is that the antifreeze solution does *not* come in direct contact with potable (drinking) water in the storage tank.

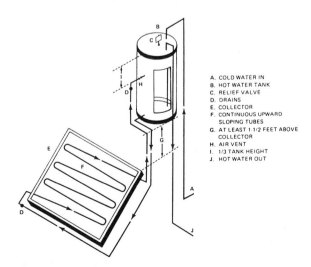

A. COLD WATER IN
B. HOT WATER TANK
C. RELIEF VALVE
D. DRAINS
E. COLLECTOR
F. CONTINUOUS UPWARD SLOPING TUBES
G. AT LEAST 1-1/2 FEET ABOVE COLLECTOR
H. AIR VENT
I. 1/3 TANK HEIGHT
J. HOT WATER OUT

Figure 3.4. Thermosiphon Solar Water Heater. (Source: Florida Solar Energy Center[7]).

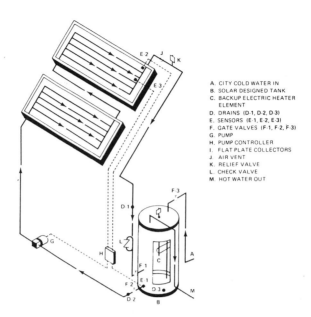

A. CITY COLD WATER IN
B. SOLAR DESIGNED TANK
C. BACKUP ELECTRIC HEATER ELEMENT
D. DRAINS (D-1, D-2, D-3)
E. SENSORS (E-1, E-2, E-3)
F. GATE VALVES (F-1, F-2, F-3)
G. PUMP
H. PUMP CONTROLLER
I. FLAT PLATE COLLECTORS
J. AIR VENT
K. RELIEF VALVE
L. CHECK VALVE
M. HOT WATER OUT

Figure 3.5. Pumped Solar Water Heater. (Source: Florida Solar Energy Center[7]).

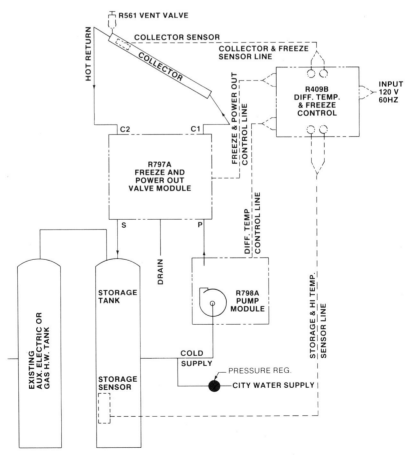

Figure 3.6. Typical Domestic Hot Water Installation. (Courtesy of Richdel, Inc. Solar Products).

SPECIFICATIONS:

INPUT: 120 VAC   60 Hz

OUTPUTS: Pump & Valve 120VAC   60 Hz
                      Protected at 2.00 Amps

TURN ON DIFFERENTIAL: 15°F±1°F

TURN OFF DIFFERENTIAL: 5°F±1°F

TANK OVERTEMP LIMIT: On 180°F
     (Typical System)   Off 176°F

COLLECTOR LOW TEMP LIMIT: On 55°F
                                                          Off 58°F

105 TO 125VAC LINE CHANGE: ½°F Maximum
                                                              Differential change

SENSOR COMPATABILITY: Thermistor type
                                                  10K@25°C

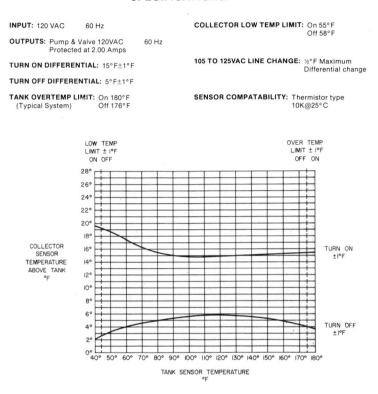

Figure 3.7. Domestic Hot Water Differential Thermostat Control. (Courtesy of Richdel, Inc. Solar Products).

3.1.2. Thermosiphon System vs. Pumped System

The thermosiphon type system relies upon convection and gravity to circulate water through the system. Figure 3.4 illustrates a thermosiphon solar water heater.

The cooler water at the bottom of the tank is heavier than the warmer water at the top. This cooler, heavier water flows through the bottom of the collector. As it is heated, the water becomes lighter and begins to rise through the collector and back into the top of the storage tank. This constant circulation alleviates the need for a pump. The main drawback is that in order for the thermosiphon system to work, the storage tank *must* be located higher than the collector. This means that the storage tank will need to be mounted in the attic or on the roof. Structural complications could arise because of the large weight of the tank being confined to a small area (a 66 gallon water tank weighs about 700 pounds). Poor access to the water heater will be another problem. One would have to go up into the attic to close the shutoff valves and open the draincocks whenever the temperature dropped below freezing. For these reasons, few thermosiphon-type water heaters are being installed today.

The pumped solar water heater is the most popular. Figure 3.5 illustrates this type. This system incorporates two sensors (E-1 and E-2) to control the

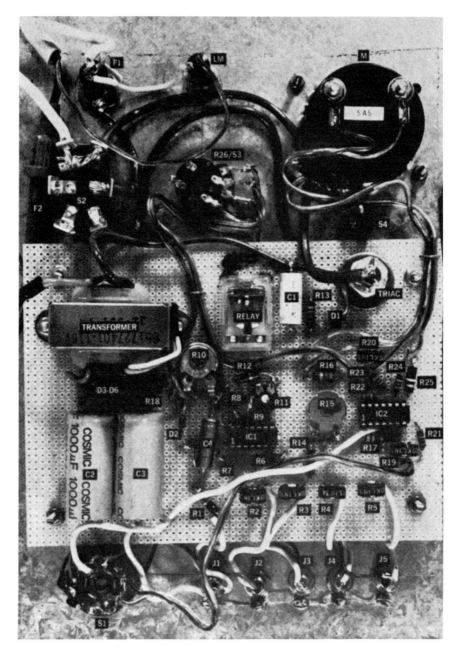

Figure 3.8. Homebuilt Solar System Differential Controller. Reprinted from Popular Science with permission © 1978, Times Mirror Magazines, Inc. (See Russell[15]).

pump. When the collector is 10° to 15° F warmer than the storage tank, the sensors switch the pump on until the differential between the collectors and the storage medium drops to 3° to 5° F. Figure 3.6 illustrates a schematic of a commercially available differential controller built by Richdel, Inc. Figure 3.7 shows specifications of the differential thermostat control. The differential controller does not have to be commercially purchased, but can be homemade.

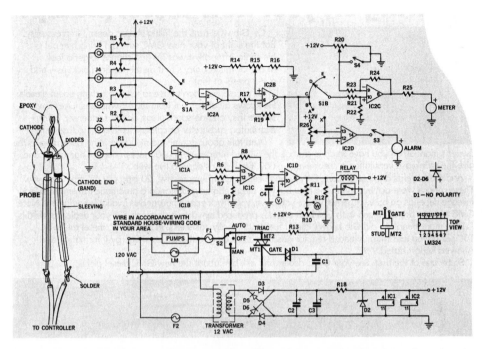

Figure 3.9. Schematic Diagram for Homebuilt Differential Controller. Reprinted from Popular Science with permission © 1978, Times Mirror Magazines, Inc. (See Russell[15]).

Figure 3.8 illustrates the parts placement of one such homemade differential controller. Also included are a parts list (Table 3.1) and a schematic diagram of the circuit (Fig. 3.9). The homemade differential controller can:

1. Control a pump
2. Display up to five separate temperature points in the system on a meter
3. Warn the user by an audible alarm circuit when excessive temperatures have built up in the system (such as in the solar collector).

3.2 MAINTENANCE CONSIDERATIONS

Corrosion generally represents the greatest maintenance problem to be dealt with in a solar water heating system. There are two main sources of corrosion:

1. The *chemicals occurring naturally in the water supply* (e.g., sulfur, iron, etc.) form scales which clog aluminum or steel collector tubing. (This is one reason why copper tubing is used in open-type systems. Even when copper tubing is used, it should never be less than 1/2 inch I.D. to keep from clogging up the system.)
2. *Galvanic corrosion* is caused when two dissimilar metals come in contact in the presence of water. The use of brass or bronze nipples to connect a steel tank to the copper water pipes will reduce the problem of galvanic

Table 3.1. Parts List for Homebuilt Differential Solar Controller. Reprinted from Popular Science with permission © 1978, Times Mirror Magazines, Inc. (Source: Russel[15]).

```
R1 -- 47,000-ohm  1/4-W resistor
R6, 7 -- 1500-ohm, 1/4-W resistor
R8, 9 -- 24,000-ohm, 1/4-W resistor
R12 -- 1200-ohm, 1/4-W resistor
R13 -- 22,000-ohm, 1/2-W resistor
R14 -- 15,000-ohm, 1/4-W resistor
R16 -- 1500-ohm, 1/4-W resistor
R17 -- 3300-ohm, 1/4-W resistor
R18 -- 100-ohm, 1/2-W resistor
R21, 22, 23, 24 -- 470,000-ohm, 1/4-W resistor
R25 -- 100,000-ohm, 1/4-W resistor
R2, 3, 4, 5, 11, 19, 20 -- 100,000-ohm, 1/4-W potentiometer
R10 -- 5000-ohm, 1/4-W potentiometer
R15 -- 500-ohm, 1/4-W potentiometer
R26/S3 -- 10,000-ohm linear taper potentiometer
          w/SPST switch
C1 -- 0.1-mfd capacitor
C2, 3 -- 1000-mfd at 25-V electrolytic capacitor
C4 -- 10-mfd at 15-V electrolytic capacitor
D1 -- ECG 6406 trigger diode
D2 -- 1N4742 12-volt, 1-watt zener diode
D3, 4, 5, 6 -- 1N4001 diode
Sensors -- 1N4148 diodes (get 15-20)
Triac -- ECG 5665 10-A at 200-VAC stud-mount triac
IC12 -- LM324N op Amp
LM -- Neon panel lamp 120 VAC
T1 -- 12 VAC secondary at 300 mA
Relay -- 12-VDC 1200-ohm coil, SPST
Meter -- 50 microamps, 1000 ohms
Alarm -- Mallory Sonalert (DO NOT SUBSTITUTE)
S1 -- 2-pole, 6-position rotary switch
S2 -- SPDT center-off toggle switch 10 A at 120 VAC
S4 -- SPST normally open momentary pushbutton
F1 -- 5A 3AG fuse and panel-mount fuse holder
F2 -- 0.25 A 3AG fuse and in-line fuse holder
J1-J5 -- Panel-mount phone jacks or equivalent
P1-P5 -- phone plugs
Misc. -- 4.5" x 6" epoxy vectarboard, 8" x 10" metal
         plate, 2 14-pin IC sockets, No. 22 solid wire

NOTE:  All components are available from Williams
       Electronic Supply, 1863 Woodbridge Avenue,
       Edison, NJ 08817
```

corrosion. Joints such as steel–copper and aluminum–copper should never be used because galvanic corrosion is inevitable.

3.3 SUMMARY

By choosing a reputable company which supplies the complete solar water heating system (e.g., solar collector, pumps, controllers, water storage tank), the user is least likely to have problems. Figure 3.10 illustrates one such system by Revere Solar and Architectural Products, Inc. This system has a five year limited warranty for the solar collectors and storage tanks and a one year limited warranty for pumps, controls and heat exchangers. Notice the 240 V resistance heating cable in Fig. 3.10. The solar water heater, like most solar water heaters, has an electrical heating element which will come on when the solar collectors cannot provide sufficient heat to warm the water. Figure 3.11 shows an installation diagram for a solar water heater manufactured by State Industries, Inc., and Fig. 3.12 shows the wiring diagram for the water heater illustrated in Fig. 3.11.

Another commercial system shown in Fig. 3.13 illustrates a unique system which uses air as the heat transfer medium. This system which is manufac-

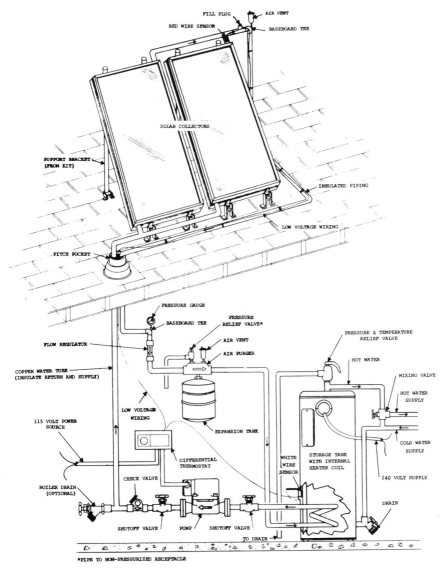

NOTES: (1) Piping insulation not shown.
(2) Some installations may require double walls between potable water and collector glycol loop.
When required, Revere can furnish water heaters with double wall heat exchangers.

Figure 3.10. Complete Solar Domestic Water Heating System. (Courtesy Revere Solar and Architectural Products, Inc., Rome, New York).

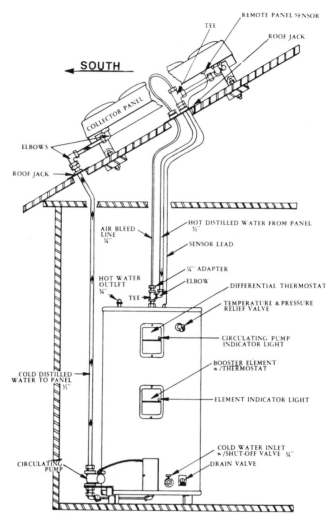

Figure 3.11. Installation Diagram for a Solar Water Heater. (Courtesy of State Industries, Inc. Solar Water Heating Manual[19]).

tured by Solaron Corporation is unique in the fact that most commercially available units use a liquid (e.g., water or antifreeze) as the heat transfer medium. Included in Fig. 3.13 is a product description.

Two other ways of obtaining a working system besides buying a completely furnished solar water heater are:

1. Buying the components separately (e.g., pump, water storage tank, controller, etc.) from different manufacturers.
2. Building one's own system (as shown in Fig. 3.14).

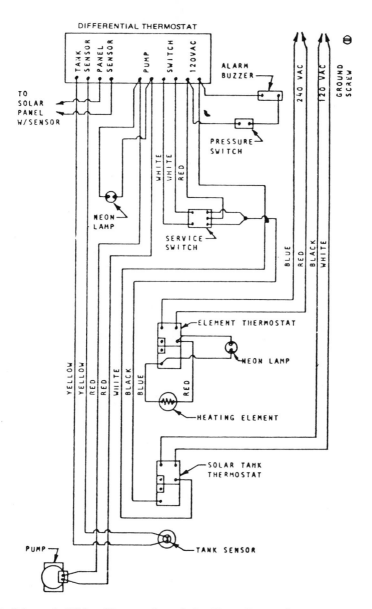

Figure 3.12. Schematic Wiring Diagram for a Solar Water Heater. (Courtesy of State Industries, Inc. Solar Water Heating Manual[19]).

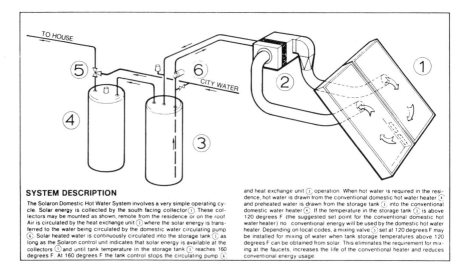

SYSTEM DESCRIPTION

The Solaron Domestic Hot Water System involves a very simple operating cycle. Solar energy is collected by the south facing collector ①. These collectors may be mounted as shown, remote from the residence or on the roof. Air is circulated by the heat exchange unit ② where the solar energy is transferred to the water being circulated by the domestic water circulating pump ⑥. Solar heated water is continuously circulated into the storage tank ③ as long as the Solaron control unit indicates that solar energy is available at the collectors ① and until tank temperature in the storage tank ③ reaches 160 degrees F. At 160 degrees F the tank control stops the circulating pump ⑥

and heat exchange unit ② operation. When hot water is required in the residence, hot water is drawn from the conventional domestic hot water heater ④ and preheated water is drawn from the storage tank ③ into the conventional domestic water heater ④. If the temperature in the storage tank ③ is above 120 degrees F (the suggested set point for the conventional domestic hot water heater) no conventional energy will be used by the domestic hot water heater. Depending on local codes, a mixing valve ⑤ set at 120 degrees F may be installed for mixing of water when tank storage temperatures above 120 degrees F can be obtained from solar. This eliminates the requirement for mixing at the faucets, increases the life of the conventional heater and reduces conventional energy usage.

Figure 3.13. Solar Air-Type Domestic Water Heating System. (Source: Solaron Corporation[18]).

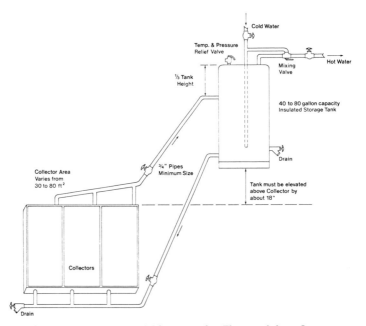

Figure 3.14. Conceptual Diagram of a Thermosiphon System.

3.4 REFERENCES AND SUGGESTED READING

1. Anderson, Bruce, *The Solar Home Book,* Harrisville, New Hampshire: Cheshire Books, 1976, p. 209.
2. Beason, Robert G., ed., "Homebuilt Solar Water Heaters," *Mechanix Illustrated,* **73,** 587:50, 1977.

3. Campbell, Stu, and Taft, Douglass, *Build Your Own Solar Water Heater,* Charlotte, Vermont: Garden Way Publishing, 1978.

4. Duffie, John A., and Beckman, William A., *Solar Energy Thermal Processes,* New York: John Wiley & Sons, Inc., 1974, pp. 252–256.

5. Environmental Information Center of the Florida Conservation Foundation, Inc., *Build Your Own Solar Water Heater,* Winter Park, Florida: Superior Minute Print, 1976, pp. 1–7.

6. Florida Solar Energy Center, *Guide to Solar Water Heating,* Cape Canaveral, Florida: State University System of Florida, October, 1978.

7. Florida Solar Energy Center, *A Guide to System Sizing and Economics of Solar Water Heating in Florida Residences,* Cape Canaveral, Florida: State University System of Florida, September, 1977.

8. Florida Solar Energy Center, *Solar Water Heating - A Question and Answer Primer,* Cape Canaveral, Florida: State University System of Florida, September, 1979.

9. Florida Solar Energy Center, *Solar Water and Pool Heating Installation and Operation,* Cape Canaveral, Florida: State University System of Florida, January, 1979.

10. International Compendium, A Division of Solar Science Industries, Inc., *Solar Energy Business.*

11. Los Alamos Scientific Laboratory, Solar Energy Group, *Pacific Regional Solar Heating Handbook,* Washington, D.C.: U. S. Government Printing Office, November, 1976, pp. 65–66.

12. Mason, Dike, "A Simple Solar-Heated Shower," *The Mother Earth News,* 46:64, (1973).

13. National Solar Heating and Cooling Information Center, Rockville, Md., *Solar Hot Water and Your Home,* 1978.

14. Rho Sigma, Inc., *Solar Controls Applications Manual,* North Hollywood, California, p. 2.

15. Russel, Thomas, "For Your Solar Water Heater, Build a Temperature Control," *Popular Science,* **212,** 3:134, (1978).

16. Sheet Metal and Air Conditioning Contractors National Association, *Fundamentals of Solar Heating,* Washington, D.C.: U. S. Government Printing Office, January, 1978, pp. 8.1–8.6.

17. Smith, A. O., *Energy Saving Conservationist Solar System,* Kankakee, Illinois, 1978, (sales brochure) pp. 2–3.

18. Solaron Corporation, Denver, Colorado, *Application Engineering Manual,* 1977, p. 706.

19. State Industries, Inc., Ashland City, Tennessee, *Solar Water Heater Manual,* 1978, p. 4.

20. U. S. Department of Commerce, *Installation Guidelines for Solar DHW Systems,* Washington, D.C.: U. S. Government Printing Office, 1979.

21. U. S. Department of Commerce, *Solar Heating and Cooling of Residential Buildings – Design of Systems,* Washington, D.C.: U. S. Government Printing Office, October, 1977, pp. 19.1–19.19.

22. U. S. Department of Commerce, *Solar Heating and Cooling of Residential Buildings – Sizing, Installation, and Operation of Systems,* Washington, D.C.: U. S. Government Printing Office, October, 1977, pp. 2.9, 7.1–7.15.

23. U. S. Department of Energy, *DOE Facilities Solar Design Handbook,* Washington, D.C.: U. S. Government Printing Office, January, 1978, pp. 24–26.

24. U. S. Department of Energy, *Introduction to Solar Heating and Cooling Design and Sizing,* Washington, D.C.: U. S. Government Printing Office, August, 1978, pp. 4.1–4.17.

25. U. S. Department of Housing and Urban Development, *Intermediate Minimum Property Standards Supplement: Solar Heating and Domestic Hot Water Systems,* Washington, D.C.: U. S. Government Printing Office, Standard 4930.2, 1977.

26. U. S. Department of Housing and Urban Development, *Solar Hot Water and Your Home,* Washington, D.C.: U. S. Government Printing Office, August, 1979.

27. Weber, William J., "A Homemade Solar Water Heater," *Mother Earth News,* 59:70–71, (1979).

CHAPTER 4
SOLAR SPACE HEATING

4.1 SOLAR SPACE HEATING OVERVIEW

A solar heating system consists of five major components:

1. Solar collector
2. Piping and ductwork
3. Storage tank
4. System controls and fluid flow devices (e.g., valves, dampers, thermostats, pumps, fans)
5. Auxiliary heating device

There are two general types of solar space heating systems, air and liquid. Both will be covered in separate sections of Chapter 4. Figure 4.1 illustrates the basic differences between the two systems.

4.2 SOLAR AIR SYSTEM

4.2.1. Description of System

The solar air system has four basic modes of operation:

1. Space heating directly from the collectors
2. Heat storage
3. Space heating from the storage bin
4. Space heating using an auxiliary heating source

Figure 4.2 illustrates the first mode of the solar air heating unit. In Fig. 4.2 the collectors can be used to supply heat directly to the conditioned space.

When there is sufficient solar energy being absorbed by the collector and no heat is required for the room, the solar heat from the collector is transported directly to the rock storage bin as shown in Fig. 4.3.

When the solar collectors cannot supply heat directly to the conditioned space inside the building due to low collection temperatures, air is blown through the rock storage bin to remove the stored heat as shown in Fig. 4.4. Notice that in Fig. 4.3 when storing collected solar energy the air is circulated from the top of the storage bin to the bottom of the bin and when using (removing) the stored heat from the storage bin as shown in Fig. 4.4, the air

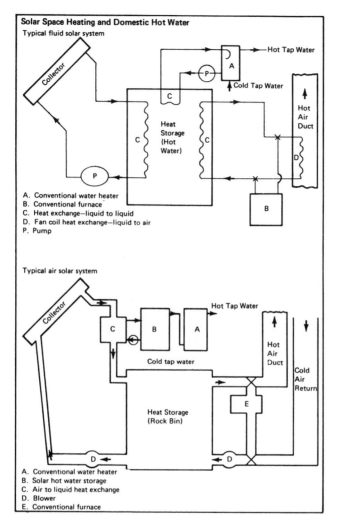

Figure 4.1. Solar Space Heating and Domestic Hot Water Systems. (Source: Federal Energy Administration[9]).

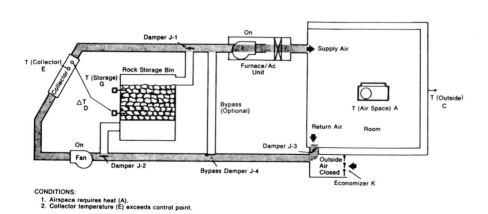

CONDITIONS:
1. Airspace requires heat (A).
2. Collector temperature (E) exceeds control point.

Figure 4.2. Heating Directly from Collectors of an All-Air System. (Source: Sheet Metal and Air Conditioning Contractors National Association[18]).

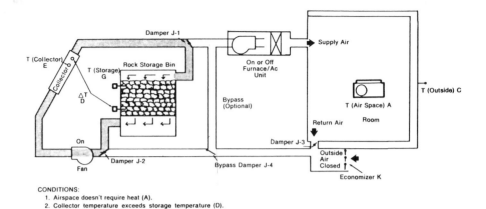

CONDITIONS:
1. Airspace doesn't require heat (A).
2. Collector temperature exceeds storage temperature (D).

Figure 4.3. Storing Heat in a Solar Space Heating System. (Source: Sheet Metal and Air Conditioning Contractors National Association[18]).

flow is reversed. This procedure takes advantage of thermal stratification which is discussed later. In the event that the rock storage bin cannot supply the necessary heat to the conditioned space, the auxiliary heating unit (e.g., furnace) will activate as shown in Fig. 4.5. In the different heating modes (of Figs. 4.2–4.5) note the positions of the dampers and conditions necessary for each mode of operation to take place.

Air-type solar systems have been used successfully for both residential and commercial applications. The following information describes the solar air heating system used in Fig. 4.6.

PROJECT INFORMATION

Owner/Builder: Air Force
Operational Date: April, 1977
Total Estimated ERDA Funds: To be determined

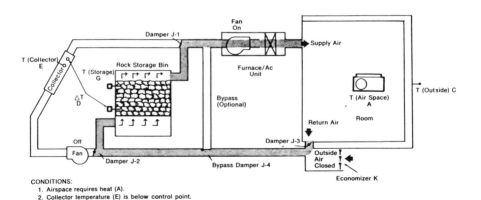

CONDITIONS:
1. Airspace requires heat (A).
2. Collector temperature (E) is below control point.
3. Storage temperature (G) is above control point.

Figure 4.4. Space Heating from Storage. (Source: Sheet Metal and Air Conditioning Contractors National Association[18]).

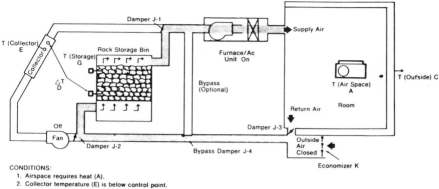

CONDITIONS:
1. Airspace requires heat (A).
2. Collector temperature (E) is below control point.
3. Storage temperature (G) is below control point.
4. Furnace is energized by 2nd thermostat stage (A).

Figure 4.5. Heating Using the Auxiliary Heating Unit. (Source: Sheet Metal and Air Conditioning Contractors National Association[18]).

Building
   Type: Residential, single family   2 units
   Area: 1487 sq.ft
Location: Andrews Air Force Base, Maryland
Latitude: 38°N
Climatic Data
   Degree Days   Heating 4333   Cooling 1237
   Avg. Temp. (°F)   Winter 46   Summer 67
   Avg. Insol. (Ly/d) Winter 260   Summer 453

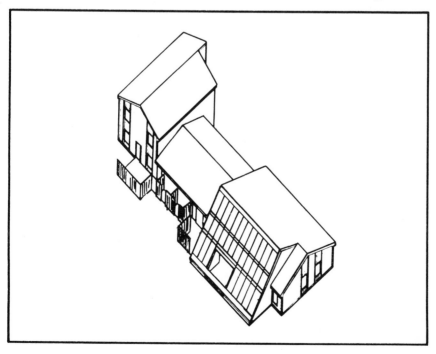

Figure 4.6. Andrews Air Force Base Residential Solar Heated Building. (Source: Energy Research and Development Administration[7]).

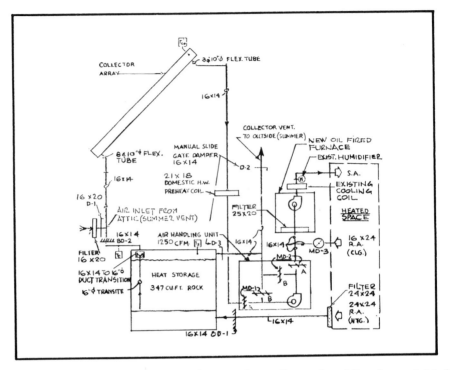

Figure 4.7. Mechanical Systems Layout. (Source: Energy Research and Development Administration[7]).

SOLAR ENERGY SYSTEM

Application   Heating 73%   Cooling 0°   Hot Water 80%
Collector
  Type: Air flat-plate
  Area: 624 sq.ft
  Manufacturer: Solaron Corp.
Storage
  Type: Rock
  Capacity: 347 sq.ft
Auxiliary System Type: Oil furnace w/humidifier

PROJECT DESCRIPTION

Objective(s): To compare liquid and air systems and higher performance collectors in a temperate climate.

Figure 4.7 shows the mechanical systems layout for the residential building shown in Fig. 4.6. Note that the storage medium is rock. Another Air Force residential building is shown in Fig. 4.8. The storage medium in this case is a change-of-phase material such as eutectic salts. The following is a description of the building.

Figure 4.8. Air Force Academy Residential Solar Heated Building. (Source: Energy Research and Development Administration[7]).

PROJECT INFORMATION

Owner/Builder: Air Force
Operational Date: April, 1977
Total Estimated ERDA Funds: To be determined
Building
  Type: Residential, single family
  Area: 1238 sq.ft
Location: Air Force Academy, Colorado
Latitude: 39°N
Climatic Data
  Degree Days   Heating 6254   Cooling 500
  Avg. Temp. (°F) Winter 41   Summer 56
  Avg. Insol. (Ly/d) Winter 301   Summer 488

SOLAR ENERGY SYSTEM

Application   Heating 76%   Cooling 0%   Hot Water 85%
Collector
  Type: Air flat-plate
  Area: 704 sq.ft
  Manufacturer: Solaron Corp.
Storage
  Type: Change of phase
  Capacity: 636,000 Btu
Auxiliary System Type: Gas furnace w/humidifier

PROJECT DESCRIPTION

Objective(s): To compare alternative methods of heat storage for air collectors utilizing existing basements.

Figure 4.9 shows the heating system used in Fig. 4.8.

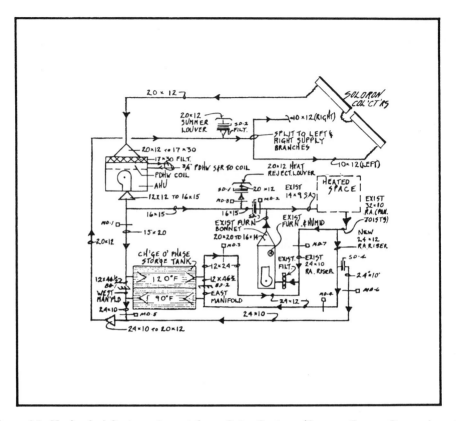

Figure 4.9. Mechanical Systems Layout for a Solar System. (Source: Energy Research and Development Administration[7]).

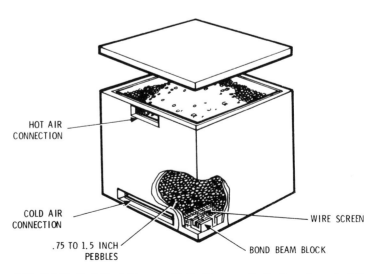

Figure 4.10. Pebble-Bed Heat Storage Unit. (Source: U. S. Department of Energy[26]).

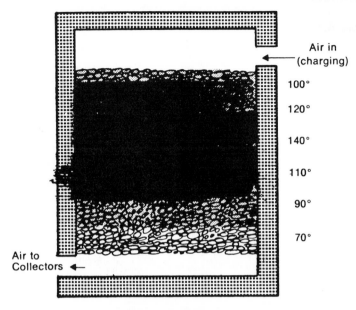

Air in
(charging)

100°

120°

140°

110°

90°

70°

Air to
Collectors ←

Thermal Stratification

Figure 4.11. Temperature "Layers" in Pebble-Bed Storage. (Source: Sheet Metal and Air Conditioning Contractors National Association[18]).

4.2.2. Storage

Generally, the storage medium for air-type collector heating systems is rock. Ideally the rocks used should be rounded and of uniform size (between 0.75 to 1.5 inches). Figure 4.10 shows a typical rock storage bin. Thermal stratification will occur in a rock storage bin. This results in layers of rock having varying temperatures such as shown in Fig. 4.11. The thermal layers will change over the course of a day as shown in Fig. 4.12. This is due to the alternating charging and discharging cycles of heat.

In determining the amount of rocks to use, the specific heat of the material

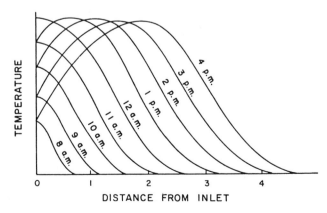

Figure 4.12. Typical Temperature Profiles in a Rock Bin Storage. (Source: U. S. Department of Commerce[24]).

Table 4.1. Sensible Heat Storage Materials. (Source: Cole[5]).

Material	Density, lbs/ft^3.	Heat Capacity, Btu/lb·°F	Volumetric Heat Capacity, Btu/ft^3·°F	
			No Voids	30% Voids
Water	62.4	1.00	62.4	–
Scrap Iron	489	0.11	53.8	37.7
Scrap Aluminum	168	0.22	36.96	25.9
Scrap Concrete	140	0.27	27.8	26.5
Rock	167	0.21	34.7	24.3
Brick	140	0.21	29.4	20.6

must be considered. Table 4.1 illustrates various properties of some common materials. Notice that rock has a volumetric heat capacity (in Btu/ft^3°F) of 24.3 with 30% voids. The voids refer to the spaces between the rocks (or other materials) piled in the storage bin. When determining the heat capacity of a material such as rock, 30% voids should be used. If there were no voids in the storage bin, no air could flow between the rocks. The 24.3 Btu/ft^3°F heat capacity of rock means that the rocks can store 24.3 Btus for every degree of temperature rise in the storage bin. Referring back to table 4.1 it can be seen that water has a heat capacity of 62.4 Btu/ft^3°F. This means that for rocks in the storage bin to be able to store the same amount of heat as water, the rock bin must be 2.6 times greater than the tank of water. Figure 4.13 illustrates

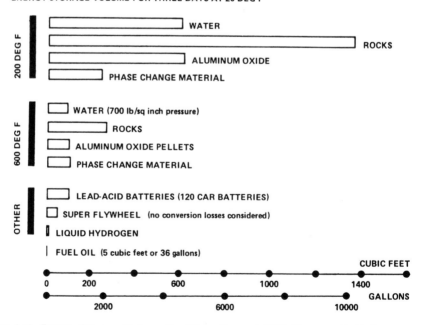

Figure 4.13. Energy Storage Volume for Three Days at 20°F. (Source: U. S. Department of Housing and Urban Development[29]).

storage volume requirements for various materials. Notice that the phase change product shown in Fig. 4.13 requires less materials than rocks to store the same amount of energy. This is one reason why phase change materials are being considered as storage mediums—such as for the building in Figs. 4.8 and 4.9. Table 4.2 illustrates some properties of phase change materials (latent heat storage).

4.2.3. Controls

Figure 4.14 illustrates a temperature control wiring diagram for the Solaron type air collector system.

The following information describes the general operation of the Solaron air collector system shown in Fig. 4.14.

GENERAL SYSTEM OPERATION

Models HC0115 and HC0116 Controllers

HC0115 controller is used in applications as follows:	HC0116 controller is used with heat pump system:
1. Heating only.	1. Energize reversing valve for heating.
2. Heating only with continuous fan operation.	2. Energize reversing valve for cooling.
3. Heating/cooling.	

Solaron Controls — Sequence of Operation
  I. *Solar Energy Available.* When a 40° F. differential is achieved between sensors T (in collector) and T_{ci} (in return air/collector inlet duct) the following events take place:
  A. Storing Heat — no demand for space heating
  Differential thermostat in HC0115 or HC0166 will energize control circuit to: Open MD-1; turn on Solaron air handler blower; turn on hot water pump (optional); close MD-2.
  B. First Stage Heating — demand for solar heat in space.
  The HC0115 or HC0116 control circuit will: Turn on auxiliary unit *fan* (auxiliary heat exchanger is off); Open MD-2.
  C. Second State Heating — demand for additional heat (auxiliary)
    1. The HC0115 control circuit will bring on the auxiliary heat souce (gas, electric, oil, etc.)
    2. The HC0116 control circuit will bring on the heat pump compressor and fully open MD-3. (Note: MD-2 closes on second stage for heat pump).
  D. Third Stage Heating (heat pump applications *only*)
  A second thermostat will bring on electric resistance strip heating elements in auxiliary heating unit.

 II. *Solar Energy Not Available.* When the differential between T_{co} and T_{ci} drops to 25° F or less the following takes place:
  A. The HC0115 control circuit will: Close MD-1; turn off air handler blower; turn off hot water pump (optional); open MD-2.

Table 4.2. Properties of Phase Change Materials (Latent Heat Storage). (Source: U. S. Department of Commerce[25]).

Material	Melting Point °F	Heat of Fusion	
		Btu/lb	Btu/cu ft
Hydrated Inorganic Salts			
Sodium chromate	67.8	78.4	7400
Manganese nitrate	79.4	60.4	6570
Ortho Phosphoric acid	84.8	61.6	7010
Lithium nitrate	85.8	127	18200
Calcium chloride	86.4	73.2	7570
Glauber Salt	90.4	102.3	9320
Disodium phosphate	94.2	121.0	11550
Manganous nitrate	95.8	50.4	6240
Zinc nitrate	97.2	56.1	7180
Calcium nitrate	108.6	61.1	6900
Thiosulfate (Hypo)	119.4	40.7	4710
Nickelous nitrate	134.0	65.6	8330
Cobaltous nitrate	134.7	54.4	8300
Cadmium nitrate	139.1	45.7	6950
Sulfur trioxide	144.0	137	16750
Magnesium nitrate	182.2	68.8	9380
Hydrazine hydrochloride	198.8	95.7	7700
Magnesium chloride	244.0	72.7	9000
Anhydrous Inorganic Salts			
Arsenic tribromide	89.4	16.0	
Meta Phosphoric Acid	108.5	46.2	
Phosphoric acid	158	67.4	
Antimony trichloride	164	24.0	
Antimony Tribromide	205.5	16.6	
Aluminum bromide	208.3	18.2	
Ammonium acid sulfate	291	53.5	
Ammonium nitrate	337	27.4	
Potassium thiocyanate	350	48	
Waxes and Organic Solids			
Anthracine	205	45.2	3480
Anthraquinone	545	67.8	6030
Naphthaline	176	64.9	4620
Naphthol	203	70.0	5280
Bees wax	143	76.2	4500
Stearic acid (tallow)	169	85.4	4500
Amorphous paraffin wax	166	99.0	4900

B. First Stage Heating — demand for stored solar heat in space.
   1. When T_s sensor (heat storage box) is above 90°F system will heat the space from heat stored in the heat storage unit. Should T_s sensor be lower than 90°F, the HC0115 control circuit will automatically bring on the auxiliary heating unit to insure a minimum supply air temperature which will avoid the sensation of drafts. (Note: the heating system will still be circulating air through the heat storage unit to insure all of the solar energy is used.)

C. Second Stage Heating (same as I.C. above)

D. Third Stage Heating (same as I.D. above) heat pump only.

III. *Domestic Water Heating.*
   A. Winter operation permits domestic water to be preheated any time the system is storing heat or heating from collector. The sensor T_w will shut off the hot water pump when set point temperature in the domestic water storage tank is reached.

   B. Summer operation permits the T_w sensor, when satisfied to shut down the hot water pump, Solaron air handler and close MD-1 until water temperature in the storage tank drops 15°F below set point of T_w.

IV. *Air Conditioning (nonsolar).* When system is switched into "cool" mode the following occurs:

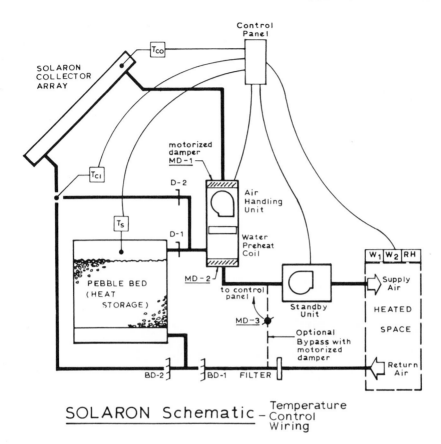

SOLARON COLLECTOR ARRAY

T_{co}

Control Panel

motorized damper MD-1

D-2

T_{ci}

T_s

D-1

Air Handling Unit

Water Preheat Coil

MD-2

PEBBLE BED (HEAT STORAGE)

to control panel

MD-3

Standby Unit

W_1 W_2 RH

Supply Air

HEATED

SPACE

Optional Bypass with motorized damper

Return Air

BD-2

BD-1

FILTER

SOLARON Schematic – Temperature Control Wiring

TEMPERATURE CONTROLS - SEQUENCE OF OPERATION

1. CONTROLS FOR THE COLLECTORS:

   When $\Delta T = T_{co} - T_{ci} =$ Set Point: The AHU will turn <u>on</u>

   When $\Delta T =$ Minimum set point the AHU will <u>shut off</u>

2. <u>CONTROLS FOR SOLAR & CONVENTIONAL HEATING</u>

   <u>Heating from collectors</u>: ΔT on and call for first stage heating, W_1

   <u>Storing heat</u>: ΔT on and no call for space heating

   <u>Heating from storage</u>: ΔT off and call for space heating, W_1 and or W_2

   <u>Heating from aux. unit</u>: Call for second stage heating and or T_s at its minimum set point.

4.14. Temperature Control Wiring Diagram for Air-Type Solar Collectors. (Source: Solaron Corporation[21]).

A. Solar available: System preheats domestic water; auxiliary system cools space as demand is dictated by thermostat; MD-3 opens fully.
B. Solar unavilable: Auxiliary system (same as IV.A. above); MD-3 opens.

4.2.4. Maintenance Considerations

The solar air system incorporating an air-type collector and using rocks as the storage medium will give almost no maintenance problems. The problems of freezing, corrosion, and pipe leaking are nonexistent in the air-type system.

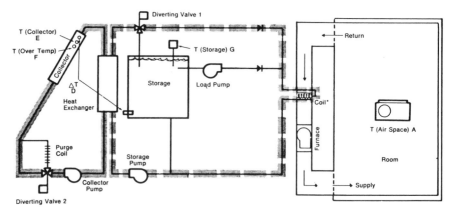

Figure 4.15. Heating Directly from Collectors of a Liquid-Type System. (Source: Sheet Metal and Air Conditioning Contractors National Association[18]).

There is no maintenance with the air-type heating system other than changing an air filter as is done in a conventional heating system.

4.3 SOLAR LIQUID SYSTEM

4.3.1. Description of System

The basic solar liquid heating system resembles the solar air heating system with respect to modes of operation. Figure 4.15 illustrates the heating mode using heat directly from the collectors. Notice that the solar energy from the collector is directed through a coil that transfers the thermal energy to the stream of air blowing through the heating duct in the building.

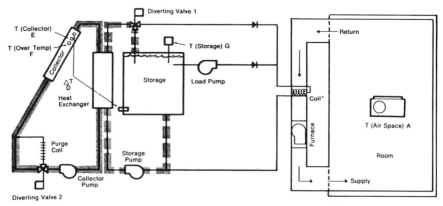

Figure 4.16. Storing Heat in a Solar Space Heating System. (Source: Sheet Metal and Air Conditioning Contractors National Association[18]).

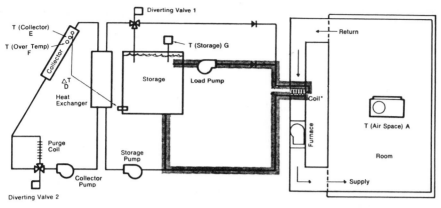

* NOTE: If heat pump is used, solar coil is placed downstream from heat pump coil.

CONDITIONS:
1. Airspace requires heat (A).
2. Collector temperature (E) is below control point.
3. Storage temperature (G) exceeds control point.

Figure 4.17. Space Heating from Storage. (Source: Sheet Metal and Air Conditioning Contractors National Association).

Figure 4.16 illustrates the heating system storage mode. When the collector temperature is too low to directly supply heat to the collector, heat is then drawn from the storage tank as shown in Fig. 4.17. Notice that the furnace is placed past the heat exchanger in the air duct. When the solar system cannot supply enough heat then the furnace activates to meet the thermal demand of the building.

In the liquid-type heating system it may be necessary to purge collected heat back into the atmosphere to prevent collectors from overheating. This is accomplished by using a purge coil such as shown in Fig. 4.18. As can be seen in Fig. 4.18, a diverting valve causes the collector fluid to flow through a heat exchanger that dissipates the collected heat back into the atmosphere.

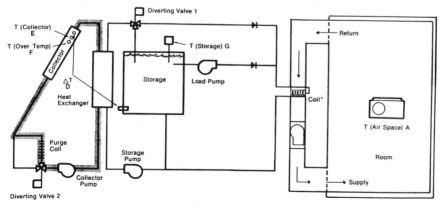

* NOTE: If heat pump is used, solar coil is placed downstream from heat pump coil.

CONDITION:
1. Over temperature condition exists in system (F).

Figure 4.18. Operation of Purge Coil on Liquid-Type Solar Space Heating System. (Source: Sheet Metal and Air Conditioning Contractors National Association[18]).

The following information describes a solar liquid-type heating system for multifamily elderly housing:

PROJECT INFORMATION

Owner/Builder: State of Connecticut
Contractor: Dept. of Community Affairs
Operational Date: December, 1976
Total Estimated ERDA Funds: $389,000
Building
  Type: Multi-family elderly housing
  Area: 16,800 sq.ft (total); 8,400 (cond.)
Location: Hamden, Connecticut
Latitude: 41.4°N
Climatic Data
  Degree Days   Heating 5461 Cooling 734
  Avg. Temp. (°F)   Winter 42.8 Summer 71.5
  Avg. Insol. (Ly/d) Winter 283   Summer 523

SOLAR ENERGY SYSTEM

Application Heating 75% Cooling 0% Hot Water
Collector
  Type: Liquid cooled flat-plate
  Area: 2600 sq.ft
Storage
  Type: Water
  Capacity: 8,000 gallons
Auxiliary System Type: Electric

PROJECT DESCRIPTION

This experiment will test the practicality and economy of using solar energy and other energy conservation techniques in housing for the elderly. The proposed project will consist of 40 housing units in low-rise structures. All will be designed for energy conservation. Half of the 40 units will be equipped with solar space and water heating and half will serve as a control in order that accurate conclusions on life cycle costs and performance may be drawn. A questionnaire will be used to determine how effective the tenants consider the solar energy system. The application will be a prototype for future state-assisted projects. The design and construction process will be integrated as closely as possible into the guidelines and procedures of Connecticut's Department of Community Affairs' (DCA) ongoing elderly housing program. As with all DCA elderly projects, the site will be acquired, the construction contract will be let, and the project will be owned and managed by a local housing authority.[7]

Figure 4.19 shows a model of the installation.
Figure 4.20 shows the mechanical layout for the elderly housing unit.
A larger commercial solar heating installation is described below:

Figure 4.19. Illustration of a Multi-Family Solar Heated Building. (Source: Energy Research and Development Administration[7]).

PROJECT INFORMATION

Owner/Builder: Louisiana State University and A&M College
Operational Date: September, 1977
Total Estimated ERDA Funds: $258,225

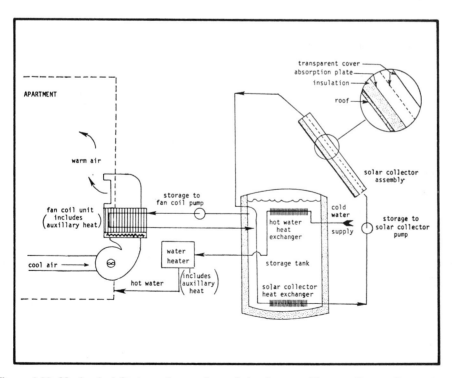

Figure 4.20. Mechanical Systems Layout for a Solar System. (Source: Energy Research and Development Administration[7]).

Building
 Type: Field House
 Area: 101,200 sq.ft (conditioned)
Location: Baton Rouge, Louisiana
Latitude: 30.7°N
Climatic Data
 Degree Days   Heating 1560   Cooling 2585
 Avg. Temp. (°F)   Winter 55.2   Summer 76.2
 Avg. Insol. (Ly/d) Winter 287   Summer 490

SOLAR ENERGY SYSTEM

Application Heating Hot Water Collector
 Type: Two cover, modular, selective absorber
 Area: 5,700 sq. ft
 Manufacturer: Lennox Industries-Honeywell, Inc.
Storage
 Type: Water
 Capacity: 20,000 gallons
Auxiliary System Type: Gas-fired Boilers

PROJECT DESCRIPTION

An existing field house used year round by students and faculty for indoor athletics will be retrofitted with the proposed solar energy system. The field house consists of a one story center wing, a two story south wing, and a two story north wing. Exterior walls are 8-in. concrete, except the upper part of the center wing, which is plywood sheathing on light gauge steel framing with glass fiber blanket insulation. Ceilings are insulated with brown cellulose acoustical-thermal insulation. Collectors are located on the roof of the south wing. The maximum domestic hot water requirement is 1600 gallons per hour. The solar system is designed to provide storage of heat equivalent to two days of peak solar radiation. The yearly sunshine received by Baton Rouge is approximately 57%.

Figure 4.21 shows a drawing of the field house. In Fig. 4.22 the mechanical system layout for the field house is shown. Notice the purge coil present as explained earlier in Fig. 4.18.

4.3.2. Storage

Heat transfer and storage medium fluids for a liquid solar heating system can consist of:

 1. Potable water
 2. Water/ethylene glycol solution
 3. Water/propylene glycol
 4. Silicone fluids
 5. Mineral oils
 6. Other (e.g., Dowtherm J, Sun-Temp, Therminol, etc.)

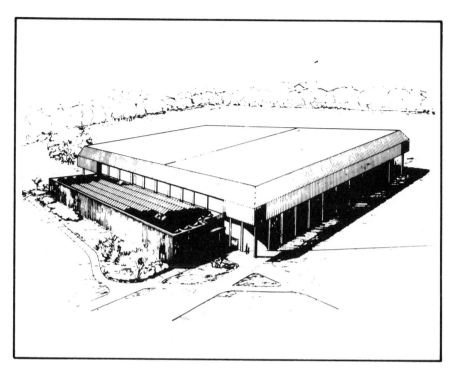

Figure 4.21. Solar Heated University Fieldhouse. (Source: Energy Research and Development Administration[7]).

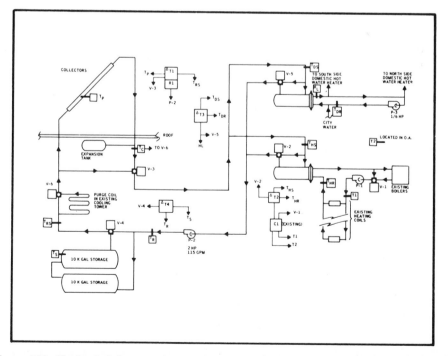

Figure 4.22. Mechanical Systems Layout for a Solar System. (Source: Energy Research and Development Administration[7]).

Table 4.3. Advantages and Disadvantages of Storage Tank Types. (Source: Cole[5]).

Steel Tank	Fiberglass Tank	Concrete Tank	Wooden Tank with Liner
ADVANTAGES			
Cost is moderate	Factory-insulated tanks are available	Cost is low	Cost is moderate
Steel tanks can be designed to withstand pressure	Considerable field experience is available	Concrete tanks may be cast in place or may be precast	Indoor installation is easy
Much field experience is available	Some tanks are designed specifically for solar energy storage		
Connections to plumbing are easy	Fiberglass does not rust or corrode		
Some steel tanks are designed specifically for solar energy storage			
DISADVANTAGES			
Complete tanks are difficult to install indoors	Maximum temperature is limited even with special resins	Careful design is required to avoid cracks and leaks	Maximum temperature is limited
Steel tanks are subject to rust and corrosion	Fiberglass tanks are relatively expensive	Concrete tanks must not be pressurized	Wooden tanks must not be pressurized
	Complete tanks are difficult to install indoors	Connections to plumbing are difficult to make leaktight	Wooden tanks are not suitable for underground installation
	Fiberglass tanks must not be pressurized		

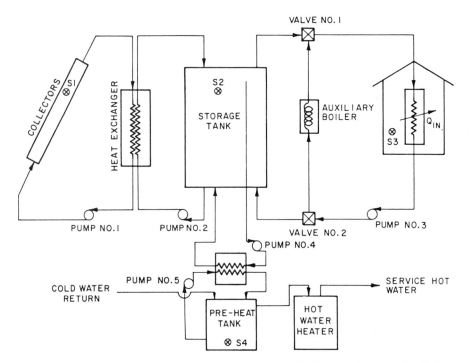

Figure 4.23. Schematic Representation of a Typical Liquid System. (Source: U. S. Department of Commerce[25]).

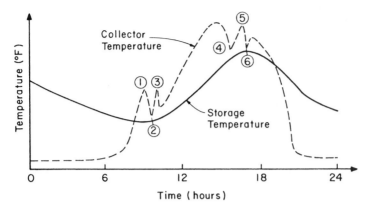

Figure 4.24. Typical Temperature Profiles of a Solar System. (Source: U. S. Department of Commerce[25]).

The selection of a storage medium may be based on such factors as economic, product availability, product life, or maintenance considerations (e.g., possible corrosion problems). The storage tanks used to hold the collected solar heat are generally made of steel, fiberglass, or concrete. Wooden tanks can also be used with appropriate liners. Table 4.3 shows some advantages and disadvantages of various tank types.

4.3.3. Controls

Figure 4.23 shows a design of a liquid solar heating system. The system shown in Fig. 4.23 utilizes four sensors:

S1 measures collector fluid temperature
S2 measures liquid storage temperature
S3 measures temperature inside building
S4 measures temperature of liquid in pre-heat tank

These four sensors, through the use of appropriate system control relays, operate the various pumps. Figure 4.24 illustrates the varying temperatures of the collector and storage tanks over a 24 hour period. In Fig. 4.24, at about 10:00 AM the temperature of the collector has increased to position 1 on the collector temperature curve. At this point, when the collector temperature (of S1) is greater than the storage temperature (of S2) by some present amount (typically 20°F), then pumps P1 and P2 of Fig. 4.23 are activated and begin pumping until the temperature difference drops to 3°F. At this point (position 2 in Fig. 4.24) pumps P1 and P2 are turned off. The ratio between the cut-on and cut-off temperatures is 20:3 or 6.7:1. In a liquid system this ratio should range from about 5:1 to 7:1. A smaller temperature ratio such as 5:1 will cause the pumps to cycle on and off more than a higher ratio such as 7:1. Higher cycling will increase the total energy that is stored in the storage tank but will lead to greater wear on the pump motors and relays. There is thus a tradeoff between higher system efficiencies and system wear. For this reason it is suggested that the on–off temperature ratio be kept between about 5:1 and 7:1. Referring back to Fig. 4.24, at position 3 of the collector temperature graph, pumps P1 and P2 begin operating again. As shown in the graph, the pumps lower the collector temperature somewhat and increase the storage temperature but the higher solar insolation at noon cause the pump to continue operating until about 3:30 PM (or position 4 on the graph). When collector temperature increases to position 5 the pumps turn on until collector temperature drops to about 3°F of the storage temperature. After the pumps cut off as shown at position 6 of the graph, the collector temperature increases but not enough to activate the pumps. At this time the sun is not able to raise the temperature of the solar collectors the required 20°F above the storage temperature until the following day.

In Fig. 4.23 thermostat (S3) located inside the building functions much as a standard residential or commercial thermostat used by the occupants in a building for setting and maintaining the desired room temperature. It is a two stage type thermostat of which the first stage controls pump number three to furnish heat to the fan coil unit located inside the building.* If the temperature of the storage tank is not high enough to provide sufficient heat to the fan coil unit, then the second stage of thermostat S3 is activated which engages an auxiliary boiler to increase the water temperature reaching the fan coil unit. Although Fig. 4.23 shows only one thermostat (e.g., S3) and a single fan coil unit used in regulating conditioned space temperature, the system can be expanded by using additional thermostat and fan coil units to provide multi-zone control.

The last temperature sensor shown in Fig. 4.23 is sensor S4 which controls pumps 4 and 5 that provide hot water to the preheat tank. The preheat tank serves to raise the temperature of the inlet water to a higher temperature. This heated water is then fed to the water heater which raises the water temperature to the desired level for household use (e.g., 60°C or 140°F).

Figure 4.25 illustrates another solar heating system control schematic. The systems control design shown in Fig. 4.25 is a further refinement of the system shown in Fig. 4.23 in that the design in Fig. 4.25 allows the heated water from the solar collector to be diverted to the storage tank for later use *or* pumped directly to the fan coil unit heat exchanger. This is an important refinement over the system shown in Fig. 4.23 which only allows the heated water from the solar collector to be transferred to the storage tank.

A heating control system that can route heated water directly from the collector to the fan coil heating unit, may realize greater efficiency. This increased efficiency derives from running the solar collectors at lower operating temperatures. In the system shown in Fig. 4.23 the solar collector pumps

---

*It should be noted that fan coil units are preferred over using baseboard heaters because fan coil units can operate at lower water temperatures (e.g., 38°C or 100°F). The baseboard radiators require temperatures of about 66°C (150°F) to operate effectively. As discussed in Chapter 2, by operating (solar flat plate) solar collectors at lower temperatures, greater collector efficiencies are obtained. In addition, by operating solar heating systems (e.g., storage tanks) at higher than necessary temperatures, greater heat losses will be incurred due to a larger ΔT.

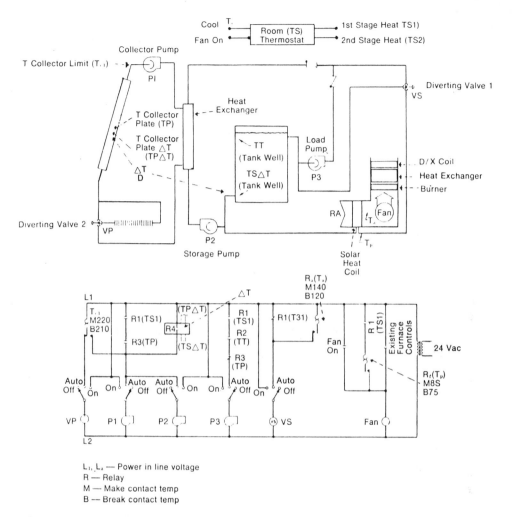

Figure 4.25. Solar Heating System Control Schematic. (Source: Sheet Metal and Air Conditioning Contractors National Association[18]).

operate only when the collectors are at a temperature 20°F greater than the storage temperature. In the system shown in Fig. 4.25, the collectors operate and supply heated water directly to the fan coil unit even if the collector temperature is lower than the storage temperature. This allows the system to operate the collectors at reduced temperatures down to about 90° F. After the thermostat inside the house signals that the proper room temperature has been reached, the solar collectors can then function much like the system shown in Fig. 4.23 and pump heated water from the collectors to the storage tank when the collector temperatures are higher than the storage tank temperature.

The basic differential controller can either be purchased from numerous manufacturers of the devices or can be homemade. Figures 4.26 and 4.27 schematically illustrate a home-made controller and the controller's power supply.

Table 4.4 is a parts list of the controller and power supply shown in Figs.

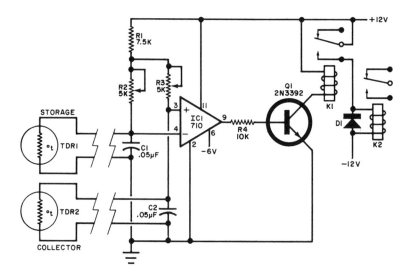

*Comparator IC1 turns on or off depending on resistances of
TDR1 and TDR2. When IC1 is on, Q1 and the relays are energized.*

Figure 4.26. Schematic Diagram of a Homebuilt Differential Controller. (Source: Cogswell[4]).

4.26 and 4.27. The total cost of the parts list will run about $45 to $50. Another home-made controller is shown in Fig. 4.28. The parts list for the controller shown in Fig. 4.28 is listed in Table 4.5.

4.3.4. Maintenance Considerations

Maintenance will be more frequent for a liquid type solar heating system than for an air type system. In the liquid system corrosion and leakage problems can occur.

It cannot be stressed enough that many maintenance problems present in liquid systems can be attenuated or eliminated by proper system planning and design before the first pipe or solar collector is installed on the building. This includes not using an open-type nondrainable solar collector system in a northern climate where freezing weather can cause the collector piping to rupture and burst due to expansion of frozen water in the pipes. In a liquid type system, attention should be paid to the materials used and how they will be connected. This means that copper tubing at least one-half inch I.D. should be used in open type systems to lessen corrosion and problems of clogged pipes. In many solar liquid heating systems a closed type system uses heat exchangers to transfer thermal energy from the solar collectors to the storage tank which decreases the chances of freezing and chemical type corrosion problems. In both the open and closed type liquid heating systems galvanic corrosion can occur. By proper system design this type of corrosion can be lessened by not joining copper fittings directly to a steel storage tank which can strip off the steel threads in the joints and cause leakage problems. Designing a liquid system using a dielectric joint such as a brass or bronze

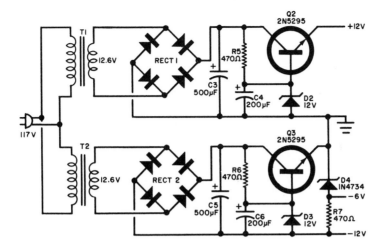

*The power supply for the solar controller is standard
design and provides regulated positive and negative outputs.*

**Figure 4.27. Schematic Diagram of the Power Supply for the Homebuilt Controller. (Source:
Cogswell[4]).**

**Table 4.4. Parts List for Homebuilt Differential Solar Collector. (Source:
Cogswell[4]).**

C1, C2 -- 0.05- F, ceramic disc capacitor

C3, C5 -- 500- F, 25-V electrolytic capacitor

C4, C6 -- 200- F, 25-V electrolytic capacitor

D1 -- General-purpose silicon rectifier diode

D2, D3 -- 12-V, 1-W zener diode (1N4742, or similar)

D4 -- 6-V, 1-W zener diode (1N4734 or similar)

IC1 -- 710 voltage comparator

K1 -- 12-V, 600-ohm coil relay

K2 -- 24V, 10-ampere contacts relay

Q1 -- 2N3392 transistor

Q2, Q3 -- 2N5295 transistor (or similar)

R1 -- 7500-ohm, 1/-W resistor

R2, R3 -- 5000-ohm multi-turn trimmer potentiometer

R4 -- 10,000-ohm, 1/-W resistor

R5, R6, R7 -- 470-ohm, 1/-W resistor

T1, T2 -- 12.6-V, 300-mA transformer (Radio-Shack 273-1385
          or similar)

TDR1, TDR2 -- TG-1/8, 100-ohm, $\pm$ 5% Sensitor

Misc. -- Suitable enclosure, perforated or pc board, socket
         for IC1, twin lead cable for sensors, heat sinks (2),
         power cord, mounting hardware.

Note: The Sensitors are avilable from Texas Instruments
          seminconductor dealers, or from Texas Instruments,
          2916 Holmes St., Kansas City, MO 64109 at $2.40 each.

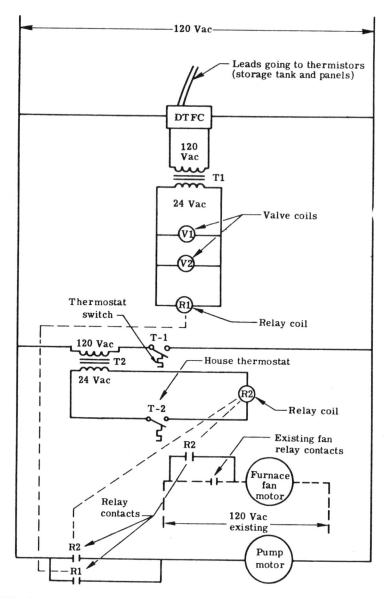

Figure 4.28. Schematic Diagram of Differential Controller Used in NASA Test House. (Source: Allred[1]).

joint to connect a copper pipe to a steel tank can reduce such galvanic corrosion problems.

The examples in the previous paragraph represent some of the design considerations which if analyzed before the heating system is installed may considerably lessen system maintenance. Basic maintenance on a liquid type heating installation includes:

1. Periodically checking the antifreeze solution (if used) for chemical breakdown due to aging—and changing if necessary.

Table 4.5. Parts List for Solar System Controller. (Source: U. S. Department of Housing and Urban Development[1]).

Component Name	Symbol	Component Description	Manufacturer (a)
Transformer	T1	120 to 24 Vac; 4 A (Type 6K80VBR or equivalent)	*Allied Electronics Corp. 401 E. 8th St. Ft. Worth, TX 76102 (1976 Engineering Manual and Purchasing Guide *760)
Transformer	T2	120 to 24 Vac; 1 A (Type 6K113HF or equivalent)	
Relay	R1	24 Vac; 10 A; double pole double throw (Type KA11AG or equivalent)	Newark Electronics Corp. 500 Pulaski Rd. Chicago, IL 60624 Allen-Bradley Co. 1201 S. 2d St. Milwaukee, WI 53204
Relay	R2	24 Vac; 10 A; double pole double throw (Type KA11AG or equivalent)	Radio Shack Div. 2617 W. 7th St. Ft. Worth, TX 76107
Thermostat	T-2	(Chromalox WR-1E30 or equivalent)	Edwin L. Weigand Div. Emerson Electric Co. 7500 Thomas Blvd. Pittsburg, PA 15208 Honeywell, Inc. 2701 Fourth Ave. S. Minneapolis, MN 55408 General Electric Co. 1 River Rd. Schenectady, NY 12345 Robertshaw Controls Co. Control Systems Div. 1701 Byrd Ave. Richmond, VA 23230
Thermostat switch	T-1	Bimetallic operating temperature, 40°F to 200°F; single pole single throw; 115 Vac; 10-A rated; well included (Mercoid type FM437-3-3516 or equivalent)	*The Mercoid Corp. 4201 Belmont Ave. Chicago, IL 60641 PSG Industries, Inc. 910 Ridge Ave. Perkasie, PA 18944 United Electric Controls 80 School St. Watertown, MA 02172 American Thermostat Corp. Box 60 South Cairo, NY 14282
Differential temperature flow controller	DTFC	120 Vac; 10°F turn-on differential; 5°F turn-off differential; two thermistors included; 300°F maximum sensor temperature (DEKO-LABS Model TC-3 or equivalent)	DEKO-LABS Box 12841 Gainesville, FL 32602 Rho Sigma 5108 Melvin Ave. Tarzana, CA 91356 Jack S. Scovel 4220 Berritt St. Fairfax, VA 22030

aAsterisk indicates manufacturer of items used in demonstration system.

2. Checking and lubricating pumps.
3. Periodically changing the air filter on the fan coil unit.
4. Checking liquid type system filters and replacing if clogged.
5. Checking fluid levels in heat exchanger loops.

4.4 REFERENCES AND SUGGESTED READING

1. Allred, Johnny W., Joseph M. Shinn, Cecil E. Kirby, and Sheridan R. Barringer, *An Inexpensive Economical Solar Heating System for Homes,* Washington, D.C.: National Aeronautics and Space Administration, July, 1976, pp. 52–53.
2. Better Heating and Cooling Bureau and the Sheet Metal and Air Conditioning Contractors National Association, Inc., *Solar Installation Standards for Heating and Air Conditioning Systems,* 1977.
3. Britton, Peter, "World's Most Advanced Solar Home," *Popular Science,* **211,** 1:92–95, 1977.
4. Cogswell, Jerald M., "Build a Solar Controller," *Popular Electronics,* **12,** 1:69, 70, 1977.
5. Cole, Roger, L., Nield, Kenneth, J., Rohde, Raymond R., and Wolosewicz, Ronald M., *Design and Installation Manual for Thermal Energy Storage,* Argonne, Illinois: Argonne National Laboratory, 1979, pp. 8, 68.
6. Copper Development Association, Inc., New York, *Solar Energy Systems Design Handbook,* 1977.
7. Energy Research and Development Administration, *Project Data Summaries Vol. I Commercial and Residential Demonstrations,* Washington, D.C.: U. S. Government Printing Office, August, 1976, pp. 127–128, 17, 26.
8. Fantel, Hans, "Our Year in NASA's Far-out House," *Popular Mechanics,* **151,** 6:77–79, 126–128, 1979.
9. Federal Energy Administration, *Buying Solar,* Washington, D.C.: U. S. Government Printing Office, 1970, p. 6.
10. Florida Solar Energy Center, *A Guide to Solar Water Heating in Florida,* Cape Canaveral, Florida: State University System of Florida, October, 1978, p. 10.
11. Florida Solar Energy Center, *Solar Water and Pool Heating Installation and Operation,* Cape Canaveral, Florida: State University System of Florida, January, 1979, pp. 5.27–5.31.
12. Gropp, Louis Oliver, "Bold Angles and a Very Special Collector," *A House and Garden Guide - Building,* Spring, 1978, pp. 73–97.
13. International Telephone and Telegraph Corporation, Fluid Handling Division, Training and Education Department, *Solar Heating Systems Design Manual,* 1976.
14. Intertechnology Corporation, *Solar Energy School Heating Augmentation Experiment,* Washington, D.C.: U. S. Government Printing Office, December, 1974.
15. Katzel, Jeanine, "Solar Energy at Work," *Plant Engineering,* 126–138, (1978).
16. Kreuter, Rodney A., "Build this Solar Controller," *Radio-Electronics,* **49,** 6:35–37, (1978).
17. Miller, Charles A., "Super Storage for Solar Heat," *Mechanix Illustrated,* **74,** 605:40–41, (1978).
18. Sheet Metal and Air Conditioning Contractors National Association, *Fundamentals of Solar Heating,* Washington, D.C.: U. S. Government Printing Office, January, 1978, pp. 4.9, 7.1–7.10, 7.14, 9.16.
19. Shuttleworth, John, ed., "Mother's Heat Grabber," *The Mother Earth News,* 47:101–103, (1977).
20. Shuttleworth, John, ed., "Mother's Heat Grabber is Back!" *The Mother Earth News,* 54:94–96, (1978).
21. Solaron Corporation, Denver, Colorado, *Application Engineering Manual,* 1977, pp. 85–85A.
22. Stepler, Richard, "Solar Heating — How to Pick the Right System for Your Home," *Popular Science,* **213,** 1:72–75, (1978).
23. Stubblefield, Richard R., "Solar System Employs Total Heat Concept," *Heating/Piping/Air Conditioning,* **47,** 7:36–40, (1975).
24. U. S. Department of Commerce, *Solar Heating and Cooling of Residential Buildings – Design of Systems,* Washington, D.C.: U. S. Government Printing Office, October, 1977, pp. 7.3, 12.13, 12.15, 15.18, 16.2–16.3.

25. U. S. Department of Commerce, *Solar Heating and Cooling of Residential Buildings – Sizing, Installation and Operation of Systems,* Washington, D.C.: U. S. Government Printing Office, October, 1977, pp. 5.7–5.11, 6.3, 11.3–11.4.
26. U. S. Department of Energy, *Introduction to Solar Heating and Cooling Design and Sizing,* Washington, D.C.: U. S. Government Printing Office, August, 1978, pp. 6.2, 4.52, 4.36, 4.33, 4.31, 4.28, 1.46–1.50.
27. U. S. Department of Energy, Solar Technology Transfer Program, *Space Heating Handbook with Service Hot Water and Heat Load Calculations: Solcost,* July, 1978.
28. U. S. Department of Housing and Urban Development, *Intermediate Minimum Property Standards Supplement: Solar Heating and Domestic Hot Water Systems,* Standard 4930.2, Washington, D.C.: U. S. Government Printing Office, 1977.
29. U. S. Department of Housing and Urban Development, *Solar Dwelling Design Concepts,* Washington, D.C.: U. S. Government Printing Office, May, 1976, p. 29.
30. Yellott, John I., "Solar Energy Update," *Heating/Piping/Air Conditioning,* **51,** 1:55–63, (1979).

CHAPTER 5
COMBINED SOLAR HEATING AND COOLING SYSTEMS

5.1 OVERVIEW OF COMBINED SYSTEMS

When solar cooling is used in a building, solar domestic water heating and/or solar space heating is usually present. Incorporating solar space and domestic water heating systems into the solar cooling system is generally done for economic reasons. In combined solar heating and cooling systems, many components are shared between the two such as solar collectors, storage tanks, pumps and piping. If a solar installation consisted of only a solar air conditioning system, then depending upon the geographical location, the system might only be used four to six months of the year. This would mean that system components such as solar collectors and storage tanks would be inactive six to eight months of the year. By adding solar space heating and/or a domestic hot water system to the solar air conditioning system in the building, these shared components are more fully utilized during the year and may therefore give a quicker monetary payback.

There are three basic methods used in space cooling. These are refrigeration, evaporative cooling, and radiative cooling. The refrigeration method is most applicable to solar applications and will be the primary method discussed. There are three basic solar systems falling under the refrigeration category:

1. Absorption refrigeration
2. Solar assisted heat pump
3. Solar Rankine-cycle engine

5.2 ABSORPTION REFRIGERATION

5.2.1. Absorption System Overview

The two basic absorption systems are the lithium bromide–water and the ammonia–water systems. It should be noted that absorption refrigeration systems have been used for a number of years in cooling commercial buildings. For these nonsolar absorption refrigeration systems, natural gas has been the main source of thermal energy to run the machines. The solar absorption refrigeration machine basically functions like a standard gas-fired absorption unit except that hot water from the solar collectors is substituted for the gas flame used in supplying heat for system operation. Presently only the lithium

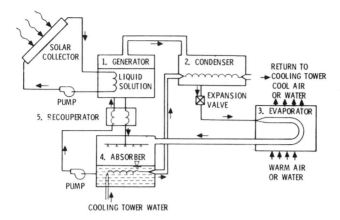

Figure 5.1. Schematic Drawing for an Absorption Air Conditioner. (Source: U. S. Department of Energy[35]).

bromide–water absorption system is suitable for solar application. This is because the lithium bromide–water absorption units will operate at lower temperatures than the presently available ammonia–water absorption units. The basic absorption system is shown in Fig. 5.1.

In the illustration water acts as the refrigerant and lithium bromide as the absorbent. The cycle begins when the lithium bromide–water solution in the generator is heated by 170°F (77°C) hot water from the solar collector. Due to low pressure in the generator, the water in the lithium bromide–water solution is superheated to a vaporous form. This water vapor is directed into the condenser where it is cooled to about 100°F (38°C) by a heat exchanger located inside the condenser. Upon being cooled, the water vapor condenses into liquid form. This water then passes through an expansion valve where part of the water is revaporized. This vapor–liquid then passes through the evaporator coils at about 40°F (4°C) when warm air from the building to be cooled passes between the evaporator coils, both heat and moisture from the room air is removed. This results in lowering both the temperature and humidity of the air in the building being cooled. The refrigerant (i.e., water) flowing through the evaporator coils, after absorbing heat from the warm air in the building, is then transported back to the absorber where the refrigerant is mixed with the concentrated lithium bromide solution from the generator at a temperature of about 100°F (38°C). While the refrigerant from the evaporator is being mixed in the absorber, heat is released. This heat in the absorber is removed by the cooling tower water. The cooled absorber liquid is then pumped back to the generator to repeat the absorption cycle. The recouperator shown in Fig. 5.1 serves to transfer heat from the liquid solution flowing from the generator into the absorber to the cooler dilute solution flowing from the absorber into the generator. This preheating serves to make the absorption system more efficient. Figure 5.2 illustrates the Arkla Industries' Solaire Model WF36 3-ton lithium bromide absorption unit. Figure 5.3 illustrates the operation of an absorption chiller in a building. Note the various inlet and outlet temperatures of the components shown in the illustration. Solar operated lithium bromide

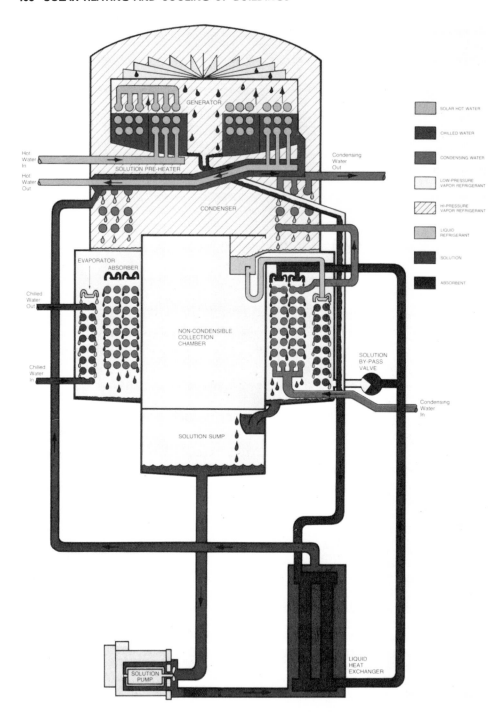

SOLAR HOT WATER

CHILLED WATER

CONDENSING WATER

LOW-PRESSURE
VAPOR REFRIGERANT

HI-PRESSURE
VAPOR REFRIGERANT

LIQUID
REFRIGERANT

SOLUTION

ABSORBENT

Figure 5.2. Arkla Industries' 3-Ton Lithium Bromide Absorption Unit. (Source: Arkla Industries, Inc.[4]).

absorption refrigeration systems have been installed in a number of commercial and residential buildings. Figure 5.4 illustrates a residential solar installation in a National Bureau of Standards test house. The absorption chiller system used in the test house is a 3-ton Arkla Industries Solaire Model WF36 unit.

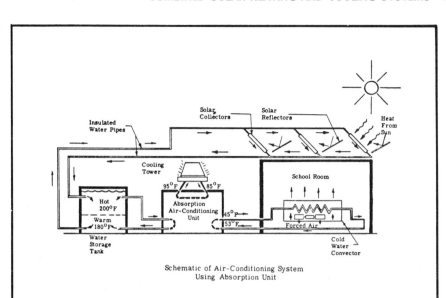

Figure 5.3. Schematic of Air Conditioning System Using Absorption Unit. (Source: Energy Research and Development Administration).

The following project information describes the solar system installed at the Florida Welcome Station.

PROJECT INFORMATION

Owner/Builder: Florida Dept. of Commerce
Contractor: Florida Dept. of General Services
Operational Date: June, 1977
Total Estimated ERDA Funds: $347,812
Building:
  Type: Welcome station
  Area: 3,300 sq.ft (conditioned)
Location: Nassau County, Florida
Latitude: 30.7°N
Climatic Data
  Degree Days   Heating 1327   Cooling 2596
  Avg. Temp. (°F)   Winter 57.7   Summer 76.0
  Avg. Insol. (Ly/d) Winter 308   Summer 496

SOLAR ENERGY SYSTEM

Application   Heating 99%   Cooling 88%   Hot Water 0%
Collector
  Type: Fresnel Lens, concentrating and tracking
  Area (sq. ft.): 2,700
  Manufacturer: Northrup, Inc.
Storage
  Type: Water
  Capacity: 10,000 gallons
Auxiliary System Type: Oil fired furnace

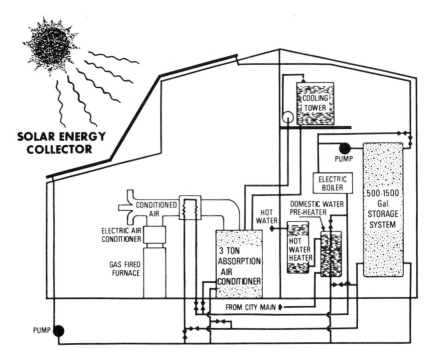

Figure 5.4. Schematic of the Solar Heating and Cooling System in the NBS Test House. (Source: Hill[16]).

PROJECT DESCRIPTION

This Florida Welcome Station, located on Interstate 95 two miles south of the Florida-Georgia line, is expected to accommodate a large number of visitors. To conserve energy, the one-story building utilizes a poured, lightweight, aggregate roof deck; a suspended insulating tile ceiling; and windows that are protected from direct sunlight. Key system components include a 25-ton ARKLA WF-300 absorption chiller, a field of concentrating collectors, a hot water storage tank and a cooling tower that will provide condensing water. An oil-fired hot water heater, a vapor compression air exchange air conditioner, and an oil-fired furnace will be auxiliary systems. The building's energy system (solar and conventional) will be controlled by a single set of valves and sensors. Automatic switching between the two systems will be possible, manual operation being necessary only for selecting the heating and cooling mode.[42]

Figure 5.5 shows the mechanical systems schematic for the welcome station. Another example of a combined solar heating and cooling system installation is described below.

PROJECT INFORMATION

Owner/Builder: Development Dynamics of Richland
Operational Date: January, 1977
Total Estimated ERDA Funds: $592,456

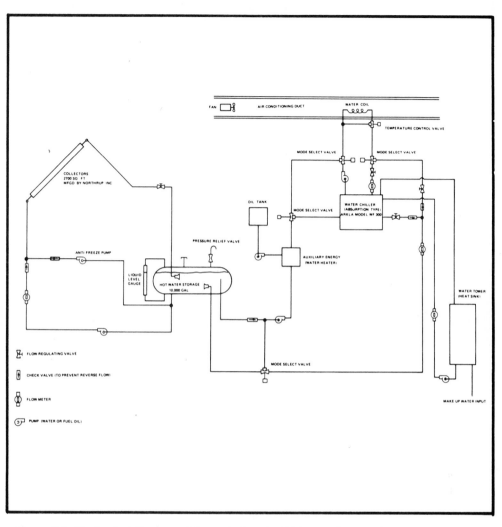

Figure 5.5. Mechanical Systems Schematic for the Welcome Station. (Source: Energy Research and Development Administration).

Building:
   Type: Office
   Area: 14,400 sq.ft
Location: Richland, Washington
Latitude: 46.3°N
Climatic Data
   Degree Days   Heating 4836   Cooling 862
   Avg. Temp. (°F)   Winter 45.8   Summer 70.5
   Avg. Insol. (Ly/d) Winter 255   Summer 553

SOLAR ENERGY SYSTEM

Application   Heating 55%   Cooling 72%   Hot Water 90%
Collector

Type: Flat-plate
Area (sq. ft.): 8,000
Manufacturer: General Electric Co.
Storage
Type: Water
Capacity: 9,000 gallons
Auxiliary System Type: Electric

PROJECT DESCRIPTION

This demonstration site, in the desert of south central Washington State, is adjacent to and readily visible from a principal north-south traffic artery in Richland. The design involves two identical single-story office buildings in a four-building complex known as Hanford Square. The buildings, each with a capacity of about 100 persons, will have the same external climatographic exposure and nearly identical internal environmental conditions. One building, heated and cooled solely by electrical systems, will act as a control, furnishing power data needed to compute solar energy input as a percentage of total power consumed. Energy conserving features incorporated into the solar building will include solar ponds, dual purpose absorption/reflection venetian blinds, solar panel insulation and double windows. Energy conserving features unrelated to the solar system will be duplicated in both buildings.[42]

Figure 5.6 shows a sketch of the Hanford Square complex and the mechanical systems layout for the complex is shown in Fig. 5.7.

Notice in Fig. 5.7 that there are eight zones shown in the schematic diagram. This type design allows for separate comfort control in eight parts of the building.

Solar heated lithium bromide–water absorption refrigeration units are now

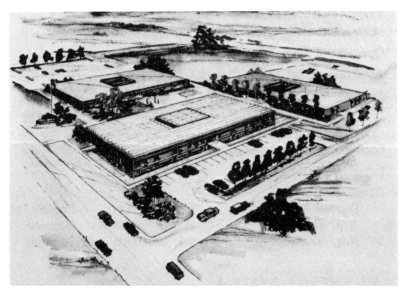

Figure 5.6. Sketch of the Hanford Square Complex. (Source: Energy Research and Development Administration).

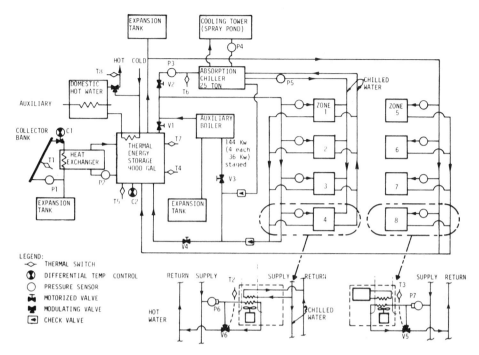

Figure 5.7. Mechanical Systems Layout for the Hanford Square Complex. (Source: Energy Research and Development Administration).

available for purchase in various cooling capacities from a number of companies. Many of these manufacturers have been producing gas-fired absorption chillers for a long time and have in recent years modified their basic (gas-fired) absorption unit to operate from heated water supplied by solar collectors. Arkla Industries, formerly the Servel Corporation, is one such company. Some readers may recall the old Servel gas-fired absorption home refrigerators manufactured by the company years ago. Arkla Industries presently sells a 3-ton solar unit for residential·application and a 25.5-ton solar unit for commercial use, along with their line of standard gas-fired absorption units. Table 5.1 provides data for the Arkla Solaire WF36 3-ton unit.

At this point two terms need to be defined. "Tons of refrigeration" relates to the cooling capacity of a refrigeration machine. The ton refers to the rate of heat removal in a cooling system and is equal to 12,000 Btu per hour (or 3514.50 W). Consequently when heat is removed from the conditioned space in a building, cooling occurs. Referring to the top of Table 5.1 it can be seen that 36,000 Btu/h equals 3 tons of refrigeration (36,000/12,000 = 3 tons). Another term to be discussed is the coefficient of performance (COP). The coefficient of performance indicates the efficiency of a refrigeration system and the formula for determining it is:

$$COP = \frac{\text{thermal energy removed}}{\text{energy supplied from external sources}} \qquad (5.1)$$

Table 5.1.   Data for Arkla 3-Ton Absorption Unit. (Source: Arkla Industries, Inc.[4]).

```
DESIGN DELIVERED CAPACITY, Btu/h....................36,000ᵃ
DESIGN DELIVERED CAPACITY, Tons I.M.E. ............  3.0ᵃ
ENERGY REQUIREMENTS
  Design Hot Water Input, Btu/h.....................50,000
  Design Hot Water Inlet Temperature, °F............ 195
  Design Hot Water Outlet Temperature, °F........... 185.9
  Permissible Range of Inlet Temp.............. 170 to 205
  Design Hot Water Flow, gpm....................... 11.0
  Pressure Drop, Feet of Water, at 11 gpm...........  9.8
  Permissible Range of Flow, gpm................. 5 to 22
  Pressure Drop, Feet of Water, at 22 gpm........... 29.9
  Maximum Working Pressure, psig................... 100
  Unit Water Volume, Gallons, Approx...............  3.0ᵇ
  Electrical Voltage, 60 Hz, 1 Phase............... 115
  Maximum Wattage Draw............................ 250
CHILLED WATER DATA
  Design Inlet Temperature, °F.....................  55
  Design Outlet Temperature, °F....................  45
  Design Flow, gpm.................................  7.2
  Pressure Drop, Feet of Water, at 7.2 gpm.........  4.6
  Permissible Range of Flow, gpm.............. 4 to 13
  Pressure Drop, Feet of Water, at 13 gpm........... 12.5
  Maximum Working Pressure, psig................... 100
  Unit Water Volume, Gallons, Approx...............  1.5
CONDENSING WATER DATA
  Design Heat Rejection, Btu/h.....................86,000
  Design Inlet Temperature, °F.....................  85
  Design Outlet Temperature, °F.................... 99.3
  Permissible Range of Inlet Temp................ 75 to 90
  Design Flow, gpm................................. 12.0
  Pressure Drop, Feet of Water, at 12 gpm...........  9.6
  Permissible Range of Flow, gpm................ 9 to 25
  Pressure Drop, Feet of Water, at 25 gpm........... 33.9
  Maximum Working Pressure, psig................... 100
  Unit Water Volume, Gallons, Approx...............  3.0
FOR COOLING TOWER SELECTION
  Maximum Heat Rejection, Btu/h....................106,000
  Range, °F........................................14 to 17ᶜ
  Minimum Permissible Sump Temperature,°F.........  75
SERVICE CONNECTIONS
  Hot Water Inlet and Outlet...................... 1" FPT
  Chilled Water Inlet and Outlet.................. 1" FPT
  Condensing Water Inlet and Outlet............... 1" FPT
PHYSICAL DATA, APPROXIMATES
  Operating Weight, Pounds........................  675ᵈ
  Shipping Weight, Pounds.........................  680ᵉ
  Crated Size, Inches..................... 36W, 34D, 75H
```

```
                        NOTES
      ᵃCapacity at design conditions.

      ᵇUnits equipped for operation on 230V-50Hz-1Ph
       available on special order.

      ᶜThermostatic switch to control tower fan MUST
       be used.  Set to "cut out" at 75°F.

      ᵈIncludes circulating water weights.

      ᵉUnits as shipped contain Lithium Bromide charge.
```

From Table 5.1 the *COP* can be determined by dividing the Design Delivered Capacity (36,000 Btu/h) by the design input hot water requirements (50,000 Btu/h) to equal a *COP* of 0.72. Most lithium bromide–water absorption machines operate in the range of 0.60 to about 0.80. From Eq. 5.1 a *COP* of less than 1.0 means that more energy is supplied to the refrigeration machine than is removed by the machine. Using Eq. 5.1, the reader can observe in Table 5.2 that the *COP* will vary somewhat as different delivery capacities are chosen.

Table 5.2 illustrates performance data for the Arkla 3-ton unit described in Table 5.1. Note that the absorption unit is rated at 3 tons when the hot water inlet temperature is 195°F (90.6°C), the outlet temperature is 185.9°F (85.5°C),

Table 5.2. Performance Data for Arkla 3-Ton Absorption Unit. (Source: Arkla Industries, Inc.[4]).

```
Hot Water Flow.....................11.0 GPM
Condensing Water Flow..............12.0 GPM
Chilled Water Flow................. 7.2 GPM
Chilled Water Leaving Temperature........45°F
```

Hot Water		Energy	Inlet Cond.	Delivered Capacity		Rejected
Inlet Temp	Outlet Temp	Input Btu/H	Water Temp	Btu/H	Tons	Heat Btu/H
	167.0	16,400	80°F	9,700	0.81	26,100
170°F	167.4	14,500	85°F	6,400	0.53	20,900
	*	*	90°F	*	*	*
	170.7	23,800	80°F	17,300	1.44	41,100
175°F	171.1	21,600	85°F	13,100	1.09	34,700
	*	*	90°F	*	*	*
	174.3	31,200	80°F	24,400	2.03	55,600
180°F	174.8	28,800	85°F	19,400	1.62	48,200
	175.7	23,800	90°F	14,200	1.18	38,000
	178.0	38,400	80°F	31,100	2.59	69,500
185°F	178.5	35,900	85°F	25,600	2.13	61,500
	179.4	30,600	90°F	19,300	1.61	49,900
	181.7	45,800	80°F	36,800	3.07	82,600
190°F	182.2	42,900	85°F	31,300	2.61	74,200
	183.2	37,500	90°F	23,800	1.98	61,300
	185.3	53,100	80°F	40,600	3.38	93,700
195°F**	185.9	50,000	85°F	36,000	3.00	86,000
	186.9	44,300	90°F	27,600	2.30	71,900
	189.3	58,800	80°F	41,800	3.48	100,600
200°F	189.8	56,000	85°F	40,200	3.35	96,200
	190.7	51,000	90°F	30,500	2.54	80,500
	193.4	63,800	80°F	42,000	3.50	105,800
205°F	193.9	60,800	85°F	42,000	3.50	102,800
	194.8	56,200	90°F	32,500	2.71	88,700

*Unit operation unstable in these areas.

**Conditions for rated capacity.

inlet condensing water temperature is 85°F (29.4°C) and when heat is rejected from the absorption unit through the condensing water at a rate of 86,000 Btu/ hr (25.19 kW). By varying these elements, the Arkla Model WF36 with a designed delivery capacity of 3 tons (or 36,000 Btu/h), can be made to operate from 0.53 to 3.50 tons capacity. The hot water inlet and outlet temperature numbers listed in Table 5.2 refer to the solar heated water entering and exiting the generator as illustrated in Fig. 5.1. The inlet condensing water temperature figures in the table refer to the water entering the absorber section of the unit. The condensing water circulating through the absorber is piped back and forth to a cooling tower located outside of the building. The rejected heat (in Btu/hr, W) rates in Table 5.2 refer to the amount of heat emitted by the absorption refrigeration unit into the condensing water which expels the heat into the atmosphere via the cooling tower.

The operating data for the 25.5-ton Arkla Solaire Model WFB 300 water chiller is shown in Table 5.3. The Arkla Solaire 300 absorption unit with a design delivered capacity of 25.5 tons can be made to supply from 5.5 to 30.0 tons of refrigeration depending upon operating conditions such as the hot water inlet temperature, etc. Another manufacturer of lithium bromide absorption chillers is Yazaki Corporation of Japan. This firm offers nine lithium bromide absorption systems varying in capacity from 1.3 tons to 49.6 tons. Table 5.4 gives specifications for the lithium bromide absorption chiller units. A third company that has entered the solar absorption air conditioning field is York

Table 5.3. Operating Data for Arkla 25.5 Ton Absorption Unit. (Source: Arkla Industries, Inc.[5]).

```
DESIGN DELIVERED CAPACITY, Btu/h................... 306,000[1]
DESIGN DELIVERED CAPACITY, Tons I.M.E. ............    25.5[1]
ENERGY REQUIREMENTS
  Design Hot Water Input, Btu/h................... 447,000
  Design Hot Water Inlet Temperature, °F..........     195
  Design Hot Water Outlet Temperature, °F.........   184.8
  Permissible Range of Inlet Temp............... 160 to 200
  Design Hot Water Flow, gpm......................      90
  Pressure Drop, Feet of Water, at 90 gpm.........    20.7
  Permissible Range of Flow, gpm................ 50 to 100
  Pressure Drop, Feet of Water, at 100 gpm........    25.6
  Maximum Working Pressure, psig..................     100
  Electrical Voltage, 60 Hz, 1 Phase..............     115[2]
  Maximum Wattage Draw............................     150
CHILLED WATER DATA
  Design Inlet Temperature, °F....................      55
  Design Outlet Temperature, °F...................      45
  Design Flow, gpm................................      60
  Pressure Drop, Feet of Water, at 60 gpm.........     9.8
  Permissible Range of Flow, gpm................ 30 to 100
  Pressure Drop, Feet of Water, at 100 gpm.......     26.9
  Maximum Working Pressure, psig..................     100
  Unit Water Volume, Gallons, Approx..............      12
  Fouling Factor..................................    0005
CONDENSING WATER DATA
  Design Heat Rejection, Btu/h.................... 753,000
  Design Inlet Temperature, °F....................      85
  Design Outlet Temperature, °F...................   101.7
  Permissible Range of Inlet Temp................ 75 to 90
  Design Flow, gpm................................      90
  Pressure Drop, Feet of Water, at 90 gpm.........    22.9
  Permissible Range of Flow, gpm................ 50 to 110
  Pressure Drop, Feet of Water, at 110 gpm........    33.5
  Maximum Working Pressure, psig..................     100
  Unit Water Volume, Gallons, Approx..............      20
  Fouling Factor..................................     001
FOR COOLING TOWER SELECTION
  Maximum Heat Rejection, Btu/h................... 853,000
  Range, °F..................................... 16 to 17
  Minimum Permissible Sump Temperature, °F........      75[3]
SERVICE CONNECTIONS
  Hot Water Inlet and Outlet......................  2" FPT
  Chilled Water Inlet and Outlet.............. 2-1/2" FPT
  Condensing Water Inlet and Outlet........... 2-1/2" FPT
PHYSICAL DATA, APPROXIMATES
  Operating Weight, Pounds........................   3,420[4]
  Shipping Weight, Pounds.........................   3,145[5]
  Crated Size, Inches..............      114W, 45D, 69H
```

NOTES:

1. Capacity at design conditions.

2. Units equipped for operation on 230V-50Hz-1Ph
   available on special order

3. Thermostatic switch to control tower fan MUST
   be used. Set to "cut out" at 75°F.

4. Includes circulating water weights.

5. Units as shipped contain Lithium Bromide charge.

Division of Borg-Warner Corporation. This company has produced large steam operated lithium bromide absorption chillers for a number of years and has shown that their steam operated Model ES can function using solar heated water. Table 5.5 gives ratings for the York Model ES absorption units. The solar versions of the York Model ES systems involve basically running the absorption unit on heated solar water rather than on steam. In doing so the solar operating water temperatures will be lower than steam and the capacity of the unit in tons will have to be lowered somewhat. Figure 5.8 illustrates the derating procedure for the York absorption units. Notice in the graph that as the entering hot water temperature is reduced, the relative tons of refrigeration

Table 5.4. Yazaki Absorption Chiller Units. (Source: Yazaki Corporation[40]).

Model		WFC-400 (A)	WFC-600 (A)	WFC-2300 (BC)	WFC-3000 (BC)	WFC-4600 (C)	WFC-6000 (C)	WFC-9000 (C)	WFC-120000 (C)	WFC-15000 (C)
	Capacity (kW) (Btu / h)	4.6(15,800)	7.0(23,800)	26.2(89,500)	34.9(119,000)	52.4(179,000)	69.8(238,000)	104.7(357,000)	139.6(476,000)	174.5(595,000)
Cooling	Chilled water temp. (°C) Outlet	9								
	Chilled water temp. (°C) Inlet	14								
Chilled water	Rated water flow (ℓ / s)	0.22	0.33	1.25	1.67	2.50	3.34	5.01	6.68	8.35
	Allowable external pressure loss (kPa)	64.6	56.8	35.3	38.2	35.3		38.2		
	Heat input (kW) (Btu / h)	7.7(26,300)	11.6(39,600)	37.4(127,000)	49.8(170,000)	74.8(259,000)	99.6(343,000)	149.4(510,000)	199.2(680,000)	249.9(850,000)
Heat medium	Hot water inlet temp. range (°C)	75-100				75-100				
	Rated hot water flow (ℓ / s)	0.31	0.46	1.78	2.38	3.56	4.76	7.14	9.52	11.9
	Generator pressure loss (kPa)	9.8	12.8	24.5	39.2	24.5		39.2		
Cooling water	Cooling water temp. (°C) Inlet	29.5				29.5				
	Cooling water temp. (°C) Outlet	34.5				34.5				
	Rated water flow (ℓ / s)	0.59	0.89	3.00	4.00	6.00	8.00	12.00	16.00	20.00
	Condenser and absorber pressure loss (kPa)	14.7	22.6	68.6	68.6			68.6		
Electrical	Voltage, 1 PH. (V)	100-240		200-240	200-240			200-240		
	Consumption (50/60 Hz) (W)	195/195		30	30			30		
Piping	Chilled water (inch)	ISO ¾ Female	ISO ¾ Female	ISO 1½ Female	ISO 1½ Female			ISO 1½ Female		
	Heat medium (inch)	ISO ¾ Female	ISO ¾ Female	ISO 1½ Female	ISO 1½ Female			ISO 1½ Female		
	Cooling water (inch)	ISO 1 Female	ISO 1¼ Female	ISO 2 Female	ISO 2 Female			ISO 2 Female		
	Cistern conn. (inch)	ISO ½ Female	ISO ½ Female							
Weight	Operating (kg)	165	175	670	780	1,340	1,460	2,190	3,120	3,900

A ♦ For heat medium inlet temperatures of 100°C increase the cooling capacities by 45%.

B ♦ For heat medium inlet temperatures of 100°C increase the cooling capacities by 46%.

C ♦ Maximum operating pressure of chilled water, heat medium and cooling water is 588 kPa.

Table 5.5. Ratings for York Model ES Absorption Units. (Source: York Division of Borg-Warner Corporation[41]).

Model ES	Nominal Tons[1]	Max. Unit Tons Capability[2]	Nominal Condenser Water GPM	Standard Chiller Passes	Standard Absorber Passes	Cond. Water GPM (Max.)	Absorber Water GPM (Min.)	Absorber Water GPM (Max.)	Standard[3] Steam Valve Size	Nominal Pounds Steam Per Hour
1A1	120	169	432	3	3	693	280	779	4C	2200
1A2	155	205	558	3	3	693	360	779	4C	2840
2A3	172	233	620	2	2	969	410	1160	4C	3140
2A4	205	264	740	2	2	969	490	1160	5C	3760
2B1	235	298	846	2	2	1018	540	1590	5C	4300
3B2	273	348	980	2	2	1415	650	1590	5C	4960
3B3	311	393	1120	2	2	1415	740	1590	6C	5650
4B4	334	422	1200	2	1	1415	800	3180	6C	6120
4C1	363	471	1308	2	2	1570	840	2285	6C	6650
5C2	410	530	1475	2	2	2175	980	2285	4B	7500
5C3	446	575	1600	2	1	2175	1050	4570	4B	8200
6C4	518	695	1870	2	1	2175	1200	4570	5B	9500
7D1	565	735	2030	2	2	2985	1350	3115	5B	10300
7D2	617	794	2220	2	1	2985	1470	6230	5B	11300
8D3	704	875	2530	2	1	2985	1670	6230	5B	12800
8E1	794	937	2860	2	1	4250	1880	8080	6B	14600
9E1	908	1042	3270	2	1	4250	2160	8080	6B	16600
10E3	960	1114	3450	1	1	4250	2280	8080	6B	17600
12F1	1148	1318	4140	2	1	5560	2720	10340	6B	21000
13F2	1235	1417	4450	1	1	5560	2900	10340	6B	22600
14F3	1377	1540	4960	1	1	5560	3270	10340	8B	25200

[1]Nominal Tons is based on cooling chilled water from 54°F to 44°F with Nominal Condenser Water GPM entering absorber at 85°F and leaving condensor at approximately 103°F when supplied with 18.3 lb./ton-hr steam at 9 psig at generator inlet flange, dry and saturated at sea level, (or when supplied with approximately 1.1 GPM/ton hot water at 260°F) and Standard Number of Evaporator & Abosrber Passes. Fouling Factor .0005 for evaporator, absorber and condenser. Refer to tables and curves for capacity factors.

[2]Maximum unit tons capability can be provided by changing evaporator passes, absorber passes, generator steam pressure, condenser water GPM, leaving chilled water temperature or entering condenser water temperature.

[3]"C" refers to caged steam valve. "B" refers to butterfly valve.

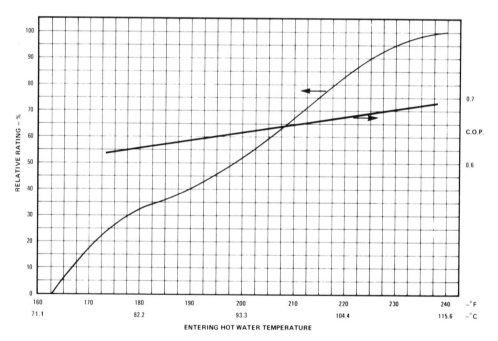

Figure 5.8. Graph of York Absorption Units Rating as a Function of Entering Hot Water Temperature. (Source: York Division of Borg-Warner Corporation[41]).

provided by the absorption unit also decreases. Also, the coefficient of performance does not drastically change as lower operating temperatures are selected.

A number of components in a solar absorption air conditioning system may be shared with the solar space heating and/or solar domestic hot water system. In many system installations the shared components may include:

1. Solar collectors
2. Back-up boiler
3. Fan coil unit
4. Storage tank
5. Some system control devices
6. Piping and duct runs

In addition to the absorption unit and the six items previously mentioned which help make up a combined solar heating and cooling installation, a cooling tower is also necessary for the absorption unit to properly operate. The cooling tower is sized to match the heat rejection requirements of the absorption air conditioning unit. Naturally the higher the tonnage capacity of the absorption unit the larger the cooling tower will have to be. The Yazaki cooling towers can range from 2.2 ft in diameter and 5.3 ft tall for the small 1.3-ton Yazaki residential absorption unit up to 18.6 ft long, 6.4 ft wide and 8.9 ft tall for their larger absorption units used in commercial applications. Some companies selling absorption units do not manufacture cooling towers but instead supply sufficient information describing the heat rejection requirements for their absorption units so that the architect/engineering firm can choose a suitable cooling tower.

5.2.2. Absorption System Maintenance

In a solar absorption cooling system, water mixed with antifreeze is generally used in the chilled water section of the absorption unit and may need to be periodically checked. The antifreeze solution keeps the cold water produced in the evaporator coils from freezing. A safety control is usually incorporated in the absorption unit to keep the chilled water from freezing but, should the control fail, freeze damage can occur and such damage to the chiller may not be covered by warranty.

Another possible problem area involves the condensing water and the cooling tower. In Fig. 5.2 the pipes leading from the condenser and from absorber go directly to the outdoor cooling tower. In almost all installations the water running through the piping between the cooling tower and the absorption air conditioning unit is simple tap water. The basic cooling tower takes heated water from the condenser piping and sprays this water from the top of the cooling toward the bottom of the tower. Generally there is also a fan at the top of the tower which aids in drawing air through the bottom louvers of the tower and expelling it through the top of the tower. The warm water droplets,

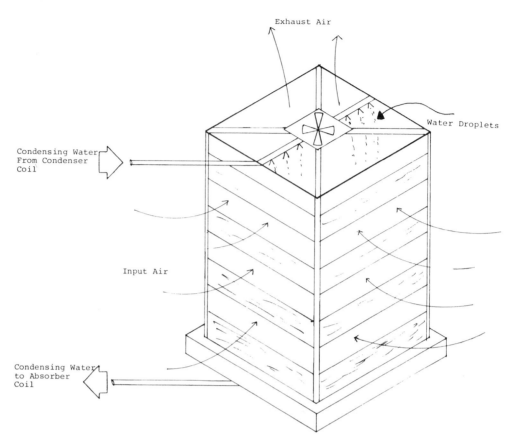

Figure 5.9. Cooling Tower Operation.

passing through the cooler air rushing up through the tower, are cooled. This cooled condensing water is collected at the bottom of the cooling tower sump and pumped back to the coil of the absorption unit. Figure 5.9 shows a typical cooling tower.

When water evaporates in the cooling tower it leaves behind scale and solid particles much like the film left in a tea kettle that has boiled off the water in it. As water evaporates from the cooling tower, the concentrations of particles present in the condenser increases. The particle concentrations can lead to scaling and corrosion problems if allowed to increase. Generally in an air conditioning system using a cooling tower, some water is allowed to drain from the system into the sewer. This expelled water is called bleed-off or blow down. As this bleed-off water is drained from the cooling tower, fresh make-up water is introduced into the system. This make-up water compensates the water loss due to blow down and cooling tower evaporation. By continually draining condensing water and adding fresh water, the concentrations of harmful materials present in the condensing water can be kept at a low level and hence corrosion and scaling problems will be minimal. Sometimes corrosion inhibitors are periodically added to the condensing water for additional corrosion and scaling control. In the cooling tower itself, algae may form in the warm water collected in the sump at the bottom of the tower. This problem can also be corrected by adding the appropriate inhibiting chemical.

In most absorption systems using a cooling tower, the manufacturer generally gives a minimum blow down or bleed-off figure in gallons per minute to be used when determining how much condensing water should be expelled into the sewer. This *minimum* figure may be calculated based only on the rate of evaporation loss in the cooling tower. The rate at which condensing water should be expelled and fresh make-up water added will be based on the composition of the tap water used. In other words, the larger the amount of dissolved solids, the more condensing water should be expelled.

The addition of chemical scale and corrosion inhibitors to the condensing water may allow for lower blow down rates than would be acceptable if no chemical additives were used. With chemical additives the minimum blow down or bleed-off rate should at least equal the rate of water evaporation from the cooling tower. When treating condensing water with chemical additives, care should be exercised because overtreatment can be almost as bad as undertreatment.

If the manufacturer does not supply recommended blow off rates, Eq. 5.2 can be used to obtain a ball park figure for the rate of blow down required.

$$B = [M_H/(S_H - M_H)] \, E \qquad (5.2)$$

where

B = blow down (or bleed-off) in gallons per minute (gpm)
M_H = make up water hardness (in parts per million)
S_H = sump water hardness (in parts per million)
E = rate of evaporation (in gpm)

The rate of evaporation (E) used in Eq. 5.2 may be estimated by Eq. 5.3.

$$E = C_R \times R_F \times 24/H_E \qquad (5.3)$$

where

C_R = Capacity of refrigeration (in tons)
R_F = Refrigeration machine heat rejection factor
H_E = Heat of evaporation for condenser water
(normal water about 1050 Btu/lb)

Corrosion can be a problem if not properly dealt with and in most cases is not covered in the system warranty issued by the manufacturer with regard to condensing water scaling and corrosion.

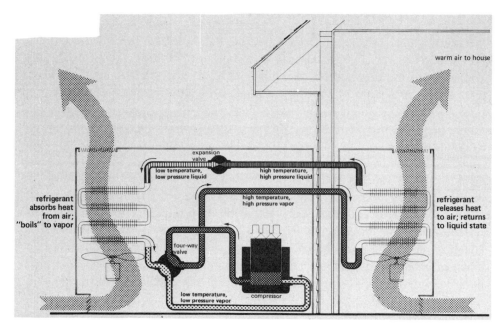

warm air to house

expansion valve

low temperature, low pressure liquid

high temperature, high pressure liquid

refrigerant absorbs heat from air; "boils" to vapor

high temperature, high pressure vapor

refrigerant releases heat to air; returns to liquid state

four-way valve

low temperature, low pressure vapor

compressor

Figure 5.10. Operation of Heat Pump in Heating Cycle. (Source: U. S. Department of Energy[34]).

5.3 SOLAR ASSISTED HEAT PUMP

5.3.1. Overview

The heat pump is basically a reversible refrigeration machine. Figure 5.10 illustrates the operation of a heat pump in the heating mode. In Fig. 5.10, as the compressor compresses the refrigerant vapor, heat is produced. The heated refrigerant vapor is then transferred through the vapor line into the room to be heated. As the warm refrigerant vapor passes through the heat exchanger inside the house and liberates heat, the warm refrigerant gas is transformed into a cooler liquid. The liquid refrigerant then passes through an expansion valve outside of the building where the liquid refrigerant is transformed into a cold gaseous state. In this cold gaseous state, the refrigerant then absorbs heat energy in the outside coil and is transported back to the compressor where the process is repeated.

In the cooling mode shown in Fig. 5.11, the cycle is reversed and heat is absorbed inside the building by the refrigerant and transported outdoors where it is expelled. As heat is removed from the building, the conditioned rooms of the building get cooler.

The heat pump, in its ability to efficiently transfer thermal energy from one point to another, has a relatively large coefficient of performance (*COP*) that can run as high as about four. This means that for every Btu of electrical energy supplied to the heat pump in the heating mode, four Btus of thermal energy will be transferred from outdoors to the conditioned space inside the building. As the outside air temperature drops, the heat pump's coefficient of performance also drops as shown in Fig. 5.12.

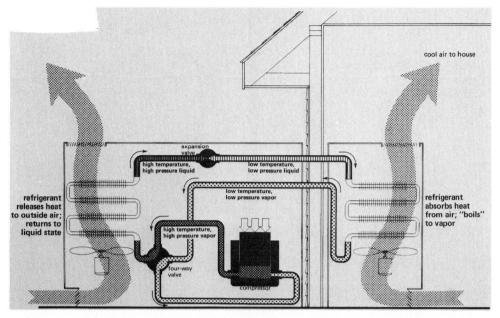

Figure 5.11. Operation of Heat Pump in Cooling Cycle. (Source: U. S. Department of Energy[34]).

The heating capacity of the heat pump decreases until a point is reached where the heat pump can no longer supply all of the heat requirements of the building being heated. At this point, an auxiliary resistance heating element will come on to supplement the heat pump. Figure 5.13 illustrates this concept. As can be seen in the diagram of Fig. 5.13, at about 35°F (2°C) the heat pump

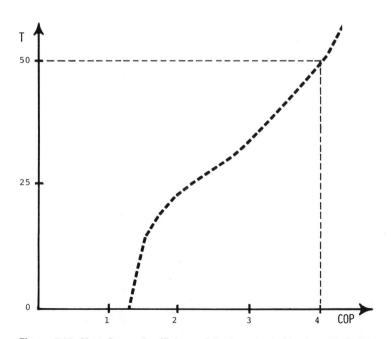

Figure 5.12. Heat Pump Coefficient of Performance. (Source: U. S. Department of Energy[35]).

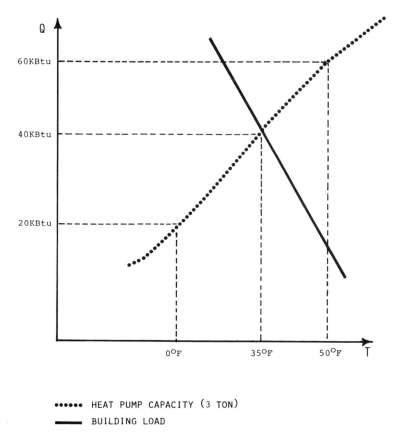

•••••• HEAT PUMP CAPACITY (3 TON)

▬▬ BUILDING LOAD

Figure 5.13. Heating with a Heat Pump. (Source: U. S. Department of Energy[35]).

alone cannot supply all of the building heat load requirements. This is where the addition of a solar collector can pay off. By using a solar collector to supply heat to the heat pump during cold outdoor temperatures, the efficiency of the heat pump is greatly increased and the heat pump will also be able to continue meeting the heating load requirements even below about 35°F (2°C). If a storage tank is also incorporated into the solar assisted heat pump system, the heat pump will be able to function efficiently at night using the stored collector heated water obtained during the day.

A number of well known air conditioning manufacturers now offer solar assisted heat pump systems for both residential and commercial application. General Electric Company offers their Weathertron solar-supplemented heat pump and Carrier offers their Weathermaker unit. Figure 5.14 shows a general piping design for the Carrier Solar Weathermaker System.

5.3.2. Maintenance

Virtually no maintenance is required for the heat pump. A number of years ago though, there were some problems with the basic heat pump system such as compressors burning out and sticking reversing valves. Today, these prob-

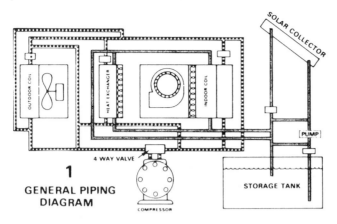

1

**GENERAL PIPING
DIAGRAM**

Figure 5.14. General Piping Design for the Carrier Solar Weathermaker System. (Courtesy of Carrier Air Conditioning, a subsidiary of United Technologies Corporation[11]).

lems have been largely eliminated and the present generation of heat pump systems is of good reliability.

The solar assisted part of the heat pump which includes the solar collector, piping, pumps, and storage tank will require periodic maintenance just as any other type of solar installation. This means lubricating the pump, checking for proper fluid circulation in the system, observing for possible leaks or corrosion problems, and checking to see that chemical additives which may have been placed in the water to retard corrosion have not been depleted due to aging or leakage.

5.4 SOLAR RANKINE-CYCLE ENGINE

The solar Rankine-cycle engine, as used in solar air conditioning, basically takes the place of an electric motor used in operating the compressor in a vapor-compression refrigeration machine. Figure 5.15 shows a detailed schematic for a Rankine-cycle air conditioning system. In the diagram, a (fluorocarbon) fluid is heated in the boiler by warm solar heated water flowing through the heat exchanger circuit. The liquid in the boiler is vaporized and forced through the turbine which causes the rotor inside to spin. The turbine rotor is coupled to a gear box which reduces its rotational speed to a rate acceptable to the air conditioning compressor. The overriding clutch coupled to an electric motor, permits running the air conditioning compressor from the turbine and/or electric motor depending upon the amount of thermal energy provided by the solar collectors to the boiler. The warm gas leaving the turbine passes through the regenerator which helps to remove heat from the gas. The gas is further cooled in the Rankine-cycle condenser by cooling tower water circulating through the condenser. As the gas is cooled it liquifies in the condenser and runs into the receiver tank. The fluid is then pumped into the regenerator and back into the boiler. As the liquid from the receiver tank passes through the regenerator, it absorbs the heat from the (gaseous) fluid leaving the turbine. The regenerator, in effect, serves to make the Rankine-

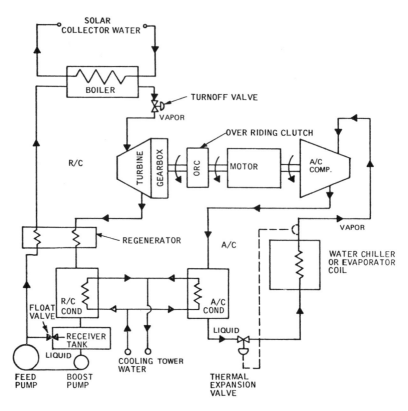

Figure 5.15. General System Schematic for Rankine-Cycle Air Conditioning System. (Source: U. S. Department of Energy[35]).

cycle engine more efficient by utilizing some of the waste heat from the exhaust turbine gas.

In the air conditioning section shown in Fig. 5.15, the air conditioning cycle begins by the compressor compressing low pressure refrigerant gas. The pressurized vapor is then passed through the air conditioning condenser where it is cooled by water flowing from the cooling tower. As heat is removed from the vapor, the high pressure gas is liquified. This high pressure liquid then passes through a thermal expansion valve whereby part of the liquid is vaporized. The combination of refrigerant liquid droplets and vapor, cooled to about 45°F (7°C) by the effects of the thermal expansion valve, pass through the water chiller or evaporator coil where heat from the inside of the building is extracted. The extracted heat then vaporizes the droplets of refrigerant liquid and the vapor travels from the water chiller or evaporator coil to the air conditioning compressor where the cycle is repeated.

The Rankine-cycle engine is still in the testing and development stage and is not presently available for purchase.

5.5 REFERENCES AND SUGGESTED READING

1. Air Conditioning and Refrigeration Institute, Arlington, Virginia, *Directory of Certified Unitary Air Conditioners, Air-Source Unitary Heat Pumps, Sound-Rated Outdoor Unitary Equipment, Central System Humidifiers*, 1978.

2. Alizadeh, S., Bahar, F., Goola, F., "Design and Optimization of an Absorption Refrigeration System Operated by Solar Energy," *Solar Energy*, **22**, 149–154, (1979).

3. Andrews, F. T., *Building Mechanical Systems*, New York: McGraw-Hill Book Company, 1977.

4. Arkla Industries, Inc., Evansville, Indiana, *Solaire 36 Model WF-36* (pamphlet), *Application Manual*.

5. Arkla Industries, Inc., Evansville, Indiana, *Solaire 300 Model WFB-300 Application Manual*, January, 1978.

6. Arkla Industries, Inc., Evansville, Indiana, *Thirty-Six P Installation Guide for PWF 36-100 Packaged Residential Solar System*, August, 1979.

7. Bartholomew, Douglas, "Solar Energy on a Large Scale," *Mechanix Illustrated*, **75**, 611:55–58, (1979).

8. Beason, Robert G., "A Giant Step for Solar Air Conditioning," *Mechanix Illustrated*, **74**, 60:45–57, 109–110, (1978).

9. Burkhardt, Charles H., *Residential and Commercial Air Conditioning*, New York: McGraw-Hill Book Company, 1959.

10. Carrier Air Conditioning Company, *Handbook of Air Conditioning System Design*, New York: McGraw-Hill Book Company, 1965, pp. 5.1–5.49.

11. Carrier Corporation, Syracuse, New York, *The Solar Weather Maker – Solar Assisted Heat Pump System*, (pamphlet).

12. Davis, John L., "Underground School Gets Its Energy From the Sun," *Heating/Piping/Air Conditioning*, **50**, 1:93–96 (1978).

13. deWinter, Francis, ed., *Solar Cooling for Buildings*, Washington, D.C.: U. S. Government Printing Office, 1974.

14. Elliott, Thomas C., (ed.), "Cooling Towers," *Power*, **117**, 3:17–40, (1973).

15. Fenner, William E., "Solar Energy Caps Unique System," *Heating/Piping/Air Conditioning*, **47**, 7:41–43, (1975).

16. Hill, James E., and Richtmyer, Thomas E., *Retrofitting a Residence for Solar Heating and Cooling: The Design and Construction of the System*, Washington, D.C.: U. S. Department of Commerce, November, 1975, pp. 20–38.

17. Hornak, John P., and Knight, Kurt, "Solar Energy System Retrofit", *Heating/Piping/Air Conditioning*, **50**, 1:87–89, (1978).

18. Hornak, John P., and Knight, Kurt, "Solar Energy Retrofit Revisited," *Heating/Piping/Air Conditioning*, 85–88, 1979.

19. Korte, Robert T., ed., "Honeywell Starts Up Largest High Temperature Solar System," *Heating/Piping/Air Conditioning*, **51**, 1:27, 35, (1979).

20. Korte, Robert T., ed., "Solar Powered Air Conditioning," *Heating/Piping/Air Conditioning*, **52**, 1:71–75, (1980).

21. Kreuter, Rodney A., "Solar Tracking System", *Radio Electronics*, **50**, 2:42–44, (1979).

22. Kuszpa, Frank J., Jr., "Solar Energy for Hospital Tower," *Heating/Piping/Air Conditioning*, **51**, 1:75–80, (1979).

23. Leitner, Gordon, "Controlling Chiller Tube Fouling," *ASHRAE Journal*, **22**, 2:40–43, (1980).

24. Mitchell, John W., Freeman, Thomas L., and Beckman, William A., "Heat Pumps — Do They Make Economic and Performance Sense with Solar?," *Solar Age*, **3**, 6:24–28, (1978).

25. Scholten, William, and Curran, Henry, "Active Cooling Update," *Solar Age*, **4**, 6:37–41, (1979).

26. Scott, David, "Tracking Solar Heater," *Popular Science*, **214**, 6:82–83, (1979).

27. Setty, Boggarm, S. V., "Solar Heat Pump Integrated Heat Recovery," *Heating/Piping/Air Conditioning*, **48**, 7:58–60, (1976).

28. Shuttleworth, John, ed., "Mother's Solar Tracking System," *The Mother Earth News*, 55:93–95, (1979).

29. Stephan, Edward, "Solar Heating a Big Building with Packaged Chillers," *Air Conditioning and Refrigeration Business*, **23**, 3:59–62, (1976).

30. Strock, Clifford, and Koral, Richard L., co-ed., *Handbook of Air Conditioning, Heating and Ventilating*, New York: The Industrial Press, 1965.

31. *Trane Air Conditioning Manual*, St. Paul, Minnesota: McGill Printing, Inc., 1965, pp. 253–257.

32. U. S. Department of Commerce, *Solar Heating and Cooling of Residential Buildings – Design of Systems*, October, 1977, pp. 17.1–17.24.

33. U. S. Department of Commerce, *Solar Heating and Cooling of Residential Buildings – Sizing, Installation and Operation of Systems*, Washington, D.C.: U. S. Government Printing Office, October, 1977, pp. 8.1–10.19.

34. U. S. Department of Energy, Fact Sheet EDM-1050, *Heat Pumps*, Oak Ridge, Tennessee: DOE Technical Information Center, pp. 1–6.

35. U. S. Department of Energy, *Introduction to Solar Heating and Cooling Design and Sizing*, Washington, D.C.: U. S. Government Printing Office, August, 1978, pp. 4.27–4.36, 4.56–4.83.

36. Ward, D. S., Duff, W. S., Ward, J. C., and Lof, G. O. G., "Integration of Evacuated Tubular Solar Collectors with Lithium Bromide Absorption Cooling Systems," *Solar Energy*, **22**, 4:335–341, (1979).

37. Ward, Dan S., "Solar Absorption Cooling Feasibility," *Solar Energy*, **22**, 3:259–268, (1979).

38. Wilbur, Paul J., and Mancini, Thomas R., "A Comparison of Solar Absorption Air Conditioning Systems," *Solar Energy*, **18**, 569–576, (1976).

39. Wilbur, Paul J., and Mitchell, Charles E., "Solar Absorption Air Conditioning Alternatives," *Solar Energy*, **17**, 193–199, (1975).

40. Yazaki Corporation, Tokyo, Japan, *Yazaki Solar Air Conditioning Systems Equipment*, (pamphlet).

41. York Division of Borg-Warner Corporation, York, Pennsylvania, *Absorption Liquid Chillers Model FS*, (pamphlet).

42. Energy Research and Development Administration, *Project Data Summaries: Commercial and Residential Demonstrations* Vol. 1, Washington, D.C.: U. S. Government Printing Office, August, 1976, pp. 21, 55.

CHAPTER 6
SOLAR SYSTEM DESIGN

6.1 DETERMINING THE SUN'S LOCATION

When designing solar heating and cooling systems, it is sometimes necessary to be able to determine exactly where the sun is for a particular latitude, day, and hour.

Two coordinates are needed to locate the sun's position: the altitude and the azimuth. The altitude is the vertical angle formed between the horizon and the sun as seen in Fig. 6.1. The azimuth, sometimes called the bearing angle, is the horizontal angle formed between a south direction and the horizontal bearing of the sun as illustrated in Fig. 6.2. The sun's altitude and azimuth may be found by using sun charts, such as those found in Appendix XI, or by Eqs. 6.1 and 6.2. The equations allow the designer to determine the altitude (α) and azimuth (γ) for an exact latitude (location) whereas the sun charts in the appendix are only for 32°N latitude to 48°N latitude in increments of 4°.

The equations for determining the sun's altitude and azimuth are:

$$\sin(\text{altitude}) = \cos(\text{declination}) \ \cos(\text{latitude}) \ \cos(15 \quad (6.1)$$
$$\text{solar hour} + \sin(\text{declination}) \ \sin(\text{latitude})$$

$$\cos(\text{azimuth}) = \cos(\text{declination}) \ \sin(15 \ \text{solar hour})/ \quad (6.2)$$
$$\cos(\text{solar altitude})$$

Before applying the equations, the variables used need to be defined and explained. The declination (δ) is the angle formed between the earth's equatorial plane facing the sun and the plane of the earth's rotation around the sun.

$$\text{declination}(\delta) = 23.45° \ \sin(360 \ (284 + \text{day of year})/365) \quad (6.3)$$

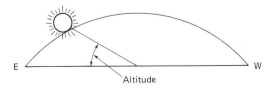

Figure 6.1. Altitude of Sun.

129

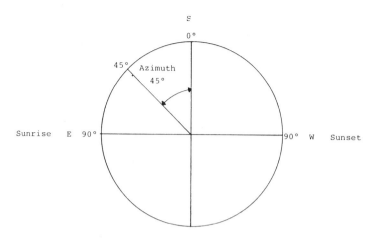

Figure 6.2. Azimuth Measurement.

In Eq. 6.3, the day of the year refers to the number of days elapsed since the first of the year. (February 17 would therefore be the 48th day of the year.)

The solar hour is found by determining the number of hours between solar noon time and the (solar) time that the sun's altitude and azimuth are to be calculated for. For example, if the sun's altitude and azimuth are to be determined for 10 AM solar time, then the solar hour would be 2.0. For 3:00 PM sun time, then the solar hour would be 3.0. In determining the solar hour, solar (or sun) time is used. Sun time differs somewhat from local standard time due to the nonuniform, eliptical orbit of the earth as illustrated in Fig. 6.3.

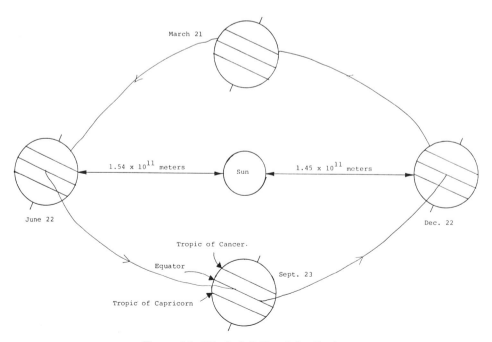

Figure 6.3. Elliptical Orbit of the Earth.

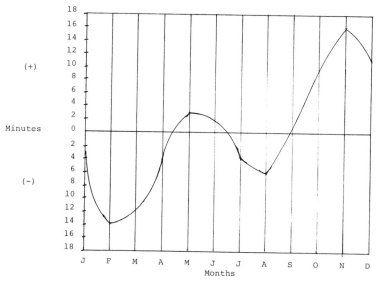

Figure 6.4. Equation of Time During Year.

As the earth moves closer to the sun it slows down and as it moves away from the sun it speeds up. Therefore, the sun time varies during the year whereas the time, as measured by someone's watch, increments time at a constant rate. Figure 6.4 illustrates how the sun time cyclically changes throughout the year. The variation of solar time throughout the year, as shown graphically in Fig. 6.4 and in tabular form in Table 6.1, is known as the "equation of time". The formula for determining the sun time using the "equation of time" is shown in Eq. 6.4.

$$\text{sun time} = \text{standard time} + \text{equation of time} + 4 \text{ (longitude} \quad (6.4)$$
$$\text{of standard meridian} - \text{longitude of observer)}$$

In Eqs. 6.4 and 6.5 the difference between the longitude of the standard meridian and the longitude of the observer is multiplied by four minutes because it takes the sun four minutes to move each degree of longitude across the sky. This is easily determined because the earth makes one 360° revolution every 24 hours. This means the earth rotates 15° per hour or 1° every four minutes.

The declination is determined by using Eq. 6.3. July 22, the 203rd day of the year, would have a declination (δ) of $+20.24°$.

The solar altitude (α) and azimuth (γ) can now be found:

$$\sin(\text{altitude}) = \cos(+20.24°) \cos(36.0) \cos(15 \times 2)$$
$$+ \sin(+20.24°) \sin(36.0) \quad (6.7)$$
$$= 59.40°$$

$$\cos(\text{azimuth}) = \cos(+20.24°) \sin(15 \times 2)/\cos(59.40°) \quad (6.8)$$
$$= 22.81°$$

Table 6.1. Equation of Time.

Date	Minutes	Date	Minutes
Jan 1	3 min slow	Jul 1	4 min slow
Jan 10	7 min slow	Jul 10	5 min slow
Jan 16	10 min slow	Jul 16	6 min slow
Jan 20	11 min slow	Jul 20	6 min slow
Feb 1	14 min slow	Aug 1	6 min slow
Feb 10	14 min slow	Aug 10	5 min slow
Feb 16	14 min slow	Aug 16	4 min slow
Feb 20	14 min slow	Aug 20	3 min slow
Mar 1	12 min slow	Sep 1	0 min slow
Mar 10	10 min slow	Sep 10	3 min fast
Mar 16	9 min slow	Sep 16	5 min fast
Mar 20	8 min slow	Sep 20	6 min fast
Apr 1	4 min slow	Oct 1	10 min fast
Apr 10	1 min slow	Oct 10	13 min fast
Apr 16	0 min slow	Oct 16	14 min fast
Apr 20	1 min fast	Oct 20	15 min fast
May 1	3 min fast	Nov 1	16 min fast
May 10	4 min fast	Nov 10	16 min fast
May 16	4 min fast	Nov 16	15 min fast
May 20	4 min fast	Nov 20	14 min fast
Jun 1	2 min fast	Dec 1	11 min fast
Jun 10	1 min fast	Dec 10	7 min fast
Jun 16	0 min fast	Dec 16	4 min fast
Jun 20	1 min slow	Dec 20	2 min fast
		Dec 30	2 min slow

Sun charts can also be used to determine the altitude and azimuth. Figure 6.5 illustrates a sun chart for 36°N latitude. The circled portion of the sun chart shows the altitude and azimuth for Durham, North Carolina. In Equation 6.4, the longitude of the standard meridian is chosen with regard to what time zone the observer is in. For the United States there are four time zones:

Eastern Time Zone	75°W Longitude
Central Time Zone	90°W Longitude
Mountain Time Zone	105°W Longitude
Pacific Time Zone	120°W Longitude

For eastern Alaska the time zone is 135°W longitude, and for the rest of Alaska and Hawaii the time zone is 150°W longitude.

To demonstrate the use of Eqs. 6.1 and 6.2 and the sun charts, it will be assumed that someone in Durham, North Carolina needs to know what the solar altitude and azimuth will be on July 22 at 11:22 AM local time. The sun time is calculated first using Eq. 6.4. Since the United States is on daylight savings time, the standard time will be 10:22 AM rather than 11:22 AM. From Table 6.1, the equation of time for July 22 is six minutes slow (−6). Since Durham is on Eastern Standard Time the longitude of the standard meridian is 75°W. The longitude for Durham is 78.92°W and the latitude is 36.0°N. Therefore, the sun time is:

$$\text{sun time} = 10:22 \text{ AM} + (-6) + 4 \ (75.00° - 78.92°) \qquad (6.5)$$

$$\text{sun time} = 10:22 \text{ AM} - 6 - 16 = 10:00 \text{ AM} \qquad (6.6)$$

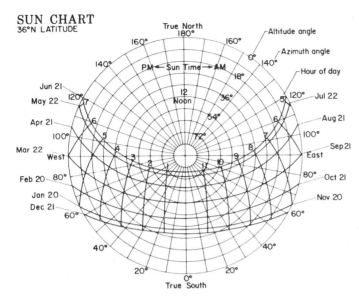

Figure 6.5. Sun Chart Diagram. (Source: Los Alamos Scientific Laboratory[68]).

Notice that in the sun chart the azimuth is shown to be about 67° east of true south or about 23° (90°-67°) south of east as predicted from Eqs. 6.2 and 6.8 and the altitude from the sun is about 60° as predicted from Eqs. 6.1 and 6.7.

6.2 INTERPRETING SOLAR INSOLATION DATA

6.2.1. Overview

Section 1.3 discussed factors affecting the reception of solar energy for various locations on the earth's surface. The problem now is to obtain solar insolation data that will be representative of the area in which the solar heated or cooled building will be located. Ideally the *best* solar insolation data obtainable would be data collected over a number of years at the exact location of the proposed solar heated or cooled building site. Unfortunately, this is not possible for the average solar system analysis. The architect/engineering (A/E) firm or home-owner must usually rely on insolation data for the city or general area where the proposed solar heated or cooled building will be located. The accuracy of the solar insolation data hinges upon a number of factors, one of which is that there are a limited number of weather stations across the country that record solar insolation. For cities that do not measure and record solar insolation, the data for that city has to be extrapolated from insolation recordings of other cities. When the solar data must be estimated, certain inaccuracies can arise. The problem can be compounded if the cities that collect solar insolation data allow incorrect data to be recorded. This can occur if the measuring instrument, such as a pyranometer, is not periodically recalibrated and if atmospheric dirt particles are allowed to accumulate on the glass lens of the pyranometer.

Solar insolation data is necessary to size a solar system when using a design methodology such as the F-chart method (described later). The reader should be aware that solar insolation data is one avenue whereby some degree of error may be introduced into the solar design analysis.

6.2.2. Detailed Estimation of Solar Radiation Falling Upon Inclined Surfaces

Generally, solar energy insolation data is available for a horizontally oriented surface and occasionally for a surface oriented normal to the sun's rays. This data will need to be adjusted to a more usable form. If the insolation figures obtained for a "horizontal surface" are used in the design analysis for a building incorporating solar collectors, low insolation values could lead to underestimation of the available solar radiation and oversizing the system. On the other hand, using insolation values obtained for a surface "normal" to the sun's rays may lead to overestimation of the amount of solar radiation available to a solar collector mounted in a permanently fixed position and underestimating the number of solar collectors needed.

Liu and Jordan [117-119] have devised a methodology for determining the amount of solar energy falling on a tilted surface such as a solar collector using solar insolation data readily obtained for a horizontal surface. The monthly average daily solar radiation falling on a tilted surface is:

$$\bar{H}_T = \bar{R} \, \bar{H} - \bar{R} \, \bar{K}_T \bar{H}_0 \qquad (6.9)$$

where

$\bar{H}_T$ = monthly average daily diffuse solar radiation

$\bar{R}$ = ratio of the monthly mean daily radiation on a tilted surface to that on a horizontal surface ($\bar{H}$)

$\bar{K}_T$ = ratio of monthly average daily solar radiation falling on a horizontal surface ($\bar{H}$) to the average daily extra terrestrial radiation ($\bar{H}_0$)

The value of $\bar{R}$ can be determined by separately considering the three components of solar radiation: beam, diffuse, and reflected. Equation 6.10 adds each component together to obtain a composite figure for $\bar{R}$.

$$\bar{R} = (1-\bar{H}_d/\bar{H})\bar{R}_b + H_d/H(1+\text{cosine } s)/2 + p(1-\text{cosine } s)/2 \quad (6.10)$$

| Beam Component | Diffuse Component | Reflective Component |

where

$\bar{H}_d$ = monthly average daily diffuse solar radiation

s = tilt of the surface (e.g., a solar collector) from a horizontal orientation.

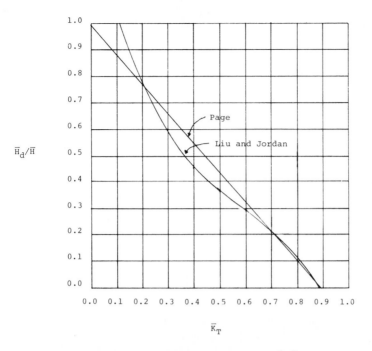

Figure 6.6. Graph Illustrating Values for $\bar{H}_d/\bar{H}$.

p = the ground reflectance which Liu and Jordan estimate varies from 0.2 to 0.7 depending on the amount of snow
    cover.

To determine the value of $\bar{H}_d/\bar{H}$, Eq. 6.11 is used.

$$\bar{H}_d/\bar{H} = 1.390 - 4.027\bar{K}_T + 5.531\ \bar{K}_T^2 - 3.108\ \bar{K}_T^3 \qquad (6.11)$$

Liu and Jordan proposed that the ratio of diffuse solar radiation ($\bar{H}_d$) to
horizontal solar radiation ($\bar{H}$) is related to $\bar{K}_T$ as shown in Eq. 6.11. Page [120]
determined that the relationship was of the form shown in Eq. 6.12.

$$\bar{H}_d/\bar{H} = 1.00 - 1.13\ \bar{K}_T \qquad (6.12)$$

Figure 6.6 graphically illustrates the difference between Eqs. 6.11 and 6.12.
Page's equation of the first order results in a linear plot whereas Liu and
Jordan's third degree equation results in a nonlinear plot. For the purposes of
this book, Liu and Jordan's interpretation of the relationship between $\bar{H}_d/\bar{H}$ and
$\bar{K}_T$ will be used. To determine $\bar{K}_T$,

$$\bar{K}_T = \bar{H}/\bar{H}_0 \qquad (6.13)$$

To obtain the value of $\bar{H}_0$ for Eq. 6.13,

$$\bar{H}_0 = 1/(m_2 - m_1) \sum_{n=m_1}^{m_2} (H_0)_n \qquad (6.14)$$

$$(H_0)_n = 24/\pi \, I_{sc}[1 + 0.033 \text{ cosine } (360 \, n/365)]$$
$$[\text{cosine}\phi \text{ cosine}\delta \text{ sine}\omega_s + \omega_s(\pi/180) \text{ sine}\phi \text{ sine}\delta]$$

where

m_1 = day of year for beginning of month (e.g., December 1 would be 334)

m_2 = day of year for end of month (e.g., December 31 would be 365)

$(H_0)_n$ = daily extraterrestrial radiation for day of the year

I_{sc} = solar constant of 4871 $\bar{K}_T/m^2$ hr

δ = solar declination (see Eq. 6.3)

ϕ = latitude of observer

ω_s = sunset hour angle which equals arcos $(-\tan\phi \tan\delta)$

The last variable to be determined for use in ultimately finding $\bar{H}_T$ is $\bar{R}_b$. Recall that $\bar{R}_b$ is necessary to compute $\bar{R}$ in Eq. 6.10. $\bar{R}_b$ is the ratio of beam solar radiation on a tilted surface (e.g., solar collector surface) to that on a horizontal surface. Equation 6.16 is used in determining $\bar{R}_b$ for each month. Equation 6.16 was derived by Liu and Jordan for use in surfaces directly facing the equator.

$$\bar{R}_b = [\cos (\phi - s) \cos\delta \sin \omega'_s + (\pi/180) \, \omega'_s \sin(\phi - s) \sin\delta] \qquad (6.16)$$
$$/[\cos\phi \cos\delta \sin \omega_s + (\pi/180)\omega_s \sin\phi \sin\delta]$$

where

ω = hour angle equaling $15° \times$(number of hours from solar noon), where the morning values are negative and afternoon values are positive

ω'_s = sunset hour angle for the tilted surface

$$\omega'_s - \text{minute}[\omega_s, \text{arcosine } (-\text{tangent } (\phi - s)\text{tangent}\delta)] \qquad (6.17)$$

Klein (Ref. 53) extended Liu and Jordan's method of finding $\bar{R}_b$. Using Klein's method, $\bar{R}_b$ can be determined for tilted surfaces not directly facing toward the equator. This is an important distinction because it allows for solar radiation falling on a solar collector oriented somewhat east or west of true south (for the northern hemisphere). Equation 6.18 illustrates Klein's extended method of determining $\bar{R}_b$.

$$\bar{R}_b = [(\text{cosine } s \text{ sine } \delta \text{ sine } \phi) \ (\pi/180) \ (\omega_{ss} - \omega_{sr}) -$$
$$(\sin \delta \cos \phi \text{ sine } s \cos \gamma) \ (\pi/180) \ (\omega_{ss} - \omega_{sr}) +$$
$$(\cos \phi \cos \delta \cos s) \ (\sin \omega_{ss} - \sin \omega_{sr}) + \qquad (6.18)$$
$$(\cos \delta \cos \gamma \sin \phi \sin s) \ (\sin \omega_{ss} - \sin \omega_{sr}) -$$
$$(\cos \delta \sin s \sin \gamma) \ (\cos \omega_{ss} - \cos \omega_{sr})]/$$
$$[2 \ (\cos \phi \cos \delta \sin \omega_s + (\pi/180) \ \omega_s \sin \phi \sin \delta)]$$

where

γ = surface azimuth angle where due south is zero and east is negative, while west is positive.

ω_{sr} and ω_{ss} = sunrise and sunset hour angles for the tilted surface

if $\gamma > 0$, then;

$$\omega_{sr} = -\text{ minute } \{\omega_s, \text{ arcosine } [(AB + \sqrt{A^2 - B^2 + 1})/A^2 + 1)]\} \qquad (6.19)$$

$$\omega_{ss} = \text{min } \{\omega_s, \text{ arcosine}[(AB + \sqrt{A^2 - B^2 - 1})/A^2 + 1)]\} \qquad (6.20)$$

if $\gamma < 0$, then;

$$\omega_{sr} = -\text{ minute } \{\omega_s, \text{ arcosine } [(AB - \sqrt{A^2 - B^2 + 1})/A^2 + 1)]\} \qquad (6.21)$$

$$\omega_{ss} = \text{minute } \{\omega_s, \text{ arcosine } [(AB + \sqrt{A^2 - B^2 + 1})/A^2 + 1)]\} \qquad (6.22)$$

$$A = \text{cosine } \phi/(\sin \gamma \tan s) + \sin \phi/\tan \gamma \qquad (6.23)$$

$$B = \tan \delta \ [\cos \phi/\tan \gamma - \sin \phi/(\sin \gamma \tan s)] \qquad (6.24)$$

6.2.3. Simplified Example of Determining Solar Radiation Falling Upon Inclined Surfaces

The angle of inclination, dependent upon the type of solar system used, must be determined before computing the amount of solar insolation hitting an inclined surface. A general guide for determining proper solar collector tilt is shown in Table 6.2.

Table 6.2.   Determination of Proper Solar Collector Tilt.

```
For domestic hot water . . . . . . . . . .   Tilt = Latitude

For combined heating and cooling . . . . .   Tilt = Latitude

For combined solar heating and
   domestic hot water  . . . . . . . . . .   Tilt = Latitude + 10°

For solar heating only . . . . . . . . .   Tilt = Latitude + 15°
```

Table 6.3. Values of $\bar{K}_T$ for Pensacola. (Source: Klein[57]).

January	= 0.50		July	= 0.56
February	= 0.53		August	= 0.57
March	= 0.55		September	= 0.55
April	= 0.59		October	= 0.61
May	= 0.59		November	= 0.53
June	= 0.58		December	= 0.48

In the following example it will be assumed that tilt will be equal to latitude $+15°$. For Pensacola, Florida, at about 30°N latitude, the collector tilt(s) will therefore be at a 45° angle. Since $\bar{K}_T$ is the ratio of monthly average daily terrestrial solar radiation falling on a horizontal surface ($\bar{H}$) to the average daily extraterrestrial radiation ($\bar{H}_0$), it gives an indication of the atmospheric condition of an area. Large average monthly values of $\bar{K}_T$ represent relatively clear and sunny skies whereas low values for $\bar{K}_T$ represent relatively cloudy and overcast skies for the month. Appendix II gives monthly values of $\bar{K}_T$ for a number of cities. Table 6.3 shows values of $\bar{K}_T$ for Pensacola. For other cities, see Appendix II. $\bar{K}_T$ can also be derived by using Table 6.4 (showing monthly average daily extraterrestrial radiation) and by using solar insolation data for a horizontal surface ($\bar{H}$).

After determining $\bar{K}_T$ for all twelve months, $\bar{H}_d/\bar{H}$ can then be determined.

$$\bar{H}_d/\bar{H} = 1.390 - 4.027 \, \bar{K}_T + 5.531 \, \bar{K}_T^2 - 3.108 \, \bar{K}_T^3 \qquad (6.24)$$

Table 6.4. Monthly Average Daily Extraterrestrial Radiation. (Source: Klein[57]).

Latitude	JAN	FEB	MAR	APR	MAY	JUN	JULY	AUG	SEP	OCT	NOV	DEC
25	23.9	28.2	33.0	37.1	39.4	40.1	39.6	37.9	34.4	29.5	24.9	22.7
30	21.1	25.7	31.3	36.5	39.6	40.7	40.1	37.6	33.1	27.3	22.1	19.7
35	18.1	23.1	29.3	35.5	39.6	41.2	40.3	37.0	31.5	24.9	19.2	16.7
40	15.1	20.3	27.2	34.3	39.3	41.4	40.3	36.2	29.7	22.3	16.3	13.6
45	12.0	17.5	24.8	32.8	38.8	41.3	40.0	35.1	27.7	19.6	13.3	10.6
50	9.0	14.5	22.3	31.2	38.1	41.2	39.6	33.8	25.4	16.7	10.3	7.6
55	6.1	11.5	19.5	29.3	37.2	40.9	39.1	32.4	23.0	13.8	7.3	4.8

Table 6.5.   Values of $\bar{H}_d/\bar{H}$.

January = 0.37	July = 0.32
February = 0.34	August = 0.31
March = 0.34	September = 0.33
April = 0.30	October = 0.28
May = 0.30	November = 0.34
June = 0.31	December = 0.39

From Eq. 6.24, the values of $\bar{H}_d/\bar{H}$ for all twelve months are shown in Table 6.5.

$\bar{R}_b$ can be determined using Tables 6.6 to 6.8.

Since the collector tilt was determined by using latitude $+15°$, Table 6.8 is used. The solar collector in the example is assumed to face due south toward the equator. Therefore, the azimuth angle would equal zero. When a solar collector is assumed to face directly south, Fig. 6.7 can be used instead of Tables 6.6 to 6.8 if desired.

Since Pensacola is exactly $30.47°$ North latitude, interpolation between monthly data for $30°$ latitude and $35°$ latitude shown in Table 6.8 yield essentially the same values as may be found from Fig. 6.7. Table 6.9 lists monthly values determined for $\bar{R}_b$.

Now that $\bar{H}_d/\bar{H}$ and $\bar{R}_b$ have been determined for each month, $\bar{R}$ can now be computed.

Table 6.6.   Values of $\bar{R}_b$ for Tilt = Latitude. (Source: Klein[57]).

Latitude	JAN	FEB	MAR	APR	MAY	JUN	JULY	AUG	SEP	OCT	NOV	DEC
				AZIMUTH ANGLE =		.0						
25	1.48	1.31	1.14	.98	.87	.83	.85	.93	1.07	1.25	1.43	1.53
30	1.66	1.43	1.20	1.00	.87	.83	.85	.93	1.07	1.25	1.43	1.53
35	1.91	1.59	1.28	1.03	.87	.81	.83	.96	1.17	1.48	1.82	2.02
40	2.26	1.79	1.38	1.06	.88	.80	.83	.98	1.24	1.64	2.12	2.42
45	2.76	2.07	1.51	1.11	.89	.80	.84	1.01	1.33	1.86	2.55	3.02
50	3.55	2.46	1.68	1.17	.90	.81	.85	1.04	1.45	2.17	3.20	4.00
55	4.94	3.06	1.92	1.25	.93	.81	.86	1.09	1.60	2.60	4.30	5.85
				AZIMUTH ANGLE =		15.0						
25	1.46	1.30	1.13	.98	.88	.84	.86	.94	1.07	1.24	1.41	1.51
30	1.63	1.41	1.19	1.00	.88	.82	.85	.95	1.11	1.33	1.57	1.71
35	1.87	1.56	1.27	1.03	.88	.82	.84	.96	1.17	1.45	1.78	1.98
40	2.21	1.76	1.37	1.07	.88	.81	.84	.98	1.24	1.61	2.07	2.36
45	2.69	2.02	1.49	1.11	.90	.81	.85	1.01	1.33	1.82	2.49	2.94
50	3.45	2.40	1.66	1.17	.91	.82	.86	1.05	1.44	2.11	3.12	3.89
55	4.79	2.97	1.88	1.25	.93	.82	.87	1.10	1.59	2.53	4.17	5.67
				AZIMUTH ANGLE =		30.0						
25	1.40	1.26	1.12	.99	.90	.86	.87	.95	1.06	1.21	1.36	1.44
30	1.56	1.36	1.17	1.01	.89	.85	.87	.96	1.10	1.29	1.50	1.62
35	1.77	1.49	1.24	1.03	.90	.84	.86	.97	1.15	1.40	1.69	1.86
40	2.06	1.66	1.33	1.07	.90	.84	.87	.99	1.22	1.54	1.94	2.20
45	2.48	1.90	1.44	1.11	.92	.84	.87	1.03	1.30	1.73	2.30	2.71
50	3.16	2.23	1.60	1.17	.93	.84	.88	1.06	1.41	1.98	2.86	3.55
55	4.36	2.73	1.80	1.25	.95	.84	.89	1.11	1.55	2.36	3.80	5.51
				AZIMUTH ANGLE =		45.0						
25	1.32	1.21	1.09	.99	.91	.88	.90	.96	1.05	1.17	1.29	1.35
30	1.44	1.29	1.14	1.00	.91	.87	.89	.96	1.08	1.24	1.40	1.49
35	1.61	1.39	1.19	1.03	.92	.87	.89	.98	1.12	1.32	1.55	1.69
40	1.84	1.54	1.27	1.06	.92	.87	.89	1.00	1.18	1.44	1.75	1.96
45	2.19	1.73	1.37	1.10	.94	.87	.90	1.03	1.25	1.59	2.05	2.36
50	2.72	2.00	1.50	1.15	.96	.87	.91	1.06	1.35	1.81	2.49	3.03
55	3.68	2.41	1.68	1.23	.98	.88	.92	1.11	1.47	2.12	3.24	4.31

Table 6.7. Values of $\bar{R}_b$ for Tilt = Latitude + 10°. (Source: Klein[57]).

Latitude	JAN	FEB	MAR	APR	MAY	JUN	JUL	AUG	SEP	OCT	NOV	DEC
					AZIMUTH ANGLE =		.0					
25	1.59	1.37	1.13	.92	.78	.72	.75	.86	1.05	1.29	1.53	1.66
30	1.79	1.49	1.20	.94	.78	.71	.74	.87	1.09	1.39	1.71	1.89
35	2.05	1.65	1.27	.97	.78	.70	.74	.88	1.14	1.52	1.94	2.19
40	2.42	1.86	1.38	1.00	.78	.70	.74	.90	1.21	1.69	2.26	2.62
45	2.95	2.15	1.51	1.04	.79	.70	.74	.93	1.30	1.91	2.71	3.25
50	3.79	2.56	1.68	1.10	.81	.70	.75	.96	1.41	2.22	3.40	4.30
55	5.25	3.17	1.91	1.17	.83	.71	.76	1.01	1.56	2.67	4.54	6.26
					AZIMUTH ANGLE =		15.0					
25	1.56	1.35	1.13	.93	.79	.73	.76	.87	1.05	1.27	1.51	1.63
30	1.75	1.47	1.19	.95	.79	.72	.75	.88	1.09	1.37	1.67	1.85
35	2.01	1.62	1.26	.97	.79	.72	.75	.89	1.14	1.49	1.90	2.14
40	2.36	1.82	1.36	1.01	.80	.71	.75	.91	1.21	1.65	2.20	2.55
45	2.87	2.10	1.48	1.05	.81	.71	.75	.94	1.29	1.87	2.64	3.16
50	3.67	2.49	1.65	1.11	.82	.71	.76	.97	1.41	2.16	3.30	4.17
55	5.09	3.08	1.87	1.18	.84	.72	.77	1.02	1.55	2.59	4.40	6.07
					AZIMUTH ANGLE =		30.0					
25	1.49	1.30	1.11	.94	.82	.77	.79	.89	1.04	1.24	1.44	1.55
30	1.65	1.40	1.16	.96	.82	.74	.78	.90	1.08	1.32	1.58	1.74
35	1.87	1.54	1.23	.98	.82	.75	.78	.91	1.13	1.43	1.78	1.99
40	2.18	1.71	1.32	1.02	.83	.75	.78	.93	1.19	1.57	2.04	2.35
45	2.63	1.95	1.43	1.06	.84	.75	.79	.96	1.27	1.76	2.42	2.90
50	3.35	2.30	1.58	1.12	.85	.75	.79	1.00	1.38	2.02	3.01	3.79
55	4.61	2.81	1.79	1.19	.87	.75	.80	1.04	1.52	2.40	3.99	5.48
					AZIMUTH ANGLE =		45.0					
25	1.38	1.24	1.09	.95	.85	.81	.83	.91	1.03	1.18	1.34	1.43
30	1.51	1.32	1.13	.96	.85	.80	.82	.91	1.06	1.25	1.46	1.58
35	1.69	1.43	1.18	.98	.85	.79	.82	.92	1.10	1.34	1.61	1.78
40	1.93	1.57	1.26	1.01	.86	.79	.82	.94	1.15	1.46	1.83	2.07
45	2.29	1.77	1.35	1.05	.87	.79	.83	.97	1.22	1.61	2.13	2.50
50	2.86	2.05	1.48	1.11	.89	.80	.84	1.01	1.32	1.83	2.60	3.20
55	3.86	2.47	1.66	1.18	.91	.80	.85	1.05	1.44	2.14	3.38	4.56

$$\bar{R} = (1-\bar{H}_d/\bar{H})\bar{R}_b + \bar{H}_d/\bar{H}(1+\cosine\ s)/2 + p(1-\cosine\ s)/2 \quad (6.25)$$

Beam	Diffuse	Reflective
Component	Component	Component

By assuming that no snow cover is present in Pensacola, p will be low and hence is assumed to be 0.2. The values of $\bar{R}$ for all twelve months are found in Table 6.10. The first three numerical columns represent each of the three components of solar radiation reaching the tilted solar collector (i.e., beam, diffuse, and reflective), while the fourth numerical column gives the total value of $\bar{R}$ used in computing $\bar{H}_T$.

The solar radiation upon a tilted (collector) surface can now be calculated using Eq. 6.26. The monthly values for solar radiation on a horizontal surface ($\bar{H}$) can be found in Appendix II.

$$\bar{H}_T = \bar{R}\ \bar{H} = \bar{R}\ \bar{K}_T\ \bar{H}_0 \quad (6.26)$$

Table 6.11 summarizes the data calculations used in determining $\bar{H}_T$.

6.2.4. Use of the Incidence Angle Modifier

Earlier in Chapter 2 some mention was given to the incident angle modifier in the discussion of collector efficiencies. Table 2.17 illustrated a summary information sheet prepared by the Florida Solar Energy Center for one manufacturer's solar collector that was tested. Table 6.12 illustrates a collector test information sheet for another manufacturer's solar collector.

Table 6.8. Values of $\bar{R}_b$ for Tilt = Latitude + 15°. (Source: Klein[57]).

Latitude	JAN	FEB	MAR	APR	MAY	JUN	JUL	AUG	SEP	OCT	NOV	DEC
					AZIMUTH ANGLE =		.0					
25	1.63	1.38	1.12	.88	.73	.66	.69	.81	1.02	1.29	1.56	1.71
30	1.83	1.50	1.18	.90	.72	.65	.68	.82	1.06	1.39	1.74	1.94
35	2.10	1.67	1.26	.92	.72	.64	.68	.83	1.11	1.52	1.98	2.25
40	2.47	1.88	1.36	.96	.73	.64	.68	.85	1.18	1.69	2.30	2.68
45	3.01	2.17	1.49	1.00	.74	.64	.68	.88	1.27	1.91	2.76	3.33
50	3.86	2.58	1.66	1.05	.75	.64	.69	.91	1.38	2.22	3.45	4.40
55	5.35	3.19	1.88	1.12	.77	.65	.70	.95	1.52	2.67	4.62	6.40
					AZIMUTH ANGLE =		15.0					
25	1.60	1.36	1.11	.89	.74	.68	.70	.82	1.02	1.27	1.53	1.68
30	1.79	1.48	1.17	.91	.74	.67	.70	.83	1.06	1.37	1.71	1.90
35	2.05	1.63	.124	.93	.74	.66	.69	.84	1.11	1.49	1.93	2.19
40	2.41	1.83	1.34	.97	.74	.66	.69	.86	1.18	1.64	2.24	2.61
45	2.93	2.11	1.46	1.01	.75	.66	.70	.89	1.26	1.87	2.68	3.24
50	3.74	2.50	1.63	1.06	.77	.66	.70	.92	1.37	2.16	3.35	4.26
55	5.18	3.10	1.85	1.13	.78	.66	.71	.96	1.52	2.59	4.47	6.19
					AZIMUTH ANGLE =		30.0					
25	1.52	1.31	1.10	.91	.77	.72	.74	.85	1.02	1.24	1.46	1.59
30	1.68	1.41	1.15	.92	.77	.71	.74	.86	1.05	1.32	1.61	1.77
35	1.91	1.54	1.22	.95	.78	.70	.73	.87	1.10	1.43	1.80	2.03
40	2.22	1.72	1.30	.98	.78	.70	.73	.89	1.16	1.57	2.07	2.40
45	2.68	1.96	1.41	1.02	.79	.70	.74	.92	1.24	1.75	2.46	2.95
50	3.40	2.30	1.56	1.08	.80	.70	.74	.95	1.35	2.01	3.05	3.86
55	4.68	2.82	1.76	1.15	.82	.70	.75	.99	1.48	2.39	4.04	5.59
					AZIMUTH ANGLE =		45.0					
25	1.40	1.24	1.07	.92	.81	.77	.79	.87	1.01	1.18	1.36	1.46
30	1.53	1.32	1.11	.93	.81	.76	.78	.88	1.04	1.25	1.47	1.60
35	1.71	1.43	1.17	.95	.81	.75	.78	.89	1.08	1.33	1.63	1.81
40	1.96	1.57	1.24	.98	.82	.75	.78	.91	1.13	1.45	1.84	2.10
45	2.32	1.77	1.34	1.02	.83	.75	.79	.93	1.20	1.61	2.15	2.53
50	2.89	2.05	1.46	1.07	.84	.75	.79	.97	1.29	1.82	2.62	3.25
55	3.90	2.47	1.64	1.14	.86	.75	.80	1.01	1.41	2.13	3.41	4.63

The incident angle modifier $(K\tau\alpha)$ equation shown in Table 6.12 is of the form shown in Eq. 6.27.

$$K\tau\alpha = (\tau\alpha)/(\tau\alpha)_n - 1.0 - b_0(1/\cosine\ \theta - 1.0) \qquad (6.27)$$

where

α = absorptance of the collector absorptance plate
τ = transmittance of the solar collector cover plate
$(\tau\alpha)$ = monthly average of the transmittance absorptance product
$(\tau\alpha)_n$ = transmittance absorptance product at normal incidence

Hence:

$(\tau\alpha/\tau\alpha)_n$ = correction factor that accounts for the collector not continuously receiving solar radiation at normal incidence throughout the month
b_0 = incident angle modifier constant
θ = monthly average incident angle for beam radiation

A solar collector is generally tested where solar radiation falls normal to the collector cover. In many solar collector rating tests, a motorized solar tracking device is used to keep the solar collector oriented directly (normal) toward the sun. This allows the collector to receive the maximum amount of solar radiation throughout the test. In a typical solar collector installation, this is not

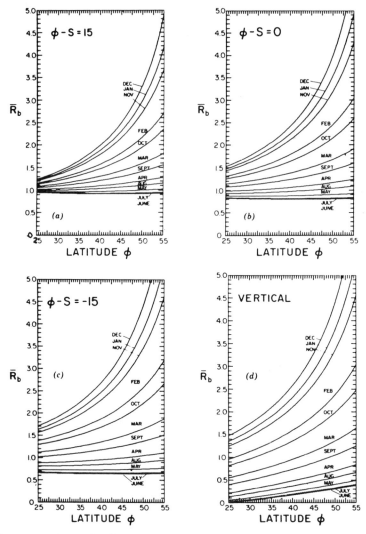

Figure 6.7. Values for $\bar{R}_b$ for South Facing Collectors. (Source: Beckman[14]).

Table 6.9.   Values of $\bar{R}_b$ for Pensacola.

January	=	1.86	July	=	0.68
February	=	1.52	August	=	0.82
March	≈	1.19	September	=	1.06
April	=	0.90	October	=	1.40
May	=	0.72	November	=	1.76
June	=	0.65	December	=	1.97

Table 6.10. Calculation of $\bar{R}$.

Month	Beam Component		Diffuse Component		Reflective Component		$\bar{R}$
Jan	1.17	+	0.32	+	0.03	=	1.52
Feb	1.00	+	0.29	+	0.03	=	1.32
Mar	0.79	+	0.29	+	0.03	=	1.11
Apr	0.63	+	0.26	+	0.03	=	0.92
May	0.50	+	0.26	+	0.03	=	0.79
Jun	0.45	+	0.26	+	0.03	=	0.74
Jul	0.46	+	0.27	+	0.03	=	0.76
Aug	0.57	+	0.26	+	0.03	=	0.86
Sep	0.71	+	0.28	+	0.03	=	1.02
Oct	1.01	+	0.24	+	0.03	=	1.28
Nov	1.16	+	0.29	+	0.03	=	1.48
Dec	1.20	+	0.33	+	0.03	=	1.56

the case. The solar collector is generally rigidly mounted in a fixed position and will receive solar energy at numerous angles of incidence throughout the year. The correction factor $(\tau\alpha)/(\tau\alpha)_n$ serves to modify the collector efficiency equations for operating conditions present for such solar collectors. When collector efficiency equations obtained at $(\tau\alpha)_n$ are multiplied by the correction factor $(\tau\alpha)/(\tau\alpha)_n$, the $(\tau\alpha)_n$ drops out leaving only $\tau\alpha$. The collector efficiency equations then represent average monthly $(\tau\alpha)$ conditions.

Before the incidence angle modifier (Eq. 6.27) can be used, the monthly values for θ must be determined.

$$\cos \theta = \cos (\phi-s)\cos \delta \cos \omega + \sin (\phi-s)\sin \delta \qquad (6.28)$$

For solar collectors facing directly toward the equator, Klein has determined that the monthly average incident angle for beam radiation (θ) is the angle formed when the sun is 2.5 hours from solar noon for a day in the middle of the month. Basically this means that the hour angle (ω) in Eq. 6.28 will be $\pm 37.5°$ since the hour angle is equal to $15°$ times the number of hours from solar noon. In deciding upon a specific day of the month to use in determining the declination (δ) in Eq. 6.28, Klein, *et al.*, presents some suggested days of the month to use as shown in Table 6.13. The average day for the month was basically determined by finding the day of the month whose day length and declination corresponded to the average day length and declination for the month. Table 6.14 shows a compilation of data used in determining $(\tau\alpha)/(\tau\alpha)_n$.

In the table, monthly values of solar declination (δ) were determined using

Table 6.11. Worksheet 1.

A1 Location = Pensacola, Florida
B1 Latitude = 30.47°
C1 Collector Tilt = (s) = 45°

D1 Ground Reflectance = (p) = 0.20
D1 Solar Reflectance Component = $p(1 - \cos s)/2$ = 0.03
F1 $(1 + \cos s)/2$ = 0.85

G1 Month	H1 $\bar{K}_T$ From Appendix II	I1 $\bar{H}_d/\bar{H}$ From Eq. 6.24	J1 $1-\bar{H}_d/\bar{H}$ 1-I1	K1 $\bar{R}_b$ From Tables 6.5 to 6.7 or Fig. 6.7	L1 Beam Component J1*K1	M1 Diffuse Component F1*I1	N1 Reflective Component E1	O1 $\bar{R}$ L1+M1+N1	P1 $\bar{H}$ MJ/day-m² From Appendix II	Q1 $\bar{H}_T$ O1*P1 10⁶ J/day-m²
JAN	0.50	0.37	0.63	1.86	1.17	0.32	0.03	1.52	10.47	15.91×10^6
FEB	0.53	0.34	0.66	1.52	1.00	0.29	0.03	1.32	13.44	17.74×10^6
MAR	0.55	0.34	0.66	1.19	0.79	0.29	0.03	1.11	16.96	18.83×10^6
APR	0.59	0.30	0.70	0.90	0.63	0.26	0.03	0.92	12.31	19.61×10^6
MAY	0.59	0.30	0.70	0.72	0.50	0.26	0.03	0.79	23.53	18.59×10^6
JUN	0.58	0.31	0.69	0.65	0.45	0.26	0.03	0.74	23.78	17.60×10^6
JUL	0.56	0.32	0.68	0.68	0.46	0.27	0.03	0.76	22.48	17.08×10^6
AUG	0.57	0.31	0.69	0.82	0.57	0.26	0.03	0.86	21.31	18.33×10^6
SEP	0.55	0.33	0.67	1.06	0.71	0.28	0.03	1.02	18.00	18.36×10^6
OCT	0.61	0.28	0.72	1.40	1.01	0.24	0.03	1.28	15.49	19.83×10^6
NOV	0.53	0.34	0.66	1.76	1.16	0.29	0.03	1.48	11.64	17.23×10^6
DEC	0.48	0.39	0.61	1.97	1.20	0.33	0.03	1.56	9.38	14.63×10^6

Table 6.12.   Collector Test Information Sheet. (Source: Florida Solar Energy Center[33]).

MANUFACTURER	
Revere Copper & Brass Inc. P.O. Box 151 Rome, New York 13440	**Collector Model** 133

This solar collector was tested by the Florida Solar Energy Center (FSEC) in accordance with prescribed methods and was found to meet the minimum standards established by FSEC. The purpose of the tests is to verify initial performance conditions and quality of construction only. The resulting certification is not a guarantee of long term performance or durability.

DESCRIPTION

Gross Length	1.956 meters	6.42 feet
Gross Width	0.889 meters	2.92 feet
Gross Depth	0.108 meters	0.35 feet
Gross Area	1.739 square meters	18.72 square feet
Transparent Frontal Area	1.605 square meters	17.28 square feet
Volumetric Capacity	2.2 liters	0.58 gallons
Weight (empty)	40.0 kilograms	88.2 pounds
Number of Cover Plates	One	
Flow Pattern	Parallel	

Incident Angle Modifier $K_{\tau\alpha}$ = $1.0 - 0.11 \ (1/\cos\theta - 1)$

Efficiency Equations   First Order η = $71.2 - 503.7 \ (Ti-Ta)/I$

Second Order η = $71.0 - 485.9 \ (Ti-Ta)/I - 292.1 \ [(Ti-Ta)/I]^2$

Tested per ASHRAE 93-77   Units of Ti-Ta/I are $^{\circ}C/Watt/m^2$

MATERIALS

Enclosure   Extruded aluminum frame, 0.22 cm thick
Glazing   Tempered water white glass, 0.32 cm thick
Absorber   Copper plate, 0.813 mm thick; black chrome coating
Insulation   Fiberglass blanket; back: 7.6 cm, sides: 2.5 cm

RATING

The collector has been rated for energy output on measured performance and an assumed standard day. Total solar energy available for the standard day is 5045 watt-hour/m^2 (1600 BTU/ft^2) distributed over a 10 hour period.

Output energy ratings for this collector based on the second-order efficiency curve are:

Collector Temperature	Energy Output			
Low Temperature, 35°C (95°F)	20,800	Kilojoules/day	19,700	BTU/day
Intermediate Temperature, 50°C (122°F)	16,500	Kilojoules/day	15,700	BTU/day
High Temperature, 100°C (212°F)	5,600	Kilojoules/day	5,300	BTU/day

Table 6.13.   Recommended Average Day for Each Month. (Source: Klein[57]).

Month	Day of the Year	Date
January	17	Jan. 17
February	47	Feb. 16
March	75	Mar. 16
April	105	Apr. 15
May	135	May 15
June	162	June 11
July	198	July 17
August	228	Aug. 16
September	258	Sept. 15
October	288	Oct. 15
November	318	Nov. 14
December	344	Dec. 10

Table 6.14. Worksheet 2.

A2. Location = Pensacola, FL   D2. Azimuth (γ) = $0°$

B2. Latitude = 30.47°   E2. bo   = 0.11

C2. Tilt   = 45°

F2. Month	G2. δ	H2. θ	I2. K_{τ} or $(\tau\alpha)/(\tau\alpha)_n$
Jan	-20.92	36.20	0.97
Feb	-12.95	36.42	0.97
Mar	-2.418	38.93	0.97
Apr	9.415	44.22	0.96
May	18.79	49.74	0.94
Jun	23.09	52.55	0.93
Jul	21.18	51.28	0.93
Aug	13.45	46.48	0.95
Sep	2.217	40.74	0.96
Oct	-9.599	36.96	0.97
Nov	-18.91	36.11	0.97
Dec	-23.05	36.40	0.97

Table 6.13 and Eq. 6.3. The monthly average incidence angle for beam radiation (θ) can then be calculated from Eq. 6.28, where

Location: Pensacola, Florida

$\phi = 30.47$

$s = 45°$

$\omega = 37.5°$

After θ is determined, $(\tau\alpha)/(\tau\alpha)_n$ is computed using Eq. 6.27 where b_0 was determined from Fig. 6.8 to be 0.11. When $(\tau\alpha)/(\tau\alpha)_n$ was determined for the example in Table 6.13, the solar collector was assumed to face due south (azimuth (γ) = 0°). The values of $(\tau\alpha)/(\tau\alpha)_n$ in Table 6.14 will not change appreciably for collector azimuth (γ) angles up to about 15° east or west of due south (0°). At greater azimuth angles (γ), Eq. 6.29 can be used.

cosine θ =sine δ sine ϕ cosine s $-$ sine δ cosine ϕ sine s
cosine γ + cosine δ cosine ϕ cosine s cosine ω +

$$(6.29)$$

cosine δ sine ϕ sine s cosine γ cosine ω + cosine δ
sine s sine γ sine ω

In using Eq. 6.29, the hour angle (ω) used will not be (37.5°) when the azimuth angle (γ) is greater than about 15° of true south. There is still some question as to what hour angle (ω) to use in calculating θ when the solar collectors are not oriented toward the equator. Therefore, if possible, the

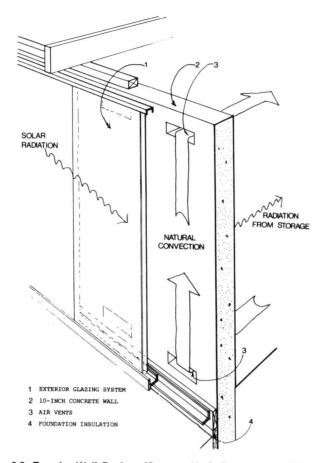

1 EXTERIOR GLAZING SYSTEM
2 10-INCH CONCRETE WALL
3 AIR VENTS
4 FOUNDATION INSULATION

Figure 6.8. Trombe Wall Design. (Source: U. S. Department of Energy[106]).

collectors should be mounted within 15° of true south and Klein's 37.5° value for ω be used.

6.3 PASSIVE SOLAR DESIGN

With passive solar design concepts, the building and surrounding land is used to help control the solar insolation. One method involves planting deciduous trees near the building to provide shade in the summer and when the leaves have fallen in the winter, to allow solar energy to reach the building unhindered. Other passive design concepts utilizing the surrounding land are planting evergreen trees some distance from the building to moderate the speed of winds reaching it in winter.

Passive design concepts that may be incorporated in a building are numerous. The Trombe wall design, shown in Fig. 6.8, is an effective means of gathering solar radiation for use in heating a building. This system uses a thick concrete wall that absorbs the solar radiation and a (glass) glazing system to retain heat. Another passive design involves utilizing a south-facing overhang or ledge on a building to allow solar radiation to pass inside during the winter

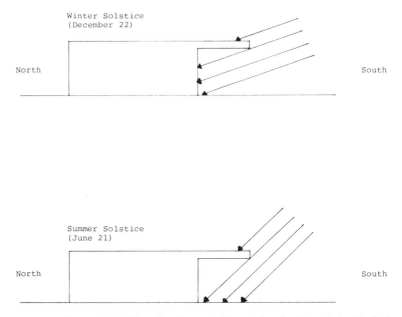

Figure 6.9. Effects of Building Overhang in Regulating Incident Solar Radiation.

but not in the summer. Figure 6.9 illustrates this concept. The solar altitude angles can be calculated using the procedure described in Section 6.1.

A very important aspect of solar design involves using insulation. Insulation helps attenuate the flow of heat. Figure 6.10 shows a heating zone map illustrating recommended values of insulation in a building. The R-value represents the thermal resistance of a material. The higher the R-value, the greater the resistance to thermal flow and therefore the greater the insulating value of the material. Table 6.15 shows an insulation materials chart that can aid in determining the type of insulation necessary based on the intended application.

There are a number of devices which can be used to detect heat losses from a building. Infrared thermography is one method whereby an infrared (infrared sensitive) camera is used to observe levels of thermal radiation. Sensitive thermometer-type devices can be used to monitor temperature changes.

6.4 ACTIVE SOLAR DESIGN

6.4.1. Overview

In an active solar system, collectors are used to trap solar energy that is then transported to a storage tank whereby the collected thermal energy is stored in either a liquid or gaseous (air) form for later use in meeting a building's thermal load requirements.

6.4.2. Reception of Solar Energy

In designing an active solar system, attention must be paid to factors which might affect the reception of solar energy. This involves determining whether

| HEATING | RECOMMENDED FOR | | |
ZONE	CEILING	WALL	FLOOR
0,1	R-26	R-13	R-11
2	R-26	R-19	R-13
3	R-30	R-19	R-19
4	R-33	R-19	R-22
5	R-38	R-19	R-22

Sources for R-value data include Federal Housing Administration, American Society of Heating, Refrigeration and Air Conditioning Engineers (ASHRAE), and insulation manufacturers.

Figure 6.10. Heating Zone Map. (Source: National Solar Heating and Cooling Information Center[78]).

trees or other buildings may block the direct solar radiation reaching the location where the proposed solar collectors are to be mounted. If there is some shading the collectors should be placed at another location on the roof or possibly on the ground. If this is not possible, the percentage of shading on the collectors should be taken into account when determining the amount of solar radiation received by the collector. One problem which can arise after the solar system has been installed in the building is when a neighbor erects a building which shadows the sunlight reaching the solar collectors. This can become a legal decision as to what rights the owner with the solar system has in regard to free access to the sunlight. Present (and potential) neighboring buildings, hills, and trees should be checked for possible shadows on a solar collector.

Figure 6.11 illustrates the altitude of the sun at solar noon for the 22nd day of each month for various latitudes. Notice that on about December 22 (the winter solstice) the sun's altitude is at its lowest point. On this date the sun will cast the longest shadows. Generally if a shadow is not cast on a solar collector during the winter solstice when the sun is at its lowest altitude, then none will appear on the solar collector when the sun is at higher altitudes. If there are deciduous trees near the proposed collector site, months in which the trees bear leaves must also be considered.

To determine whether objects will cast shadows on a solar collector and to

Table 6.15. Insulation Materials Chart. (Source: National Solar Heating and Cooling Information Center[78]).

Type	Comments	Application
BATTS/BLANKETS Preformed glass fiber or rock wool with or without vapor barrier backing.	Fire resistant, moisture resistant, easy to handle for do-it-yourself installation, least expensive. Fiberglass may cause skin irritation.	Unfinished attic floor; rafters; underside of floors; between studs.
FOAMED IN PLACE Plastic installed as a foam under pressure. Hardens to form insulation. 1.) Urethane 2.) Ureaformaldehyde 3.) Polystyrene	1.) Has highest R-value. If ignited burns explosively and emits toxic fumes. Should be covered with 1/2" gypsum wallboard to assure fire safety. 2.) Fire resistant, high R-value, first choice of many experts. Requires installation by reliable experienced contractor (as all foams do). May shrink or cause odor problems if improperly mixed or cured, and is light sensitive. 3.) Lacks fire resistance as does urethane and has lower R-value than urethane.	Finished frame walls; floors; ceilings. Ureaformaldehyde can be used as masonry or block insulation in place of pealite, vermiculite, or styrene beads. Foams should not be used in hot, humid climate.
RIGID BOARD 1.) Extruded polystyrene bead 2.) Extruded polystyrene 3.) Urethane 4.) Glass fiber	All have high R-values for relatively small thickness. 1.2.3.) Are not fire resistant, require installation by contractor with 1/2" gypsum board to insure fire safety. 2.) Is its own vapor barrier. 3.) Is its own vapor barrier, however, when in contact with water, it should have a skin to prevent degrading. 1.4.) Requires addition of vapor barrier.	Basement walls; new construction frame walls; commonly used as an outer sheathing between siding and studs.
LOOSE FILL (POURED-IN) 1.) Glass fiber 2.) Rock wool 3.) Treated cellulosic fiber	All easy to install; require vapor barrier bought and applied separately. Vapor barrier may be impossible to install in existing walls. 1.2.) Fire resistant, moisture resistant. 3.) Check label to make sure material meets federal specifications for fire, and moisture resistance and R-value.	Unfinished attic floor; uninsulated existing walls.
LOOSE FILL (BLOWN-IN) 1.) Glass fiber 2.) Rock wool 3.) Treated cellulosic fiber	All require vapor barrier bought separately; all require space to be filled completely. Vapor barrier may be impossible to install in existing walls. 1.2.) Fire resistant, moisture resistant. 3.) Fills up spaces most consistently. When blown into closed spaces, has slightly higher R-value; check label for fire and moisture resistance and R-value. Chemicals added to cullulose may reduce its thermal resistance; a potential for corroding pipes exists if too many sulfate chemicals are added.	Unfinished attic floor; finished attic floor; finished frame walls; underside of floors.
CAULKING COMPOUNDS 1.) Oil or resin base 2.) Latex, butyl, polyvinyl base 3.) Elastomeric base: silicones, polysulfides, polyurethanes	1.) Lowest cost, least durable; replacement time approximately 2 years. 2.) Medium priced, more durable; look for guarantees on time of durability. 3.) Most expensive, most durable. Recommended by some experts as the best buy. Note: Lead base caulk is not recommended because it is toxic.	At stationary joints; exterior window and door frames; whenever different materials or parts of building meet.

Table 6.15.   (Continued)

WEATHERSRIPPING		

1.) Felt or foam strip
2.) Rolled vinyl; with
   or without metal
   backing
3.) Thin spring metal
4.) Interlocking metal
   channels
	1.) Inexpensive, easy to install,
	not very durable.
	2.) Medium priced, easy to install,
	durable, visible when installed.
	3.) More expensive, somewhat
	difficult to install. Very
	durable, invisible when installed.
	4.) Most expensive, difficult to
	install, durable, excellent
	weather seal.
		At moving joints; peri-
		meter of exterior doors;
		inside of window sashes.

what extent, the sun's altitude should be plotted for the daylight hours. Figure 6.12 illustrates a plot of the sun's path for Pensacola. (Both Figs. 6.11 and 6.12 were determined using the methodology described in Section 6.1.) After the sun's path is plotted, the skyline is then plotted. The easiest way to plot the skyline is with a transit such as the one shown in Fig. 6.13. If one is not available, a hand level and a compass can be used instead.

An isogonic chart of the United States is used in determining magnetic declination. Basically, the isogonic chart is used to determine true north based on the magnetic north read from the compass needle. For example, Pensacola, located at 87.22° west longitude, has a magnetic declination of 4° east—as shown in Fig. 6.14. Figure 6.14 illustrates how the 4° east declination would be interpreted. The true south orientation is for both the transit and later, for placement of the solar collector toward true south.

Figure 6.14 also shows how a magnetic declination of 4° west would be interpreted. Once true south has been determined, the transit can then be properly oriented. The next step involves sighting the transit on objects in the

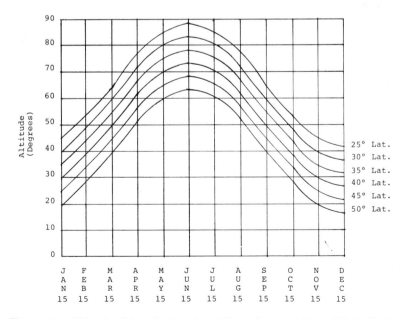

Figure 6.11. Altitude of Sun During Solar Noon for 22nd Day of Each Month.

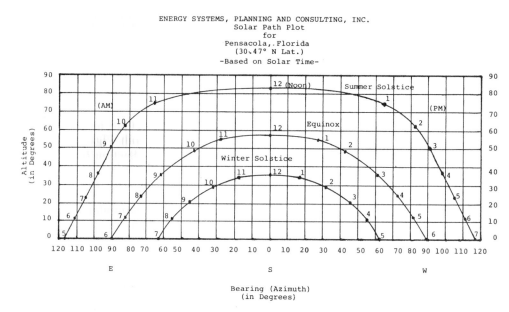

ENERGY SYSTEMS, PLANNING AND CONSULTING, INC.
Solar Path Plot
for
Pensacola,.Florida
(30.47° N Lat.)
-Based on Solar Time-

Figure 6.12. Sun Chart for Pensacola, Florida.

field of view and plotting the azimuth and altitude readings on a sun chart such as shown in Fig. 6.12. Once the skyline has been plotted, the solar designer or homeowner can easily determine what object near the solar collector installation may cause problems and the chart will also aid in determining the amount of time that the solar collectors may be shielded from direct sunlight. If a sun chart is constructed using only one end of a long row of collectors as a reference point, the results of the skyline plot may not necessarily hold true for the solar collectors located at the furthest point away in a long row of solar collectors. Therefore, several reference points should be used.

When interpreting and using the sun charts such as the one shown in Fig. 6.12, the type of solar system used in the building is important. For example, if strictly a solar heating system is used in a building where the collectors would only operate during the winter months, then a tall deciduous tree plotted in Fig. 6.12 may not necessarily be reason for concern if it were plotted as having a greater altitude angle than the sun had for a particular hour of the day in winter. On the other hand, if the solar collectors were to operate year-round, the deciduous tree could have an effect when it grows leaves in the warmer months.

It is not always possible to obtain the ideal collector mounting location where there is absolutely no shading of the collectors from nearby objects. Sometime a compromise position may have to be accepted. The majority of solar energy received by a solar collector occurs between 9:00 AM and 3:00 PM. Therefore, collector shading during this time period will cause a greater loss of potentially available solar energy than if shading were to occur early in the morning or late in the afternoon.

The design or architect/engineering firm can easily plot a chart for their solar installation locale which would be based on an exact altitude. Then by

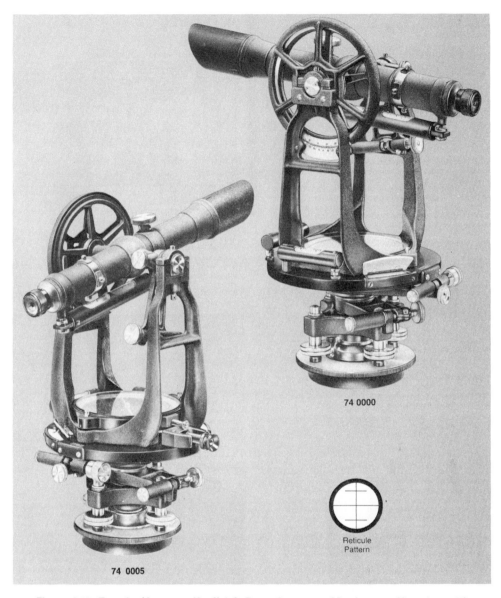

74 0000

74 0005

Reticule
Pattern

Figure 6.13. Transit. (Courtesy Keuffel & Esser Company, Morristown, New Jersey[50]).

the use of a thermofax machine, plastic transparencies can be produced. The user can then use an erasable grease pencil to plot the skyline for a particular solar job site and can then erase and reuse the chart for another site. There are numerous design aids available to the design firm or homeowner which aid in solar site selection. One design aid is shown in Fig. 6.15. This "sun locator" appears in the *Solar Energy Handbook, 1978* and is available free from the National Solar Heating and Cooling Information Center.

To use the design aid shown in Fig. 6.15 involves cutting the "sun locator" to the latitude of the proposed solar installation site. The locator is then oriented to true south and placed where the proposed collectors will be

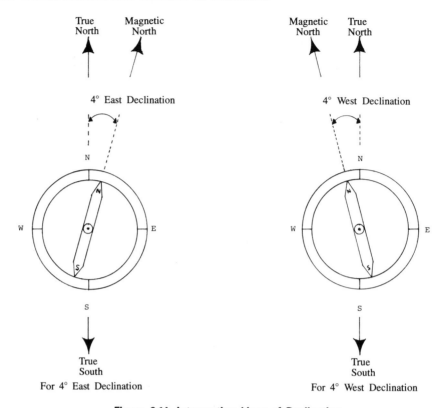

Figure 6.14. Interpreting Lines of Declination.

located. By siting from the rear sight aperture through the hourly notches in the front of the locator, the observer can determine what obstructions may block and shade the solar collectors.

6.4.3. Solar Heating and Cooling Load Determination

6.4.3.1. Domestic Hot Water Load Determination. The basic equation used in determining the monthly domestic hot water load requirement (Q_w) is shown in Eq. 6.30

$$Q_w - D \ N \ C_p P(T_s - T_m) \qquad (6.30)$$

where

$$
\begin{aligned}
D &= \text{daily hot water use (liters or gallons)} \\
N &= \text{number of days in month} \\
C_p &= \text{specific heat of water (4190 J/kg–°C or 1 Btu/lb°F)} \\
P &= \text{density of water (1 kg/l or 8.33 lb/gal)} \\
T_s &= \text{hot water temperature (°C or °F)} \\
T_m &= \text{water main temperature (°C or °F)}
\end{aligned}
$$

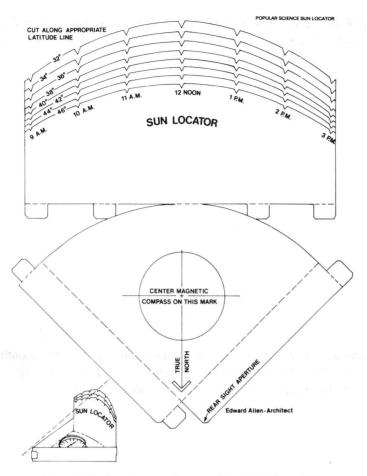

Figure 6.15. Sun Locator. (Source: Popular Science[90,78]).

Table 6.16 illustrates some daily hot water usage rates (D).

The hot water temperature (T_s) is generally assumed to be 60°C (140°F). The monthly average water main temperatures can be obtained from the local city utility (water) department. The U. S. Department of Housing and Urban Development (HUD) advised using 13°C (55°F) as an average source water temperature if the local water main temperature is not known.

For example, to determine the domestic hot water load requirement for a family of four in Pensacola, Florida, check Table 6.16. A family of four uses 265 liters of hot water per day. The average water main temperature for Pensacola is assumed to be 20°C (68°F) year round. Equation 6.31 shows the calculation procedure for determining January's monthly domestic hot water load requirement.

$$Q_w = 265 \text{ liters/day} \times 31 \text{ days} \times 4190 \text{ J/kg-°C} \qquad (6.31)$$
$$\times \ 1 \text{ kg/liter} \times (60°C - 201°C) = 1.38 \times 10^9 \text{ J.}$$

Table 6.17 summarizes the twelve monthly domestic hot water load requirements for the example problem.

Table 6.16.   Daily Hot Water Usage 60°C (140°F) for Solar System Design. (Source: Adapted from U. S. Department of Housing and Urban Development[108]).

Category	One and Two Family Units 1/ and Apts. up to 20 Units					Apts. of 2/ 20-200 Units	Apts. of 2/ over 200 Units
No. of People	2	3	4	5	6	---	---
No. of Bedrooms	1	2	3	4	5	---	---
Hot Water/Unit							
liters/day	151	208	265	322	378	151	132
(gal/day)	(40)	(55)	(70)	(85)	(100)	(40)	(35)

Assumes 76 liters (20 gal.) per person for first 2 people and 57 liters (15 gal.) per person for additional family members.

From: R. G. Werden and L. G. Spielvogel: "Part II Sizing of Service Water Heating Equipment in Commercial and Institutional Buildings", ASHRAE Transactions, Vol. 75 PII , 1969 p. iv.1.1.

6.4.3.2. Heating and Cooling Load Determination. To find monthly space heating loads, first determine the heat (transmission) loss rate (Q). The heat loss rate for the building is of the form shown in Eq. 6.32.

$$Q = U A (T_i - T_o) \tag{6.32}$$

where

U = overall heat transfer coefficient

A = surface area

T_i = design dry bulb (inside) building temperature

T_o = design dry bulb outside winter design temperature

The easiest method of calculating the building heat loss rate (Q) is by the use of load calculation worksheets such as shown in Table 6.18. Worksheets for calculating both commercial and residential heating and cooling loads can be obtained from ASHRAE and ACCA (formerly NESCA). Both of these organizations also publish load calculation manuals which explain in great detail procedures for computing loads in both commercial and residential buildings.

To demonstrate the calculations used in determining monthly space heating loads, an example problem will be used. The heat loss rate (Q) for a building located in Pensacola is found to be 21,966 W (or 75,000 Btu/hr). After the overall building heat loss rate has been found, it is divided by the design temperature difference (i.e., T_i-T_o). Since the winter design temperature (T_o) for Pensacola is −1.7°C (29°F) and the indoor design temperature (T_i) is 22°C (72°F), the design temperature difference is 23.7°C [22 − (−1.7)]. Equation

Table 6.17. Monthly Domestic Hot Water Load Calculation.

Month	Monthly Domestice Hot Water Usage (in liters) (D*N) (from Table 6.17 and Eq. 6.30)	Monthly Domestic Hot Water Lead Requirement (Q_w) (in Joules) $D*N*C_p*p*(T_S-T_m)$ (from Eq. 6.30)
Jan	8215	1.38×10^9
Feb	7420	1.24×10^9
Mar	8215	1.38×10^9
Apr	7950	1.33×10^9
May	8215	1.38×10^9
Jun	7950	1.33×10^9
Jul	8215	1.38×10^9
Aug	8215	1.38×10^9
Sep	7950	1.33×10^9
Oct	8215	1.38×10^9
Nov	7950	1.33×10^9
Dec	8215	1.38×10^9

6.33 shows the computation of the overall heat transfer coefficient − surface area product (UA).

$$UA = 21,966 \text{ watts}/23.7°\text{Centigrade} = 927 \text{ watts}/°\text{Centigrade} \quad (6.33)$$

Equation 6.34 is used to compute the monthly space heating load (Q_s).

$$Q_s = (UA)(24 \text{ hours/day})(DD)(3600 \text{ J/1W-hr}) \quad (6.34)$$

where

$$DD = \text{degree days}$$

The degree day concept is based on the assumption that when the outside temperature is at least 18°C (65°F), internal heat losses will be offset by heat gains from solar radiation, internal lighting, human body warmth, etc. When the outside building temperature is below 18°C (65°F) supplemental space heating will be required to maintain the temperature inside the building. The number of degree days is the difference between the average temperature of the day and 18°C (65°F). Appendix II contains the number of degree days per month for all 12 months of the year for a number of cities. Equation 6.35 shows the calculations for determining Q_s for January.

Table 6.18. Load Calculation Worksheet. Reprinted with permission from Load Calculation Forms, SPLCMFT, American Society of Heating, Refrigerating and Air Conditioning Engineers, Inc.[6]).

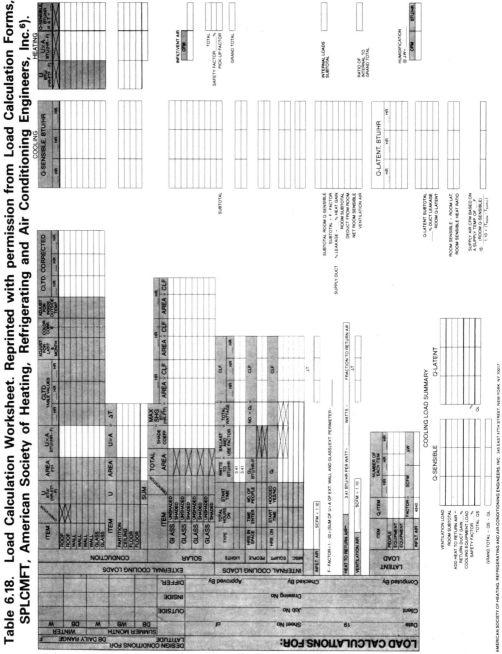

Table 6.19. Summary Building Load Worksheet 3.

A3. Month	B3 Heating Degree Days (DD)	C3. Monthly Space Heating Loads (Q_s) (J/Mo.) (from Eq. 6.34)	D3. Domestic Water Heating Load (Q_w) (J/mo.) (from Eq. 6.30 and Table 6.18)	E3. Total Building Load (Q_t) (J/mo.) (C3.+D3.)
Jan	237	1.90×10^{10}	1.38×10^9	2.04×10^{10}
Feb	179	1.43×10^{10}	1.24×10^9	1.55×10^{10}
Mar	117	9.37×10^9	1.38×10^9	1.08×10^{10}
Apr	21	1.68×10^9	1.33×10^9	3.01×10^9
May	0	0	1.38×10^9	1.38×10^9
Jun	0	0	1.33×10^9	1.33×10^9
Jul	0	0	1.38×10^9	1.38×10^9
Aug	0	0	1.38×10^9	1.38×10^9
Sep	0	0	1.33×10^9	1.33×10^9
Oct	18	1.44×10^9	1.38×10^9	2.82×10^9
Nov	105	8.41×10^9	1.33×10^9	9.74×10^9
Dec	199	1.59×10^{10}	1.38×10^9	1.73×10^{10}
TOTALS	876	7.01×10^{10}	1.62×10^{10}	8.64×10^{10}

$$Q_s = (927 \text{ W/°C}) \ (24 \text{ hours/day}) \ (237°\text{C-day})(3600 \text{ J/1W-hr}) \quad (6.35)$$
$$= 1.90 \times 10^{10} \text{ J}$$

Table 6.19 illustrates a summary of the building load requirements including both the monthly domestic water heating loads (Q_w) and the monthly space heating loads (Q_s) for the example.

6.4.4. Design Guidelines

Table 6.20 illustrates some suggested rules of thumb to use when designing a solar system. These guidelines are the result of computer simulations, practical experience and testing (experiments) by many people in the field. These recommendations are not necessarily hard and fast rules. Some manufacturers producing solar (e.g., heating) systems may recommend different operating design parameters which should be heeded.

Take for example the recommended rock depth for an air type storage bin shown in Table 6.20. This figure is based on such things as the performance characteristics of rock beds with respect to temperature stratification, pressure drop, etc., while the recommended collector slope is based on the months of heating and what the sun's declination and solar insolation will be during these heating months. If so inclined, one can derive some of these guidelines. Using the methodology found in Sections 6.1 and 6.2, one can plot the month (as x axis) with respect to solar insolation received per day (as y axis) for various collector tilts at a certain design location (i.e., a specific city) to visualize which angle allows the collector maximum solar energy reception for the

Table 6.20. Rules of Thumb for Solar System Design.

Solar Liquid Heating Systems

Collector Flow Rate	0.015 $1/s-m^2$ (i.e., for 50/50 ethylene glycol water)
	$0.02*(\dfrac{C_p \text{ of storage liquid}}{C_p \text{ of collector liquid}})$ gpm/ft^2 of collector
Water Storage Tank Size	50 to 100 $1/m^2$ 1.25 - 2.0 gal/ft^2
Pressure Drop Across Collector	.035 to 0.703 Kg/cm^2 0.5 - 10 PSI/collector module
Load Heat Exchanger	$1 < E_L (\dot{m}Cp)\text{min}/UA < 5$
Collector Heat Exchanger	$F_R'/F_R > 0.9$

Solar Air Heating Systems

Collector Flow Rate	5 - 10 $1/s-m^2$ of collector 1 - 2 CFM/ft^2 of collector
Pebble Bed Storage Size	0.15 to 0.30 m^3 of rock/m^2 of collector 0.5 - 1.0 ft^3 of rock/ft^2 of collector
Rock Depth	1.25 to 2.5 m in air flow direction 4 to 8 ft in air flow direction
Pebble Size	1 - 2 cm dia. round washed river rock screened to uniform size 0.5 - 1.0 inch dia.
Pressure Drops: Pebble-bed	3.75 - 7.5 mm H$_2$O 0.15 - 0.3 in W.G.
Flat Plate Collector (12-14 ft length)	5.0 - 7.5 mm H$_2$O 0.2 - 0.3 in W.G.
Flat Plate Collector (18-20 ft length)	17.5 - 20 mm H$_2$O 0.7 - 0.8 in W.G.
Ductwork	1 mm H$_2$O/15 m 0.08 in W.G./100 ft duct length
Insulation on duct exposed to unconditioned spaces	2.5 cm fiberglass minimun 1 in fiberglass minimum

Solar Domestic Hot-Water Heating Systems

Pre-Heat Tank Size	1.5 - 2.0 times conventional water heater size
Air-Water Coil Size	$0.2 < E < 0.5$
Water-Water Coil Size	$0.5 < E < 0.5$

specific months the solar collectors will operate. The derived collector tilt angle should come close to the recommended collector tilt guidelines shown in table 6.2 and 6.20.

6.4.5. The *F*-Chart Method

6.4.5.1. Design Procedure. The F-chart method allows the solar designer or homeowner to calculate the heating contribution fraction that solar energy supplies the thermal load requirements of the building. An auxiliary heater is used to make up the difference between the percentage of thermal energy supplied by solar and the total energy required by the building.

Two variables (x and y) are used in the calculation procedure for ultimately determining the fraction (F) of the building thermal load supplied by solar energy. The equations used in calculating x and y are shown.

$$x = F_R U_L \ (F_R'/F_R) \ (T_{ref} - \bar{T}_a) \ \Delta t \ A/Q_t \qquad (6.36)$$

$$y = F_R(\tau\alpha)_n \ (F_R'/F_R) \ (\bar{\tau}\alpha)/(\tau\alpha)_n \ \bar{H}_T A \ N/Q_t \qquad (6.37)$$

where

U_L = overall collector energy loss coefficient (W/°C, Btu/hr°F)
F_R' = efficiency of the collector heat exchanger
F_R = efficiency of the collector in removing heat
T_{ref} = reference temperature 100°C (212°F)
$\bar{T}_a$ = average monthly outdoor (ambient) temperature
Δt = number of seconds in each month
A = solar collector area (m², ft²)
N = number of days in the month
Q_t = building heating load for each month

In Eq. 6.36, x represents the ratio of collector losses to the building heating load, while in Eq. 6.37, y represents the ratio of absorbed solar energy (collector gain) to the building heating load. For calculation purposes, Eqs. 6.36 and 6.37 are modified as shown.

$$x/A = F_R U_L \ (F_R'/F_R) \ T_{ref} - \bar{T}_a) \ \Delta t \ /Q_t \qquad (6.38)$$

$$y/A = F_R(\tau\alpha)_n \ (F_R'/F_R) \ (\tau\alpha)/(\tau\alpha)_n \ \bar{H}_T \ N/Q_t \qquad (6.39)$$

The factor (F_R'/F_R) accounts for the effects of a heat exchanger used between the collector and the water storage tank. Figure 6.16 illustrates three basic types of solar systems. If no heat exchanger is used between the solar collector and the storage tank, then F_R'/F_R will be 1.0 because there is no energy (efficiency) loss there. In Fig. 6.16 the heat exchangers shown are used to

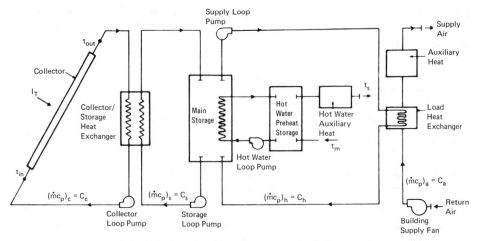

a) Liquid System: Space Heating and Domestic Hot Water

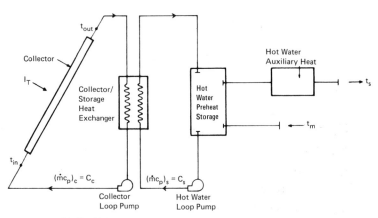

b) Liquid System: Domestic Hot Water Only

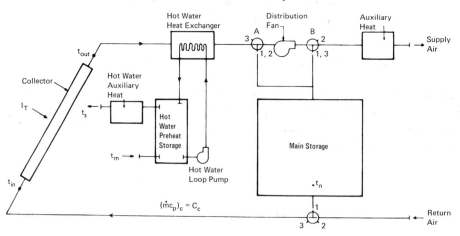

c) Air System: Space Heating and Domestic Hot Water

Figure 6.16. Three Basic Types of Solar Systems. (Source: U. S. Department of Housing and Urban Development[108]).

transfer thermal energy from one circulating liquid to another. Since a heat exchanger is not 100% efficient, F'_R/F_R is used to account for the heat exchanger energy loss penalty. Equation 6.40 can be used in determining F'_R/F_R.

$$F'_R/F_R = 1/[1 + (A\ F_R U_1/Cc)\ (Cc/\epsilon_c\ C_{min} - 1)] \qquad (6.40)$$

where

$$CC = \text{collector loop capacitance rate} = (\dot{m}cp)c$$
$$\dot{m} = \text{mass flow rate of fluid (kg/s, lb/hr)}$$
$$Cp = \text{specific heat of fluid (J/kg°C, Btu/lb°F)}$$
$$Cs = \text{storage loop capacitance rate} = (\dot{m}cp)$$
$$C_{min} = \text{minimum capacitance rate} = \text{lesser value of Cc or Cs}$$
$$\epsilon_c = \text{effectiveness of the collector/storage heat exchanger}$$

The heat exchanger effectiveness (ϵ_c) is a factor of the type of heat exchanger used and the operating conditions (e.g., capacitance rate of each loop). The value for ϵ_c can be determined from the heat exchanger manufacturer's technical sales literature. ϵ_c is basically a ratio of actual heat transfer to the greatest amount of heat transfer possible. Equation 6.41 can be used in calculating ϵ_c.

$$\epsilon_c = (\dot{m}cp)c\ (T_o - T_i)/(\dot{m}cp)_{min}\ (T_o - T) \qquad (6.41)$$

where

$$T_o = \text{outlet temperature of solar collector (°C, °F)}$$
$$T_i = \text{inlet temperature of collector (°C, °F)}$$
$$T = \text{storage temperature (°C, °F)}$$

Figure 6.17 illustrates three basic types of heat exchangers that may be used in a solar system.

The shell and tube heat exchanger should not be used for transferring heat from a toxic fluid to potable water because of the possibility of the toxic liquid mixing with the potable water if corrosion caused a break in the piping. The shell and double tube heat exchanger and the double wall heat exchanger are both suitable for transferring heat from a toxic liquid to potable water. The shell and double tube type heat exchanger can be purchased with a visual sight indicator (as shown in the diagram) that will alert the observer when there is a leak. This is generally accomplished by using a dye in the toxic liquid as a signal.

In the following example the collector/storage heat exchanger (ϵ_c) is assumed to be 0.7, the mass flow rate ($\dot{m}$) of the heat exchanger is assumed to be 0.0150 kg/s-m² for the storage loop and 0.0160 kg/s-m² for the collector loop. These flow rates are chosen based on the design recommendations shown

1. SHELL AND TUBE   This type of heat exchanger is used to transfer heat from a circulating transfer medium to another medium used in storage or in distribution. Shell and tube heat exchangers consist of an outer casing or shell surrounding a bundle of tubes. The water to be heated is normally circulated in the tubes and the hot liquid is circulated in the shell. Tubes are usually metal such as steel, copper or stainless steel.   A single shell and tube heat exchanger cannot be used for heat transfer from a toxic liquid to potable water because double separation is not provided and the toxic liquid may enter the potable water supply, in the case of tube failure.

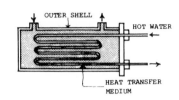

2. SHELL AND DOUBLE TUBE   This type of heat exchanger is similar to the previous one except that a secondary chamber is located within the shell to surround the potable water tube. The heated toxic liquid then circulates inside the shell but around this second tube. An intermediary non-toxic heat transfer liquid is then located between the two tube circuits. As the toxic heat transfer medium circulates through the shell, the intermediary liquid is heated, which in turn heats the potable water supply circulating through the innermost tube.   This heat exchanger can be equipped with a sight glass to detect leaks by a change in color - toxic liquid often contains a dye - or by a change in the liquid level in the intermediary chamber, which would indicate a failure in either the outer shell or intermediary tube lining.

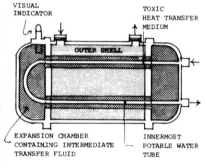

3. DOUBLE WALL   Another method of providing a double separation between the transfer medium and the potable water supply consists of tubing or a plate coil wrapped around and bonded to a tank. The potable water is heated as it circulates through the coil or through the tank. When this method is used, the tubing coil must be adequately insulated to reduce heat losses.

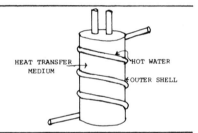

Figure 6.17. Three Basic Types of Heat Exchangers. (Source: U. S. Department of Housing and Urban Development[108]).

in Table 6.20. For example, the recommended collector flow rate is 0.015 l/s-m². Since the density of the 50/50 ethylene glycol–water solution in the collector loop in the example is about 1065 kg/m³, the mass flow rate is assumed to be 0.0160 kg/s-m² as shown in Eq. 6.42. The storage loop and tank in the example is assumed to contain potable water with a density of 1000 kg/m³. Since the liquid flow rate in the storage loop is assumed to be 0.015 l/s-m³, the mass flow rate will be 0.0150 kg/s-m² as shown in Eq. 6.43.

collector loop
$$\text{mass flow rate } (\dot{m}) = 0.015 \text{ l/s-m}^2 \ 1065 \text{ kg/m}^3 \ 0.001 \text{ m}^3/\text{l} \qquad (6.42)$$
$$= 0.0160 \text{ kg/s-m}^2$$

storage loop
$$\text{mass flow rate } (\dot{m}) = 0.015 \text{ l/s-m}^2 \ 1000 \text{ kg/m}^3 \ 0.001 \text{ m}^3/\text{l} \qquad (6.43)$$
$$= 0.0150 \text{ kg/s-m}^2$$

After the mass flow rates have been determined, both the collector loop capacitance rate (Cc) and storage loop capacitance rates (Cs) can be calculated. Since the specific heat of the 50/50 ethylene glycol–water solution is 3350 J/kg-°C (or 0.80 Btu/lb-°F), Cc will be 53.60 J/s-m² °C as shown in Eq. 6.44. The value of Cs will be 62.85 J/s-m² °C since the specific heat of water is assumed to be 4190 J/kg-c.

collector loop
capacitance rate (Cc) = (0.0160 kg/s-m²) (3350 J/kg-c) (6.44)
$$= 53.60 \text{ J/s-m}^2 \text{ °C}$$

storage loop
capacitance rate (Cs) = (0.0150 kg/s-m²) (4190 J/kg-c) (6.45)
$$= 62.85 \text{ J/s-m}^2 \text{ °C}$$

Since Cc is smaller than Cs, C_{min} will be 53.60 J/s-m² °C (or 53.60 W/m² °C since a watt is a Joule per second). The collector used in this example is assumed to be the one described in Table 6.12. Using Eq. 6.40, F'_R/F_R can be determined as shown.

$$F'_R/F_R = 1/[1 + (A \times 5.037 \text{ W/c-m}^2/53.60 \text{ J/s-m}^2\text{°C})$$
$$(53.60 \text{ J/s-m}^2\text{°C}/0.7 \times 53.60 \text{ J/s-m}^2\text{°C-1})] \qquad (6.46)$$
$$= 0.96$$

Notice that the calculated value of F'_R/F_R falls within the suggested design guidelines shown in Table 6.20.

After F'_R/F_R has been found, Eqs. 6.38 and 6.29 can be used to obtain monthly values for x/A and y/A, respectively. The values for the average monthly temperatures $(\bar{T}_a)$ can be found in Appendix II. The values for $(\tau\alpha)/(\tau\alpha)_n$ are found in Table 6.13. The heating load (L) is found in Column E1 of Table 6.17. Since both monthly space heating (Q_s) and domestic water heating (Q_w) loads will be partially satisfied by collected solar energy, the total building load (Q_t) will be the sum of Q_s and Q_w. If only a (solar) domestic water heating system was used, then the load (Q_t) used in Eqs. 6.36, 6.37, 6.38 and 6.39 would be found in Column D1 of Table 6.17. Values for $\bar{H}_T$ are found from previous calculations in Table 6.11.

The calculations of x/A and y/A for the month of January are shown in Eqs. 6.47 and 6.48, respectively.

$$x/A = (5.037 \text{ W/c-m}^2) (0.96) (100\text{-}11\text{°C}) (2.68 \times 10^6 \text{ seconds/} \qquad (6.47)$$
$$\text{month})/(2.04 \times 10^{10} \text{ J/m}^2\text{-day}) = 0.0117$$

$$y/A = (0.712) (0.96) (0.97) (15.91 \times 10^6 \text{ J/day-m}^2) (31 \text{ days})/ \qquad (6.48)$$
$$(2.04 \times 10^{10} \text{ J/m}^2\text{-day}) = 0.0160$$

The other eleven months of the year are now calculated from Eqs. 6.47 and 6.48. Table 6.21 summarizes the calculations used in determining x/A and y/A.

Table 6.21. x/A and y/A Calculation Worksheet 4.

A4 $-$ $F_R (\iota a)_n (F'_R/F_R)$ = 0.684   B4 $-$ $F_R U_L (F'_R/F_R)$ = 4.836

C4	D4	E4	F4	G4 $(\bar{H}_T)$	H4 (Q_T)	I4	J4 (x/A)	K4 (y/A)
Month	N	ΔT	$\iota a/(\iota a)_n$	(J/day-m²)	(J/mo)	(100-$\bar{T}a$)	B4*I4*E4/H4	A4*F4*G4*D4/H4
Jan	31	2.68×10^6	0.97	15.91×10^6	2.04×10^{10}	89	0.0565	0.0160
Feb	28	2.42×10^6	0.97	17.74×10^6	1.55×10^{10}	88	0.0664	0.0213
Mar	31	2.68×10^6	0.97	18.83×10^6	1.08×10^{10}	85	0.102	0.0359
Apr	30	2.59×10^6	0.96	19.61×10^6	3.01×10^9	80	0.333	0.128
May	31	2.68×10^6	0.94	18.59×10^6	1.38×10^9	77	0.723	0.269
Jun	30	2.59×10^6	0.93	17.60×10^6	1.33×10^9	74	0.697	0.253
Jul	31	2.68×10^6	0.93	17.08×10^6	1.38×10^9	73	0.686	0.244
Aug	31	2.68×10^6	0.95	18.33×10^6	1.38×10^9	73	0.686	0.268
Sep	30	2.59×10^6	0.96	18.36×10^6	1.33×10^9	75	0.706	0.272
Oct	31	2.68×10^6	0.97	19.83×10^6	2.82×10^9	79	0.363	0.145
Nov	30	2.59×10^6	0.97	17.23×10^6	9.74×10^9	85	0.109	0.0352
Dec	31	2.68×10^6	0.97	14.63×10^6	1.73×10^{10}	88	0.0659	0.0174

Before x/A and y/A can be used in determining the fraction (f) of the total building thermal load (Q_t) supplied by the solar collectors, three correction factors need to be discussed. Referring back to Fig. 6.16, for diagram "a" there are two heat exchangers used in the heating mode circuit:

1. the collector/storage heat exchanger
2. the load heat exchanger

Earlier, the correction factor F'_R/F_R was used to account for the collector/storage heat exchanger. A correction factor for the load heat exchanger must now be computed. The load heat exchanger is basically a water-to-air type heat exchanger used to transfer the thermal energy from the liquid storage fluid to the room. From Table 6.20, the recommended range of $[\epsilon_L (mcp)_{min}]/UA$ is between one and five. The optimum thermal value would be infinitely large, but the heat exchanger costs could be astronomical. The range given in the table is for a realistic design range based on a tradeoff between economics and efficiency for the heat exchanger. The load heat exchanger correction factor (Y_c/Y) is determined from Eq. 6.49.

$$(Y_c/Y) = [0.39 + 0.65 \text{ exponent } (-0.139/\epsilon_L C_{min}/UA)] \qquad (6.49)$$
$$\text{for } 0.5 < \epsilon_L C_{min}/UA < 50$$

where

ϵ_L = effectiveness of the load heat exchanger

C_{min} = smallest capacitance rate of either the air or liquid medium flowing in the load heat exchanger

In the example, the value of $\epsilon_L C_{min}/UA$ is assumed to be 2.0. Therefore, using either Eq. 6.49 or Fig. 6.18, Y_c/y is found to be 1.0. If the solar system

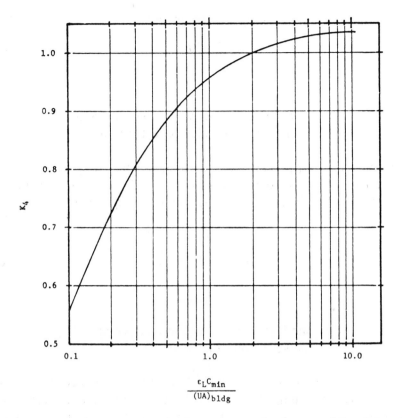

$$\frac{\varepsilon_L C_{min}}{(UA)_{bldg}}$$

Figure 6.18. Load Heat Exchanger Factor. (Source: U. S. Department of Housing and Urban Development[108]).

in the example had been an air type system rather than a liquid type system, the value of y_c/y would automatically have become 1.0.

The storage size correction factor (x_c/x) is used to adjust the F-chart calculation method for storage sizes other than 74 liters of stored water per square meter of collector area. (The F-chart method is based on using 75 l/m².) The storage size correction factor (x_c/x) can be calculated using Eq. 6.50.

$$(x_c/x) = (s/75)^{-0.25}$$
$$\text{for } 37.5 < s < 300 \tag{6.50}$$

where

s = storage size (in liters of water per square meter of solar collector area)

In the example, x_c/x is assumed to be 1.0 since the storage size (s) is assumed to be 75 liters of storage fluid per square meter of solar collector area. If an air type solar collector system had been used in the example, the storage size correction factor (x_c/x) would have been calculated from Eq. 6.51.

$$(x_c/x) = P/0.25)^{-0.3}$$

$$\text{for } 0.215 < P < 1.0 \qquad (6.51)$$

where

$$P = \text{rock storage capacity in cubic meters per}$$
$$\text{square meter of solar collector area.}$$

When using an air type solar collector, an additional correction factor is also used to modify x. This additional modifier is the collector air flow rate correction factor, termed (x_c/x). It is determined using Eq. 6.52.

$$(x_c/x) = (F/10.1)^{0.28}$$

$$\text{for } 5 < F < 20 \qquad (6.52)$$

where

$$F = \text{air flow rate (in one/second per meter}^2$$
$$\text{of solar collector area)}$$

Each monthly value of (x/A) is now multiplied by the storage size correction factor (x_c/x) and then multiplied by the collector air flow rate correction factor (x_c/x). Since the storage size correction factor and the collector air flow rate correction factor are both 1.0, the corrected value of x/A is the same as the original value of x/A as shown in Eq. 6.53 for January.

$$\text{corrected } x/A = (0.0565 \text{ l/m}^2) \underbrace{(1.0)}_{\substack{\text{storage} \\ \text{size} \\ \text{correc-} \\ \text{tion} \\ \text{factor}}} \underbrace{(1.0)}_{\substack{\text{coll.} \\ \text{air flow} \\ \text{rate cor-} \\ \text{rection} \\ \text{factor}}} = 0.0565 \text{ l/m}^2 \qquad (6.53)$$

Next, y/A is corrected. This involves multiplying y/A by the load heat exchanger correction factor (y_c/y). Since y_c/y is determined in the example to be 1.0, the original monthly values of y/A remain unchanged as shown in Eq. 6.54 for the month of January.

$$\text{corrected } y/A = (0.0160 \text{ l/m}^2) \underbrace{(1.0)}_{\substack{\text{load heat} \\ \text{exchanger} \\ \text{correction factor}}} = 0.0160 \text{l/m}^2 \qquad (6.54)$$

After all twelve monthly values of x/A and y/A have been corrected, they are multiplied by the proposed solar collector areas which for this example will be

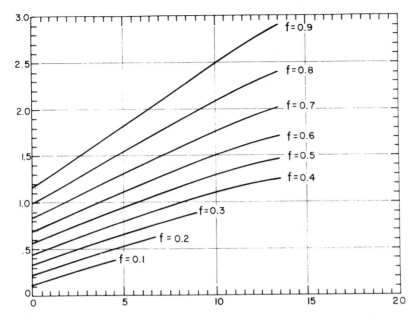

Figure 6.19. F-Chart For Liquid-Based Solar Heating Systems. (Source: U. S. Department of Housing and Urban Development[108]).

25 m², 50 m², and 100 m². These collector areas are arbitrarily chosen. In Chapter 7 an economic analysis will be done for each of the three collector areas to determine which collector size is most economical. After the monthly x/A and y/A values have been multiplied by the three proposed collector areas, the denominator drops out leaving x and y. Once values for x and y have been determined for each month of the year, the building thermal load fraction that the solar system satisfies can be determined from either Fig. 6.19 or Eq. 6.55.

$$F = 1.029 \, y - 0.065 \, x - 0.245 \, y^2 + 0.0018 \, x^2 + 0.0215 \, y^3 \qquad (6.55)$$

where

$$0 < y < 3 \text{ and } 0 < x < 18$$

Tables 6.22–6.25 summarize the calculations of the solar load fraction (F) for solar collectors 10 m², 25 m², and 40 m² in size. The annual solar load fractions for the three collectors are 0.28, 0.50, and 0.62, respectively. If an air type solar system had been chosen, the calculation of F from the monthly values of x and y would be determined from Fig. 6.20 or Eq. 6.56.

$$F = 1.040 \, y - 0.065 \, x - 0.159 \, y^2 + 0.00187 \, x^2 + 0.0095 \, y^3 \qquad (6.56)$$

where

$$0 < y < 3 \text{ and } 0 < x < 18$$

Table 6.22. Determination of Annual Solar Load Fraction (F).

A5 Collector Area (a) = 10m^2
B5 Storage Size Correction Factor (x/xo) = 1.0
C5 Collector Air Flow Rate Correction Factor (x/xo) = 1.0 (if liquid system, x/xo = 1.0)
D5 Load Heat Exchanger Correction Factor (y/yo) = 1.0 (if air system, y/yo = 1.0)

E5 Month	F5 Corrected x/A (J4)(B5)(C5)	G5 Corrected y/A (K4)(D5)	H5 x	I5 y	J5 F	K5 Monthly Solar Load (J5)(E3)
January	0.0565	0.0160	0.565	0.160	0.12	2.45x10^9
February	0.0664	0.0213	0.664	0.213	0.17	2.64x10^9
March	0.102	0.0359	1.02	0.359	0.27	2.92x10^9
April	0.333	0.128	3.33	1.28	0.76	2.29x10^9
May	0.723	0.269	7.23	2.69	1.00	1.38x10^9
June	0.697	0.253	6.97	2.53	1.00	1.33x10^9
July	0.686	0.244	6.86	2.44	1.00	1.38x10^9
August	0.686	0.268	6.86	2.68	1.00	1.38x10^9
September	0.706	0.272	7.06	2.72	1.00	1.33x10^9
October	0.393	0.145	3.63	1.45	0.83	2.34x10^9
November	0.109	0.0352	1.09	0.352	0.26	2.53x10^9
December	0.0659	0.0174	0.66	0.174	0.13	2.25x10^9

Annual Fraction of Total Building Thermal
Load Supplied by Solar Energy (L5 Total/E3 Total) = 0.28   L5 TOTAL = 2.42x10^{10}

Table 6.23. Determination of Annual Solar Load Fraction (F).

A5 Collector Area (a) = 25m^2
B5 Storage Size Correction Factor (x/xo) = 1.0
C5 Collector Air Flow Rate Correction Factor (x/xo) = 1.0 (if liquid system, x/xo - 1.0)
D5 Load Heat Exchanger Correction Factor (y/yo) = 1.0 (if air system, y/yo = 1.0)

E5 Month	F5 Corrected x/A (J4)(B5)(C5)	G5 Corrected y/A (K4)(D5)	H5 x	I5 y	J5 F	K5 Monthly Solar Load (J5)(E3)
January	0.0565	0.0160	1.41	0.40	0.29	5.92x10^9
February	0.0664	0.0213	1.66	0.53	0.38	5.89x10^9
March	0.102	0.0359	2.55	0.90	0.59	6.37x10^9
April	0.333	0.128	8.33	3.20	1.00	3.01x10^9
May	0.723	0.269	18.08	6.73	1.00	1.38x10^9
June	0.697	0.253	17.43	6.33	1.00	1.33x10^9
July	0.686	0.244	17.15	6.10	1.00	1.38x10^9
August	0.686	0.268	17.15	6.70	1.00	1.38x10^9
September	0.706	0.272	17.65	6.80	1.00	1.33x10^9
October	0.363	0.145	9.08	3.63	1.00	2.82x10^9
November	0.109	0.0352	2.73	0.88	0.57	5.55x10^9
December	0.0659	0.0174	1.65	0.44	0.40	6.92x10^9

Annual Fraction of Total Building Thermal
Load Supplied by Solar Energy (L5 Total/E3) = 0.50   L5 Total = 4.33x10^{10}

If the solar system had consisted of only a domestic water heating system, rather than a combined solar space heating and domestic water heating system as in the previous example calculations, a correction would be necessary. This factor is defined in Eq. 6.57.

Table 6.24. Determination of Annual Solar Load Fraction (F).

A5 Collector Areas (a) = 40m^2
B5 Storage Size Correction Factor (x/xo) = 1.0
C5 Collector Air Flow Rate Correction Factor (x/xo) = 1.0 (if liquid system, x/xo = 1.0)
D5 Load Heat Exchanger Correction Factor (y/yo) = 1.0 (if air system, y/yo = 1.0)

E5 Month	F5 Corrected x/A (J4)(B5)(C5)	G5 Corrected y/A (K4)(D5)	H5 x	I5 y	J5 F	K5 Monthly Solar Load (J5)(E3)
January	0.0565	0.0160	2.26	0.640	0.43	8.77x10^9
February	0.0664	0.0213	2.66	0.852	0.55	8.53x10^9
March	0.102	0.0359	4.08	1.436	0.80	8.64x10^9
April	0.333	0.128	13.32	5.12	1.00	3.01x10^9
May	0.723	0.269	28.92	10.76	1.00	1.38x10^9
June	0.697	0.253	27.88	10.12	1.00	1.33x10^9
July	0.686	0.244	27.44	9.76	1.00	1.38x10^9
August	0.686	0.268	27.44	10.72	1.00	1.38x10^9
September	0.706	0.272	28.24	10.88	1.00	1.33x10^9
October	0.393	0.145	15.72	5.80	1.00	2.82x10^9
November	0.109	0.0352	4.36	1.41	0.77	7.50x10^9
December	0.0659	0.0174	2.64	0.696	0.45	7.79x10^9

Annual Fraction of Total Building Thermal
Load Supplied by Solar Energy (L5 Total/E3 Total) = 0.62

L5 Total = 5.39x10^{10}

$$(x_c/x) = [11.6 + 1.18\ T_w + 3.86\ T_m - 2.32\ \bar{T}_a]/(100 - \bar{T}_a) \qquad (6.57)$$

For a solar system using only a domestic water heating system, the correction factor in Eq. 6.57 would be multiplied by the twelve monthly values of x. These corrected values for x would also be multiplied by the collector air flow rate correction factor (x_c/x) which for liquid systems would be 1.0.

6.4.5.2. Interpretation of Design Results. The three annual solar fractions determined in Tables 6.22–6.25 mean that the 10 m^2 solar collector area will supply 28% of the buildings' total annual thermal load requirements. For 25 m^2, it will be 50% and for 40 m^2 collector areas, the solar fraction will be 62%. These percentages give the designer or homeowner an idea of how a certain size collector array area can be expected to perform based on the design factors assumed in the design (i.e., values for $\bar{H}_T$, $F_R U_L$, Q_w, etc.). There can be no guarantee that any solar system will supply exactly x percent of the annual total buildings' thermal load requirements for a particular year. This is because the meterological data used in the design calculations is based on *average* weather conditions for a number of years. For one particular year the weather data collected could vary far from the yearly average and the solar fraction (F) could be off considerably. Additionally, there may be some inaccuracies within the meteorological data itself as explained earlier in Section 6.2. The determination of domestic hot water and space heating loads will inject some inaccuracies because weather conditions and building load conditions may not exactly correspond year after year to what was assumed to be the case when the original load calculations were made. The values assumed

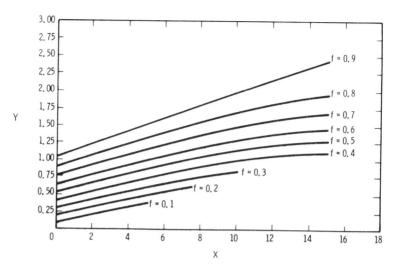

Figure 6.20. F-Chart for Solar Air Heating Systems. (Source: U. S. Department of Energy[106]).

for certain components in the solar system design may change due to age or different operating conditions which may also have an effect on the yearly value of the solar fraction (F). (See Figs. 6.19 and 6.20.)

In almost any type of solar design analysis some mathematical models are used to predict how a solar system will perform in the real world. For example, mathematical equations (models) may be used to compute performance based on such things as solar insolation ($\bar{H}_T$), collector tilt(s), types of construction materials, efficiency of heat exchangers, storage tanks, etc. Mathematical models are used to predict everything from how a particular shape will affect the aerodynamics of an airplane to the effects upon consumer spending as a factor of the rate of inflation in the economy. Many of these models come very close to estimating how the real world functions but they are never able to predict *exactly* what will happen. Some models are more accurate, but each may introduce a small amount of error which can affect the accuracy of final calculations.

In discussing possible sources of error which can change the computed value of say, the solar fraction (F), one must be aware that when interpreting the results of solar, or any type of design analysis, the end result figures should be interpreted based on the realization that some inaccuracies may have been introduced during various phases of the design calculations. With possible sources of error in mind, designers (or homeowners) can then temper their interpretation of the results.

6.5 REFERENCES AND SUGGESTED READING

1. Afgan, N. H. and Schlunder, E. U., eds., *Heat Exchangers: Design and Theory Sourcebook*, Washington, D.C.: Scripta Book Company, 1974.
2. AIA Research Corporation, *A Survey of Passive Solar Buildings*, Washington, D.C.: U. S. Government Printing Office, 1979.

3. Air Conditioning Contractors of America (ACCA), 1228 17th St., N.W.: Washington, D.C. 20036.

4. Allen, Edward, "Solar Siter Charts the Sun," *Popular Science,* **213,** 3:152–156, (1978).

5. American Society of Heating, Refrigerating and Air Conditioning Engineers, Inc., *ASHRAE Handbook and Product Directory 1977 Fundamentals,* New York: ASHRAE, 1977.

6. American Society of Heating, Refrigerating and Air Conditioning Engineers, Inc., *Cooling and Heating Load Calculation Manual,* New York: ASHRAE, 1979.

7. American Society of Heating, Refrigerating, and Air Conditioning Engineers, Inc., *Energy Conservation in New Building Design,* New York: ASHRAE, 1977.

8. American Society of Heating, Refrigerating, and Air Conditioning Engineers, Inc., 345 East 47th Street, New York, N.Y. 10017.

9. Andrews, F. T., *Building Mechanical Systems,* New York: McGraw-Hill Book Co., 1977.

10. Appelbaum, J., and Bany, J., "Shadow Effect of Adjacent Solar Collectors in Large Scale Systems," *Solar Energy,* Vol. 23, New York: Pergamon Press, 1979, pp. 497–507.

11. Arens, Edward A., and Carrol, William L., *Geographical Variation in the Heating and Cooling Requirements of a Typical Single-Family House, and Correlation of These Requirements to Degree Days,* Washington, D.C.: U. S. Government Printing Office, November, 1978.

12. Babbitt, Harold E., *Plumbing,* New York: McGraw-Hill Book Co., 1960.

13. Beck, E. J., and Field, R. L., *Solar Heating of Buildings and Domestic Hot Water,* Springfield, Virginia: National Technical Information Service, November, 1977.

14. Beckman, William A., Klein, Sanford A., and Duffie, John A., *Solar Heating Design by the F-Chart Method,* New York: John Wiley & Sons, Inc., 1977.

15. Bell and Gossett Company, *Engineering Design Manual For: Centrifugal Pumps, Heat Exchangers, Refrigeration, Air Conditioning, Hydronic Systems, Domestic Water Systems,* Morton Grove, Illinois: Fluid Handling Division, International Telephone and Telegraph Corporation, 1970.

16. Bell and Gossett, *System Syzer,* 8200 N. Austin Ave., Morton Grove, Illinois 60053.

17. Bennett, R. T., *Bennett Sun Angle Charts,* Bala Cynwyd, Pa.

18. Better Heating & Cooling Bureau and the Sheet Metal and Air Conditioning Contractors' National Association, Inc., *Solar Installation Standards for Heating and Air Conditioning Systems,* 1977.

19. Brookhaven National Laboratory with the Assistance of Dynatech R/D Co., *An Assessment of Thermal Insulation Materials and Systems for Building Applications,* Washington, D.C.: U. S. Government Printing Office, June, 1978.

20. Carrier Air Conditioning Co., *Handbook of Air Conditioning System Design,* New York: McGraw-Hill Book Co., 1965.

21. Cinquemani, V. Jr., Owenby, J. R. Jr., and Baldwin, R. G., *Input Data for Solar Systems,* Asheville, North Carolina: National Oceanic and Atmospheric Administration, August, 1979.

22. Code Development Team, *State of Florida Model Energy Efficiency Building Code – Code for Energy Conservation in New Building Construction-With Florida Amendments,* Tallahassee, Florida: State Energy Office, November, 1978.

23. Cole, Roger, L., *Design and Installation Manual for Thermal Energy Storage,* Springfield, Virginia: National Technical Information Service, February, 1979.

24. Copper Development Association, Inc., *Solar Energy Systems Design Handbook,* 1977.

25. Cousins, Frank W., *Sundials: The Art and Science of Gnomonics,* New York: Pica Press, 1969.

26. Drew, M. S., and Selvage, R. B. G., "Correspondence Between Solar Load Ratio Method for Passive Water Wall Systems and F-Chart Performance Estimates," *Solar Energy,* Vol. 23, New York: Pergamon Press, 1979, pp. 327–331.

27. Dubin-Bloome Associates, *Building the Solar Home,* Washington, D.C.: U. S. Government Printing Office, June, 1978.

28. Duffie, John A., and Beckman, William A., *Solar Energy Thermal Processes,* New York: John Wiley & Sons, Inc., 1974.

29. Federal Energy Administration, Energy Resource Development, *Buying Solar,* Washington, D.C.: U. S. Government Printing Office, 1976.
30. Field, Richard L., *Design Manual for Solar Heating of Buildings and Domestic Hot Water,* Camarillo, California: Solpub Co., 1976.
31. Florida Solar Energy Center, *Mean Solar Radiation for Florida Cities,* Cape Canaveral, Florida, 1977.
32. Florida Solar Energy Center, *Solar Water and Pool Heating - Installation and Operation,* Cape Canaveral, Florida, January, 1979.
33. Florida Solar Energy Center, *Summary Test Package,* **II,** (1), Cape Canaveral, Fl., June 1979.
34. Fraas, Arthur P., and Ozisik, M. Necati, *Heat Exchanger Design,* New York: John Wiley & Sons, Inc., 1965.
35. General Electric, Philadelphia, Pennsylvania, *Solartron TC-100 Vacuum Tube Solar Collector Commercial and Industrial Application Guide,* August, 1979.
36. Gibson, Paul D., "Design and Installation of Solar Mechanical Systems," *Heating/Piping/Air Conditioning,* **50,** 7, 1978.
37. Giles, Ronald V., *Fluid Mechanics & Hydraulics,* New York: McGraw-Hill Book Co., 1962.
38. Gonk, Rodney L., Jones, Dennie E., and Cole-Appel, Bruce E., "Calculation of Performance of N Collectors in Series from Test Data on a Single Collector," *Solar Energy,* Vol. 23, New York: Pergamon Press, 1979, pp. 535–536.
39. Hastings, S. Robert, and Crenshaw, Richard W., *Window Design Strategies to Conserve Energy,* Washington, D.C.: U. S. Government Printing Office, June, 1977.
40. Hay, John E., "Calculation of Monthly Mean Solar Radiation For Horizontal and Inclined Surfaces," *Solar Energy,* Vol. 23, New York: Pergamon Press, 1979, pp. 301–307.
41. Hill, James E., and Richtmyer, Thomas E., *Retrofitting a Residence for Solar Heating and Cooling: The Design and Construction of the System,* Washington, D.C.: U. S. Government Printing Office, November, 1974.
42. Holman, S. P., *Heat Transfer,* New York: McGraw-Hill Book Co., 1976.
43. Hunn, B. D., and Calafell, D. O. II, "Determination of Average Ground Reflectivity for Solar Collectors," *Solar Energy,* Vol. 19, New York: Pergamon Press, 1977, pp. 87–89.
44. Hurley, Charles W., and Kreider, Kenneth G., *Applications of Thermography for Energy Conservation in Industry,* Washington, D.C.: U. S. Government Printing Office, October, 1976.
45. Inter Technology Corporation, *Solar Energy School Heating Augmentation Experiment - Design, Construction and Initial Operation,* Washington, D.C.: U. S. Government Printing Office, December, 1974.
46. ITT Fluid Handling Division, *Solar Heating Systems Design Manual,* Morton Grove, Illinois: International Telephone and Telegraph Corp., 1977.
47. Jaffe, Martin, and Erley, Duncan, *Protecting Solar Access for Residential Development: A Guidebook for Planning Officials,* Washington, D.C.: U. S. Government Printing Office, May, 1979.
48. Janz, George J., Allen, Carolyn B., Downey, Joseph R. Jr., and Tomkins, R. P. T., *Physical Properties Data Compilations Relevant to Energy Storage,* Washington, D.C.: U. S. Government Printing Office, March, 1978.
49. Keuffel & Esser Company, *K & E Solar Ephemeris 1980,* Morristown, New Jersey, 1979.
50. Keuffel & Esser Company, *K & E Surveying Instruments,* Catalog 10a, Morristown, New Jersey, 1979.
51. Kern, Donald Q., and Kraus, Allan D., *Extended Surface Heat Transfer,* New York: McGraw-Hill Book Co., 1972.
52. Khashab, A. M., *Heating Ventilating and Air-Conditioning Systems Estimating Manual,* Washington, D.C.: McGraw-Hill Book Co., 1977.
53. Klein, S. A., "Calculation of Monthly Average Insolation on Tilted Surfaces," *Solar Energy,* Vol. 19, New York: Pergamon Press, 1977, pp. 325–329.
54. Klein, S. A., Beckman, W. A., and Duffie, J. A., "A Design Procedure for Solar Heating Systems," *Solar Energy,* Vol. 18, New York: Pergamon Press, 1976, pp. 113–127.

55. Klein, S. A., Beckman, W. A., and Duffie, J. A., "A Design Procedure for Solar Air Heating Systems," *Solar Energy,* Vol. 19, New York: Pergamon Press, 1977, pp. 509–512.

56. Klein, S. A., and Beckman, W. A., "A General Design Method for Closed-Loop Solar Energy Systems," *Solar Energy,* Vol. 19, New York: Pergamon Press, 1979, pp. 269–282.

57. Klein, S. A., Beckman, W. A., and Duffie, J. A., *Monthly Average Solar Radiation on Inclined Surfaces for 261 North American Cities,* Madison, Wisconsin: University of Wisconsin Press, August, 1978.

58. Klucher, T. M., "Evaluation of Models to Predict Insolation on Tilted Surfaces," *Solar Energy,* Vol. 23, New York: Pergamon Press, 1979, pp. 111–114.

59. Kondratyer, K. Ya., *Radiation in the Atmosphere,* New York: Academic Press, 1969.

60. Krueger, P. D., "Energy Storage for a High Temperature Solar Heating and Cooling System," *ASHRAE Journal,* **21,** 8, (1979).

61. Kusuda, T., and Ishii, K., *Hourly Solar Radiation Days for Vertical and Horizontal Surfaces on Average Days in the United States and Canada,* Washington, D.C.: U. S. Government Printing Office, April, 1977.

62. Langley, Billy C., *Comfort Heating,* Restin, Virginia: Restin Publishing Company, Inc., 1975.

63. Lewis & Associates, *Solar Site Selector – Instant Visual Solar Calculator,* 105 Rockwood Drive, Grass Valley, Ca., 95945.

64. Lewis, Dan, "Using F-Chart to Select Between Competing Systems," *Solar Age,* **4,** 2:18, 1979.

65. Libbey-Owens-Ford Co., Toledo, Ohio, *Designing with the LOF Sun Angle Calculator,* 1974.

66. Liddle, Peggy, "The Passive Approach," *Hudson Home Guide,* Building and Remodeling, 111:58–61, 124, 126, (1979).

67. Löf, George O. G., Duffie, John A., and Smith, Clayton, O., *World Distribution of Solar Radiation,* Madison, Wisconsin: The University of Wisconsin Press, July, 1966.

68. Los Alamos Scientific Laboratory, *Pacific Regional Solar Heating Handbook,* Washington, D.C.: U. S. Government Printing Office, November, 1976.

69. Lunde, Peter J., "Prediction of the Performance of Solar Heating Systems Over a Range of Storage Capacities," *Solar Energy,* Vol. 23, New York: Pergamon Press, 1977, pp. 115–121.

70. The Marley Co., Mission, Kansas, *Engineering Weather Data,* (pamphlet).

71. Mazria, Edward, *The Passive Solar Energy Book,* Emmaus, Pennsylvania: Rodale Press, 1979.

72. McGuinness, William J., and Stein, Benjamin, *Mechanical and Electrical Equipment for Buildings,* New York: John Wiley & Sons, Inc., 1971.

73. Merrill, Richard, and Gage, Thomas, eds., *Energy Primer – Solar, Water, Wind, and Biofuels,* New York: Dell Publishing Co., Inc., 1978.

74. Motz, Lloyd, and Duveen, Anneta, *Essentials of Astronomy,* New York: Columbia University Press, 1977.

75. NAHB Research Foundation, Inc., *Insulation Manual – Homes Apartments,* Scranton, Pennsylvania: Haddon Craftsmen, Inc., 1979.

76. National Association of Home Builders, *Designing, Building and Selling Energy Conserving Homes,* Washington, D.C.: 1978.

77. National Lumber Manufacturers Association, *Insulation of Wood - Frame Structures,* Washington, D.C.: National Forest Products Association, 1964.

78. National Solar Heating and Cooling Information Center, *Factsheet,* Rockville, Maryland.

79. Olivieri, Joseph B., *How to Design Heating – Cooling Comfort System,* Birmingham, Michigan: Business News Publishing Co., 1971.

80. Pasch, Roger M., "Solar Process and Make-Up Air Heating," *Building Systems Design,* June–July, 1979, pp. 23–31.

81. Pitts, Donald R., and Sissom, Leighton E., *Heat Transfer,* New York: McGraw-Hill Book Co., 1977.

82. Ripka, L. Z., *Plumbing Installation and Design,* Chicago, Illinois: American Technical Society, 1978.

83. Rohsenow, Warren M., and Hartnett, James P., *Handbook of Heat Transfer,* New York: McGraw-Hill Book Co., 1973.

84. Rossiter, Walter J. Jr., and Mathey, Robert G., *Criteria for Retrofit Materials and Products for Weatherization of Residences,* Washington, D.C.: U. S. Government Printing Office, September, 1978.

85. Rossiter, Walter J. Jr., Mathey, Robert G., Burch, Douglas M., and Pierce, E. Thomas, *Urea–Formaldehyde Based Foam Insulations: An Assessment of Their Properties and Performance,* Washington, D.C.: U. S. Government Printing Office, July, 1977.

86. Sandler, Jeffrey, "Build Our Cold Sleuth to Keep Heating Costs Down," *Popular Mechanics,* **151,** 2:194–196 (1979).

87. Schwolsky, Rick, "Liquid Storage Tanks," *Solar Age,* **4,** 8:36–37, (1979).

88. Sheet Metal and Air-Conditioning Contractors National Association, *Fundamentals of Solar Heating,* Washington, D.C.: U. S. Government Printing Office, January, 1978.

89. Shuttleworth, John, ed., "The Amazing $30 Solar Site Selector," *The Mother Earth News,* 52:75, (1978).

90. Smay, V. Elaine, "For Solar Designers: An Inexpensive, Cut-and-Paste Shadow Plotter," *Popular Science,* **215,** 5:144, (1979).

91. Solar Energy Laboratory, *F-Chart Users Manual,* Madison, Wisconsin: University of Wisconsin Press, June, 1978.

92. Solar Energy Laboratory, *TRNSYS A Transient Simulation Program,* Madison, Wisconsin: University of Wisconsin, February, 1978.

93. Solaron Corporation, Denver, Colorado, *Application Engineering Manual,* 1977.

94. State Energy Office, Tallahassee, Florida, *A Planners Handbook on Energy,* November, 1975.

95. Strahler, Arthur N., *The Earth Sciences,* New York: Harper & Row, Publishers, Inc., 1963.

96. Strock and Koral, eds., *Handbook of Air Conditioning, Heating and Ventilating,* New York: Industrial Press, Inc., 1965.

97. Sun, Tseng-Yao, *Time Integrated Solar Heat Gain Factors,* Chicago, Illinois: Reinhold Publishing Corp., 1968.

98. Tenny, Ralph, "Energy Leak Detector Reveals Home Heat and Cooling Losses," *Popular Electronics,* **14,** 4:59–61, (1978).

99. Threlkeld, James L., *Thermal Environmental Engineering,* Englewood Cliffs, New Jersey: Prentice-Hall, Inc., 1970.

100. The Trane Co., *Trane Air-Conditioning Manual,* St. Paul, Minnesota: McGill Printing, Inc., 1965.

101. The Trane Co., La Crosse, Wisconsin, *Ductulator.*

102. Trefil, James, "For Your Home — Easy, No-Cost Way to Measure Heat Loss," *Popular Science,* **213,** 4:62, 64, 66, 68, 72, (1978).

103. U. S. Department of Commerce, *Solar Heating and Cooling of Residential Buildings - Sizing, Installation and Design of Systems,* Washington, D.C.: U. S. Government Printing Office, October, 1977.

104. U. S. Department of Commerce, *Solar Heating and Cooling of Residential Buildings - Design of Systems,* Washington, D.C.: U. S. Government Printing Office, October, 1977.

105. U. S. Department of Energy, *DOE Facility Solar Design Handbook,* Washington, D.C.: U. S. Government Printing Office, January, 1978.

106. U. S. Department of Energy, *Introduction to Solar Heating and Cooling Design and Sizing,* Washington, D.C.: U. S. Government Printing Office, August, 1978.

107. U. S. Department of Energy, Solar Technology Transfer Program, *Space Heating Handbook with Service Hot Water and Heat Load Calculations: Solcost,* July, 1978.

108. U. S. Department of Housing and Urban Development, *HUD Intermediate Minimum Property Standards Supplement - Solar Heating and Domestic Hot Water Systems,* Washington, D.C.: U. S. Government Printing Office, 1977.

109. U. S. Department of Housing and Urban Development, *Installation Guidelines for Solar DHW Systems - In One- and Two-Family Dwellings,* Washington, D.C.: U. S. Government Printing Office, April, 1979.
110. U. S. Department of Housing and Urban Development, *Regional Guidelines for Building Passive Energy Conserving Homes,* Washington, D.C.: U. S. Government Printing Office, November, 1978.
111. U. S. Department of Housing and Urban Development, *Solar Dwelling Design Concepts,* Washington, D.C.: U. S. Government Printing Office, May, 1976.
112. U. S. Department of Interior, *Plants/People/and Environmental Quality,* Washington, D.C.: U. S. Government Printing Office, 1972.
113. U. S. Geological Survey, Arlington, Virginia, *Isoconic Map of the United States, 1975.*
114. Ward, Dan S., and Ward, John C., "Design Considerations for Residential Solar Heating and Cooling Systems Utilizing Evacuated Tube Solar Collectors," *Solar Energy,* Vol. 22, New York: Pergamon Press, 1979, pp. 113–118.
115. Watt Engineering Ltd., *On the Nature and Distribution of Solar Radiation,* Washington, D.C.: U. S. Government Printing Office, March, 1978.
116. Wright, David, *Natural Solar Architecture – A Passive Primer,* New York: Van Nostrand Reinhold Co., 1978.
117. Liu, B. Y. H., and Jordon, R. C., "Daily Insolation on Surfaces Tilted Toward the Equator," *Trans. ASHRAE,* 526–541, (1962).
118. Liu, B. Y. H., and Jordon, R. C., "The Interrelationship and Characteristic Distribution of Direct, Diffuse, and Total Solar Radiation," *Solar Energy,* **4,** 3, (1960).
117. Liu, B. Y. H., and Jordon, R. C., "The Long-Term Average Performance of Flat-Plate Solar Energy Collectors," *Solar Energy,* **7,** 2:53–74, (1963).
120. Page, S. K., "The Estimation of Monthly Mean Values of Daily Total Short-Wave Radiation on Vertical and Inclined Surfaces from Sunshine Records for Latitudes 40°N–40°S," *Proceedings of the UN Conference on New Sources of Energy,* Paper No. 35/5/98, (1961).

CHAPTER 7
SOLAR ECONOMICS

7.1 INTRODUCTION TO SOLAR ECONOMICS

In Chapter 6, a solar design example was presented with three proposed solar collector areas: 10 m², 25 m², and 40 m². In the design analysis the fraction of the building thermal load that the solar system could carry was determined to be 28%, 50%, and 62%, respectively. In Chapter 7, a methodology will be employed to determine which of the three collector sizes is the most economical.

7.2 LIFE CYCLE COSTING

Many times the idea of incorporating a solar system in a new building will be viewed with skepticism after comparing the initial costs of the solar system to conventional systems. Using initial costs alone is an unsound method of determining whether to use a solar system. Life cycle costing is the most meaningful method of comparison because the life cycle costing method considers all future costs (and savings) over the life of the system. The life cycle costing methodology allows for all future cash flows (i.e., costs) and through present value analysis to discount these cash flows to the present. In a sense one can think of present value (also called discounting) as the opposite of compound value. For example, suppose a person has $500.00 and deposits it in a savings account with 5% interest compounded annually. To find out how much there will be at the end of four years use Eq. 7.1.

$$Cv = P(1 + i)^N \tag{7.1}$$

where

$$Cv = \text{Compounded value}$$
$$P = \text{Principal}$$
$$i = \text{Interest rate (discount rate)}$$
$$N = \text{Number of interest periods}$$

When the values are plugged into Eq. 7.1 to yield Eq. 7.2, the individual will realize a $107.75 increase over the original principal of $500.00.

$$Cv = \$500.00 \times (1.0 + 0.05)^4 = \$607.75 \qquad (7.2)$$

To illustrate the concept of present value it will be assumed that an individual is offered the choice of receiving either $525.00 now or $607.75 four years from now. Assuming that the discount rate (i.e., the interest rate) is 5%, which is the best choice? The present value formula shown in Eq. 7.3 can be used to determine this.

$$Pv = P_N [1/(1 + i)^N] \qquad (7.3)$$

where

Pv = Present value or present worth

P_N = Value of the principal at the end of N periods.

Notice that Eqs. 7.1 and 7.3 are similar but have opposite functions. Equation 7.1 compounds a principal where Eq. 7.3 discounts a principal. Equation 7.4 shows the numerical values used in Eq. 7.3.

$$Pv = \$607.75 [1/1 + 0.05)^4] = \$500.00 \qquad (7.4)$$

From Eq. 7.4 it can be seen that $607.75 in four years is only worth $500.00 right now (at a discount rate of 5%). Therefore, the individual would be better off taking the $525 today rather than taking the $607.75 at the end of four years. Hence, inflation notwithstanding, money does have a "time value" because $1.00 today is worth more than $1.00 in the future because the $1.00 received today can be invested to obtain a greater yield.

The concept of present value analysis is applicable to solar economic costing because an individual can take the projected future costs such as increases in fuel prices and determine whether a solar installation is economically feasible. In life cycle costing using present value analysis, future costs (and savings) can be seen in terms of present values to determine whether a solar system will yield a greater return than a conventional system. The main drawback of life cycle costing analysis is that a number of variables must be estimated. This includes such things as predicting future rates of general inflation and fuel cost increases.

7.3 SOLAR SYSTEM FINANCING

Financing has an effect upon the economics of using solar systems for space heating and/or domestic water heating. Table 7.1 illustrates how solar system finance costs vary. Since many lending agencies have not had much long-term experience with solar installations, they may voice some concern over the high initial costs of some systems. Figure 7.1 illustrates some issues of concern to lenders when making loans for homes utilizing solar systems. Notice that in Fig. 7.1, the number one concern to lenders is generally the uncertainty of the

Table 7.1. How Financing Affects Solar Costs. (Source: Barrett[3]).

	Loan Type							
	First Mortgage					Second Mortgage	Home Improvement	
	Conventional			FHA	VA	Conventional	Title 1	Conventional
Loan/Value Ratio	70%	80%	90%	93%	100%	75%	100%	100%
Interest Rate	8.5%	8.75%	9.0%	8.25%	9.0%	13.5%	11.5%	12.5%
Maturity (years)	27	27	27	30	30	10	12	5
Mortgage Insurance Insurance	--	15%	25%	5%	--	--	5%	--
Monthly Cost per $1000 of Loan	$7.88	$8.16	$8.41	$8.06	$8.23	$16.23	$13.13	$22.50
Down Payment For an $8000 Solar Energy System	$2400	$1600	$800	$506	0	$2000	0	0
Monthly Cost For an $8000 Solary Energy System	$44.15	$52.23	$60.53	$58.53	$64.37	$91.36	$105.07	$179.98

Source: Based on loan terms in the Boston area in March, 1976. Assumes that appraised value of system is same as full cost of $8000.

market value. Tables 7.2 and 7.3 list further considerations of lenders with regard to solar system financing.

7.4 SOLAR TAX INCENTIVES

Solar systems that would have been impractical without tax incentives may become feasible when tax incentives are considered in the economic analysis. Generally, incentives for using solar energy are on both the federal and state level. The solar tax credits now (1980) amount to 40% on the first $10,000 of a solar domestic hot water, space heating, or cooling system. Previously, this tax credit was 30% of the first $2,000 and 20% of the next $8,000 spent on a solar system. In effect the federal government plays a major role in making many solar systems feasible as a direct result of the tax incentives offered. Appendix X lists various states' legislation boosting the economic feasibility of solar systems. When deciding to install a solar system, readers should contact their state energy office and the National Solar Heating and Cooling Information Center for up-to-date state solar legislation.

7.5 PRESENT ECONOMIC SITUATION OF SOLAR ENERGY SYSTEMS

Highlights of a U. S. Department of Energy study[25] are presented in the following paragraphs:

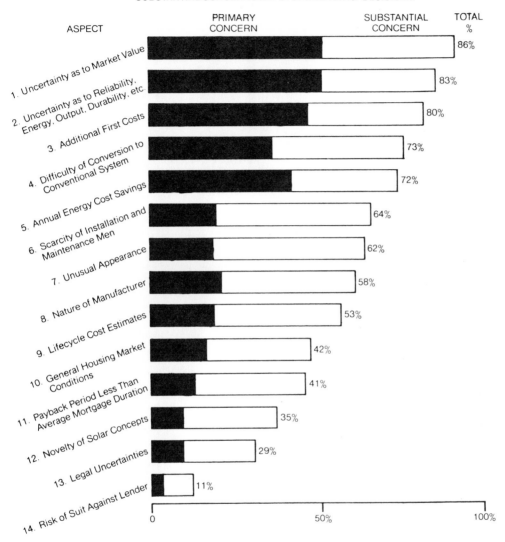

PERCENTAGE OF LENDERS IDENTIFYING
SELECTED ASPECTS OF SOLAR ENERGY
HEATING SYSTEMS AS PRIMARY OR
SUBSTANTIAL CONCERNS IN FUTURE LENDING DECISIONS

Figure 7.1. Issues of Concern to Lenders in Making Solar Loans. (Source: Barrett[3]).

OBJECTIVES OF THE STUDY

- Assess the current (1977) economic feasibility of solar water and space heating for single family and multifamily residences.
- Determine the impact of the proposed solar income tax credit on economic feasibility.
- Determine whether solar water and space heating can compete with electricity, fuel oil, and natural gas at $3 to $5 per million BTU's.

Table 7.2. Lending Considerations for Construction Finance. (Source: U. S. Department of Housing and Urban Development[29]).

Solar Construction Loans	Non-Solar Construction Loans
Site Location	Type, size and price or rent of proposed unit
Site design and layout	Materials/construction
Unit design and floor plans	Site location
Type, size and price or rent of proposed unit	Site design and layout
Materials/Construction	Unit design and floor plans
Construction methods/ technology	Project amenities
Solar energy system	Construction methods/ technology
Unit options/amenities	HVAC system
HVAC system	Unit options amenities
Energy efficiency	Energy efficiency
Project amenities	

METHODOLOGY

Utilized actual 1977 solar cost and fuel price data and a computerized solar economic performance model to determine solar economic feasibility for the owners of single family homes and the owners of multifamily rental units.

DATA BASE AND MAJOR ASSUMPTIONS

- Installed solar system costs range between $22/ft² and $30/ft² depending on size and application.
- No incremental property tax assessment.
- Solar water heating carries 70% of the load, solar water and space heating carries 50% of the load.
- Homeowners in 30% tax bracket, apartment owners in 50% tax bracket.
- Solar system maintenance costs are 1.5% of installed costs annually.
- Fuel Prices:
  - Fuel oil is $.50/gal ($3.54/MMBtu).
  - Natural gas ranges between $.15 and $.40/therm.
  - Electricity ranges between $.03 and $.05/kWh.
- Inflation:
  - CPI, 5%/yr
  - Electricity and oil, 7%/yr
  - Gas, 10%/yr
- System lives of 20 yrs.
- 10% discount rate (OMB recommendation).
- No utility rate discrimination.

**Table 7.3. Lending Considerations for Permanent Finance.
(Source: U. S. Department of Housing and Urban Development[29]).**

Non-Solar Permanent Loans	Solar Loans: Permanent Financing Arranged by Builder/Developer	Solar Loans: Permanent Financing Arranged by Purchaser
Type, size and price or rent of proposed unit	Site location	Site location
	Site design and layout	Site design and layout
Site design and layout	Type, size and price or rent of proposed unit	Type, size and price or rent of proposed unit
Site location		
Unit design and floor plans	Project amenities	Unit design and floor plans
	HVAC system	
Project amenities	Solar energy system	HVAC system
		Project amenities
Materials/ construction	Unit design and floor plans	Unit options/ amenities
HVAC system	Unit option/ amenities	Materials/ construction
Construction methods/ technology	Materials/ construction	Construction methods/ technology
Energy efficiency	Construction methods/ technology	Solar energy system
	Energy efficiency	Energy efficiency

MAJOR FINDINGS

- Solar water and space heating is, at present, economically competitive, on a life cycle cost basis, with electricity and fuel oil. Enactment of the solar tax credit make solar marginally competitive, on the same basis, with natural gas. However, most purchasing decisions are not made on a life cycle cost basis. Utilizing 1977 actual (average) fuel prices and cash flow decision criteria, solar water heating and combined water and space heating are competitive against electricity and fuel oil in some regions.
- The solar tax credit has a dramatic positive impact on solar economic feasibility for detached residences.
- Utilizing the replacement (marginal) costs of fuel oil and natural gas makes solar water and space heating competitive against both of these conventional fuels; enactment of the solar tax credit enhances this advantage.
- Due to the business tax deduction for the expense of fuels, it is difficult for solar water and space heating to become economically feasible in commercial multi-family buildings, even with the tax credit.
- These solar applications do not compete with distillate at $3/MMBtu but can compete (with adoption of the tax credit) with distillate at $5/MMBtu.
- Average cost pricing (rather than replacement cost pricing) of fossil fuels and the allowance of commercial expensing of utility costs are two of the most important institutional barriers to the economic feasibility of solar energy.

POLICY IMPLICATIONS

From a national or societal decision criteria which must recognize the true marginal and environmental costs of new energy sources, solar water and space heating is economically feasible today. This basic fact suggests consideration of the following policy options to enhance the economic feasibility of these solar energy applications to private decision makers:

- Adoption of the solar tax credit program recommended in the NEP.
- Replacement (marginal) cost pricing of fuel oil and natural gas.
- Increases in the real costs of conventional energy sources (for example, by a tax on fossil fuel usage).
- Encouragement of life cycle costing decision criteria.
- Elimination of the business tax deduction for the costs of fuels.
- Federally supported research, development and demonstration programs aimed at achieving a minimum 25% reduction in real system costs by 1985.

Although the study is somewhat dated, some of the general conclusions can be applied today. (In the study the federal tax credits referred to were for credits of 30% on the first $1,500.00 and 20% of the next $8,500.00 for residential solar installations.)

Tables 7.4 and 7.5 give relative costs for various system components for liquid and air type solar installations, respectively. Table 7.6 illustrates some relative solar system costs for commercial buildings.

Table 7.4, 7.5, and 7.6 are presented to give the reader a feel for *relative* solar system costs. Individuals in solar design or consulting positions, such as the architect and mechanical engineer, should obtain current cost estimates and do *their own* general economic analyses to determine solar system feasibility and paybacks for their specific geographic area. This general feasibility analysis will give them a feel for *relative* solar system feasibility for *their* particular city. This is necessary because each geographic location will probably have different fuel costs (e.g., electric rates), state solar legislation incentives, equipment costs, labor installation charges, etc. (Naturally an in-depth analysis should be performed for each individual client based on the client's particular situation such as specific tax bracket, cash flow position, etc.)

7.6 COMBINED SOLAR SPACE AND WATER HEATING SYSTEM

The following sections (Sections 7.6.1 and 7.6.2) utilize collector data from the design example presented in Chapter 6. The basic worksheet formats in Tables 7.7, 7.9, 7.10, 7.11, 7.14, and 7.15 are from Ref. 4, Beckman, *et al.* (The numerical data presented in the worksheets are furnished by this author.)

7.6.1. Economic Analysis for Residential Buildings

Table 7.7 illustrates the first worksheet used in determining which of the three solar collector sizes (i.e., 10 m², 25 m², 40 m²) is the most economical. In Table 7.7 items which are self explanatory will not be discussed. In item K, the collector area dependent costs refer to costs which increase as the collector

Table 7.4. Representative Costs for a Water System with 800 ft² Collector. (Source: U. S. Department of Commerce[21]).

House Design Heat Load = 110,000 Btu/hr

ITEM		COST	
1.	Collectors	$4,080	
2.	1200 gallon steel storage tank (lined)	1,400	
3.	Pumps - 350 80 185 130	745	
4.	Heat exchangers - 275 185	460	$ 6,685
5.	Plumbing - 1-1/4" pipe Misc. fittings Valves Flow regulators Manifolds Cu pump Air vents	230 220 536 160 180 100 20	1,446
6.	Controls and display	1,148	
7.	Insulation	1,481	
8.	Preheat tank	160	
9.	Expansion tank	80	
10.	Labor	3,600	
11.	Testing, balancing, adjusting and periodic checks	1,000	
	Subtotal		15,600
12.	Profit and overhead (20% of $15,600)	3,120	
	TOTAL COST		$18,720

Installed cost per unit area of collector $23.40.

area increases. This would include such items as collector costs, partial increases in cost of storage tank, etc. The area independent costs referred to in item L of Table 7.7 do not increase as the collector area is expanded. This includes costs for fans, pumps, system controls, heat exchangers, the rest of the storage unit costs not covered under collector dependent costs, etc. Items M and N represent the estimated fuel costs for the first year of operation. After the first year, item U accounts for the future estimated fuel costs. In the example it is assumed that electricity cost is $0.04 per kilowatt-hour (1 kW = 3.6 million Joules, 1 gigaJoule = 277.78 kW. Therefore, $0.04/kW × 277.78 kW = $11.11/gJ), or $11.11/gJ. In lines O and P it is assumed that electricity is used to power the backup (electric) furnace. Therefore, since the efficiency of electricity utilization is assumed to be 100%, the values of O and P will be 1.0. If, say a gas furnace had been used, then the efficiency would be assumed to be less than 100%. Item Q is the property tax rate. The tax rate is assumed to be 2.5% of the assessed value. In this example, the ratio of assessed value to actual cost is 0.80. Therefore, the tax rate as a fraction of investment is 0.020. Item V is the discount rate which represents the rate with which the extra money invested in the solar system could have been otherwise invested,

Table 7.5. Representative Costs for an Air System with 390 ft² of Collector. (Source: U. S. Department of Commerce[21]).

House Design Heat Load = 42,000 Btu/hr

ITEM		COST
1.	Solar equipment collectors, cap strips, end caps, air handler, controller, domestic water package, butyl sealant	$ 6,315.59
2.	Engineering charge	400.00
3.	Storage materials labor	500.00 500.00
3.	Gas hookup	80.00
5.	Installation collectors, duckwork, controls, air handling unit, etc.	3,101.54
	TOTAL COST	$10,897.23

Installed cost per unit area of collector $27.94.

Table 7.6. Solar System Cost Estimates for Commercial Buildings. (Source: U. S. Department of Energy[26]).

A.	Subsystem Costs	Fraction of Total Solar Cost
1.	Collectors and supports	35%
2.	Storage and heat exchangers	20%
3.	Piping, controls, electrical, and installation	45%

B.	Component Costs	Cost/SF_C
1.	Collectors	$5.00-20.00
	a. Nonselective	$5.00-10.00
	b. Selective	$10.00-20.00
	c. Collector support structure	$3.00-10.00
2.	Heat exchangers	$0.40-0.80
3.	Collector fluid	$0.15-0.20
4.	Storage tank and insulation	$2.00-5.00
5.	Piping, insulation, expansion tanks, valves	$3.00-6.00
6.	Pumps	$0.40-1.00
7.	Controls and electrical =$3000-5000	

C.	Systems Costs by Type	Installed Cost/SF_C
1.	BSHW only	$20-35
2.	Space and BSHW heating	$25-50
3.	Space heating and cooling	$35-65

Notes: 1. These costs do not include auxiliary energy equipment,
engineering design and inspection, or contingency
costs.

2. Costs for retrofit projects are approximately 10 to
25% higher than the above.

Table 7.7.   Residential Example. Form reprinted with permission from Beckman, William A., Klein, Sanford A., and Duffie, John A., Solar Heating Design by the F-Chart Method, New York: John Wiley & Sons, Inc., 1977.

```
H. Annual mortgage interest rate                        0.10%/100
I. Term of mortgage                                      30 Yrs.
J. Down payment (as fraction of investment)             0.10%/100
K. Collector area dependent costs                        270 $/m²
L. Area independent costs                                1500 $
M. Present cost of solar backup system fuel             11.11 $/GJ
N. Present cost of conventional system fuel             11.11 $/GJ
O. Efficiency of solar backup furnace                    1.0%/100
P. Efficiency of conventional system furnace             1.0%/100
Q. Property tax rate (as fraction of investment)        0.02%/100
R. Effective income tax bracket (state+federal-          .30%/100
     state x federal)
S. Extra ins. & maint. costs (as fraction of             .01%/100
     investment)
T. General inflation rate per year                      0.10%/100
U. Fuel inflation rate per year                         0.12%/100
V. Discount rate (after tax return on best              0.09%/100
     alternative investment)
W. Term of economic analysis                             20 Yrs.
X. First year non-solar fuel expense (total,             960 $
     E3)(N.)/(P.)÷10⁹
*Y. Depreciation lifetime                                 -- Yrs.
Z. Salvage value (as fraction of investment)             0 %/100
AA. Table 7.8 with Yr=(W.), Column=(U.) and Row=(V.)    24.040
BB.       "            (W.)    "    (T.)    "   (V.)     20.039
CC.       "         MIN(I.,W.)  "   (H.)    "   (V.)     20.039
+*DD.     "         MIN(W.,Y.)  "   (Zero)  "   (V.)       --
EE.       "            (I.)    "    (Zero)  "   (H.)      9.427
FF.       "         MIN(I.,W.)  "   (Zero)  "   (V.)      9.129
GG. (FF.)/(EE.), Loan payment                            0.968
HH. (GG.)+(CC.)[(H.)-1/(EE.)], Loan interest             0.846
II. (J.)+(1-J.)[(GG.)-(HH.)(R.)], Capital cost           0.743
JJ. (S.)(BB.), I&M cost                                  0.200
KK. (Q.)(BB)(1-R.), Property tax                         0.281
LL. (Z.)/(1+V.)(W.), Salvage value                        --
*MM. (R.)(DD.)(1-Z.)/(Y.), Depreciation                   --
NN. Other costs                                           --
OO. (II.)+(JJ.)+KK)-(LL.)+(NN.), Residential costs       1.224
PP. (II.)+(JJ.)(1-R.)+(KK.)-(LL.)-(MM.)+(NN.)(1-R.),      --
     Commercial costs

* Commercial only
+ Straight line only.  Use Tables 7.12 and 7.13 for other
  depreciation methods.
```

such as in bonds, certificates of deposit, etc. For items AA–FF, Table 7.8 (Sections A through C) is used. (Equation 7.5 can be used to develop tables for other values of N, i, and d.)

$$\text{Present worth factor} = [1/(d - i)][1 - (1 + i)/(1 + d)]^N = i = d \quad (7.5)$$
$$= N/(1 + i)$$

where

$$d = \text{Discount rate}$$
$$i = \text{Annual inflation rate}$$
$$N = \text{Number of periods}$$

Table 7.9 is completed after Table 7.7 has been filled in. Notice that item R8 of Table 7.9 shows the residential savings. The 25 m² collector is shown to yield the greatest savings over the other two collector sizes examined. Once the optimum collector size has been chosen, Table 7.10 can be used to give a

Table 7.8. Present Worth Factor Table. Reprinted with permission from Beckman, William A., Klein, Sanford A., and Duffie, John A., Solar Heating Design by the F-Chart Method, New York: John Wiley & Sons, Inc., 1977.

d, MARKET DISCOUNT RATE (%) — i, ANNUAL INFLATION RATE (%)

d	0	1	2	3	4	5	6	7	8	9	10	11	12
0	5.000	5.101	5.204	5.309	5.416	5.526	5.637	5.751	5.867	5.985	6.105	6.228	6.353
1	4.853	4.950	5.049	5.150	5.253	5.358	5.466	5.575	5.686	5.799	5.915	6.033	6.153
2	4.713	4.807	4.902	4.999	5.098	5.199	5.302	5.407	5.514	5.623	5.734	5.847	5.962
3	4.580	4.669	4.761	4.854	4.950	5.047	5.146	5.246	5.349	5.454	5.561	5.669	5.780
4	4.452	4.538	4.626	4.716	4.808	4.901	4.996	5.093	5.192	5.293	5.395	5.500	5.606
5	4.329	4.413	4.497	4.584	4.672	4.762	4.853	4.947	5.042	5.139	5.238	5.338	5.441
6	4.212	4.292	4.374	4.457	4.542	4.629	4.717	4.807	4.898	4.992	5.087	5.183	5.282
7	4.100	4.177	4.256	4.336	4.418	4.501	4.586	4.673	4.761	4.851	4.942	5.036	5.131
8	3.993	4.067	4.143	4.220	4.299	4.379	4.461	4.545	4.630	4.716	4.804	4.894	4.986
9	3.890	3.961	4.035	4.109	4.185	4.263	4.342	4.422	4.504	4.587	4.672	4.759	4.847
10	3.791	3.860	3.931	4.003	4.076	4.151	4.227	4.304	4.383	4.464	4.545	4.629	4.714
11	3.696	3.763	3.831	3.900	3.971	4.043	4.117	4.191	4.268	4.345	4.424	4.505	4.586
12	3.605	3.669	3.735	3.802	3.870	3.940	4.011	4.083	4.157	4.231	4.308	4.385	4.464
13	3.517	3.580	3.643	3.708	3.774	3.841	3.909	3.979	4.050	4.122	4.196	4.271	4.347
14	3.433	3.493	3.555	3.617	3.681	3.746	3.812	3.879	3.948	4.018	4.089	4.161	4.235
15	3.352	3.410	3.470	3.530	3.592	3.655	3.719	3.784	3.850	3.917	3.986	4.056	4.127
16	3.274	3.331	3.388	3.447	3.506	3.567	3.629	3.691	3.755	3.821	3.887	3.954	4.023
17	3.199	3.254	3.309	3.366	3.424	3.482	3.542	3.603	3.665	3.728	3.792	3.857	3.924
18	3.127	3.180	3.234	3.288	3.344	3.401	3.459	3.518	3.577	3.638	3.700	3.764	3.828
19	3.058	3.109	3.161	3.214	3.268	3.323	3.379	3.436	3.493	3.552	3.612	3.673	3.736
20	2.991	3.040	3.091	3.142	3.194	3.247	3.301	3.357	3.413	3.470	3.528	3.587	3.647

INFLATION-DISCOUNT FUNCTION FOR N = 10

d, MARKET DISCOUNT RATE (%) — i, ANNUAL INFLATION RATE (%)

d	0	1	2	3	4	5	6	7	8	9	10	11	12
0	10.000	10.462	10.950	11.464	12.006	12.578	13.181	13.816	14.487	15.193	15.937	16.722	17.549
1	9.471	9.901	10.354	10.831	11.335	11.865	12.425	13.014	13.635	14.289	14.979	15.705	16.470
2	8.983	9.383	9.804	10.248	10.716	11.209	11.728	12.275	12.851	13.458	14.097	14.770	15.479
3	8.530	8.903	9.295	9.709	10.144	10.603	11.085	11.594	12.129	12.692	13.286	13.910	14.567
4	8.111	8.459	8.825	9.210	9.615	10.042	10.492	10.965	11.462	11.986	12.537	13.117	13.727
5	7.722	8.046	8.388	8.748	9.126	9.524	9.942	10.383	10.846	11.334	11.847	12.386	12.953
6	7.360	7.664	7.983	8.319	8.672	9.043	9.434	9.845	10.277	10.731	11.208	11.710	12.238
7	7.024	7.308	7.607	7.921	8.251	8.598	8.962	9.346	9.749	10.172	10.618	11.085	11.577
8	6.710	6.976	7.256	7.550	7.859	8.184	8.525	8.883	9.259	9.655	10.070	10.507	10.965
9	6.418	6.667	6.930	7.205	7.495	7.798	8.118	8.453	8.805	9.174	9.562	9.970	10.398
10	6.145	6.379	6.625	6.884	7.155	7.440	7.739	8.053	8.382	8.728	9.091	9.472	9.872
11	5.889	6.110	6.341	6.584	6.838	7.105	7.386	7.680	7.989	8.313	8.652	9.009	9.383
12	5.650	5.858	6.075	6.303	6.543	6.793	7.057	7.333	7.622	7.926	8.244	8.578	8.929
13	5.426	5.622	5.826	6.041	6.266	6.502	6.749	7.008	7.280	7.565	7.864	8.177	8.505
14	5.216	5.400	5.593	5.795	6.007	6.229	6.462	6.705	6.961	7.228	7.509	7.803	8.111
15	5.019	5.193	5.374	5.565	5.765	5.974	6.193	6.422	6.662	6.914	7.177	7.453	7.743
16	4.833	4.997	5.169	5.349	5.537	5.734	5.940	6.156	6.383	6.619	6.867	7.127	7.399
17	4.659	4.814	4.976	5.146	5.323	5.509	5.704	5.908	6.121	6.344	6.577	6.822	7.077
18	4.494	4.641	4.794	4.955	5.123	5.298	5.482	5.674	5.875	6.085	6.305	6.536	6.776
19	4.339	4.478	4.623	4.775	4.934	5.100	5.273	5.455	5.644	5.843	6.050	6.267	6.494
20	4.192	4.324	4.462	4.606	4.756	4.913	5.077	5.248	5.428	5.615	5.811	6.016	6.230

d, MARKET DISCOUNT RATE (%) — i, ANNUAL INFLATION RATE (%)

d	0	1	2	3	4	5	6	7	8	9	10	11	12
0	25.000	28.243	32.030	36.459	41.646	47.727	54.864	63.249	73.106	84.701	98.347	114.413	133.334
1	22.023	24.752	27.929	31.633	35.958	41.014	46.933	53.869	62.003	71.550	82.762	95.935	111.419
2	19.523	21.832	24.510	27.622	31.245	35.470	40.401	46.164	52.906	60.800	70.051	80.897	93.621
3	17.413	19.375	21.644	24.272	27.322	30.867	34.994	39.804	45.417	51.974	59.659	68.606	79.104
4	15.622	17.298	19.229	21.459	24.038	27.028	30.498	34.531	39.224	44.693	51.071	58.516	67.213
5	14.094	15.532	17.184	19.085	21.277	23.810	26.740	30.137	34.079	38.660	43.990	50.197	57.431
6	12.783	14.024	15.444	17.072	18.943	21.098	23.585	26.458	29.784	33.639	38.112	43.308	49.350
7	11.654	12.729	13.954	15.356	16.961	18.803	20.923	23.364	26.183	29.440	33.210	37.578	42.645
8	10.675	11.611	12.674	13.885	15.269	16.851	18.666	20.750	23.148	25.912	29.103	32.791	37.058
9	9.823	10.641	11.568	12.620	13.817	15.182	16.743	18.530	20.580	22.936	25.648	28.774	32.382
10	9.077	9.796	10.607	11.525	12.566	13.749	15.097	16.636	18.396	20.412	22.727	25.388	28.452
11	8.422	9.056	9.769	10.574	11.482	12.512	13.682	15.012	16.530	18.264	20.248	22.523	25.134
12	7.843	8.405	9.035	9.743	10.540	11.440	12.459	13.615	14.929	16.425	18.133	20.086	22.321
13	7.330	7.830	8.388	9.014	9.716	10.506	11.398	12.406	13.548	14.846	16.322	18.005	19.926
14	6.873	7.320	7.817	8.372	8.993	9.689	10.473	11.356	12.353	13.483	14.764	16.220	17.878
15	6.464	6.865	7.309	7.803	8.355	8.971	9.662	10.439	11.314	12.301	13.417	14.683	16.119
16	6.097	6.457	6.856	7.298	7.790	8.338	8.950	9.636	10.406	11.272	12.249	13.353	14.602
17	5.766	6.092	6.451	6.848	7.288	7.776	8.321	8.929	9.609	10.372	11.230	12.197	13.288
18	5.467	5.762	6.086	6.444	6.839	7.277	7.763	8.304	8.907	9.582	10.339	11.189	12.146
19	5.195	5.463	5.758	6.081	6.437	6.830	7.266	7.749	8.286	8.886	9.556	10.306	11.148
20	4.948	5.192	5.460	5.753	6.075	6.430	6.822	7.255	7.735	8.269	8.864	9.529	10.272

Table 7.8. (Continued)

INFLATION-DISCOUNT FUNCTION FOR N = 30

d, MARKET DISCOUNT RATE (%)	i, ANNUAL INFLATION RATE (%)											
	0	1	2	3	4	5	6	7	8	9	10	11	12
0	30.000	34.785	40.568	47.575	56.085	66.439	79.058	94.461	113.283	136.307	164.494	199.021	241.333
1	25.808	29.703	34.389	40.042	46.878	55.164	65.225	77.462	92.367	110.545	132.735	159.843	192.981
2	22.396	25.589	29.412	34.002	39.529	46.201	54.270	64.050	75.922	90.353	107.916	129.313	155.400
3	19.600	22.235	25.374	29.126	33.624	39.029	45.541	53.404	62.914	74.435	88.413	105.392	126.034
4	17.292	19.481	22.076	25.163	28.846	33.254	38.541	44.900	52.563	61.813	73.000	86.545	102.965
5	15.372	17.203	19.363	21.919	24.955	28.571	32.891	38.065	44.276	51.746	60.748	71.613	84.744
6	13.765	15.307	17.116	19.246	21.765	24.751	28.302	32.537	37.601	43.668	50.953	59.716	70.272
7	12.409	13.716	15.241	17.028	19.131	21.612	24.549	28.037	32.190	37.147	43.076	50.182	58.715
8	11.258	12.372	13.667	15.176	16.942	19.017	21.461	24.351	27.778	31.851	36.704	42.499	49.433
9	10.274	11.230	12.335	13.618	15.111	16.856	18.904	21.313	24.157	27.523	31.518	36.271	41.937
10	9.427	10.253	11.202	12.299	13.569	15.046	16.771	18.792	21.166	23.965	27.273	31.192	35.848
11	8.694	9.411	10.232	11.175	12.262	13.520	14.982	16.687	18.681	21.022	23.776	27.027	30.873
12	8.055	8.682	9.395	10.211	11.147	12.225	13.472	14.918	16.603	18.572	20.879	23.590	26.786
13	7.496	8.046	8.670	9.379	10.190	11.119	12.188	13.423	14.855	16.520	18.464	20.738	23.407
14	7.003	7.489	8.037	8.658	9.363	10.169	11.091	12.151	13.375	14.792	16.438	18.356	20.599
15	6.566	6.997	7.482	8.028	8.646	9.347	10.147	11.063	12.115	13.327	14.729	16.356	18.250
16	6.177	6.562	6.992	7.475	8.019	8.633	9.331	10.126	11.035	12.078	13.279	14.667	16.275
17	5.829	6.174	6.558	6.987	7.468	8.009	8.621	9.315	10.104	11.007	12.041	13.231	14.605
18	5.517	5.827	6.171	6.554	6.981	7.460	7.999	8.608	9.298	10.083	10.979	12.005	13.184
19	5.235	5.515	5.825	6.168	6.550	6.976	7.453	7.990	8.596	9.282	10.061	10.951	11.968
20	4.979	5.233	5.513	5.822	6.165	6.545	6.970	7.446	7.980	8.583	9.265	10.040	10.922

INFLATION-DISCOUNT FUNCTION FOR N = 15

d, MARKET DISCOUNT RATE (%)	i, ANNUAL INFLATION RATE (%)											
	0	1	2	3	4	5	6	7	8	9	10	11	12
0	15.000	16.097	17.293	18.599	20.024	21.579	23.276	25.129	27.152	29.361	31.772	34.405	37.280
1	13.865	14.851	15.926	17.098	18.375	19.767	21.285	22.942	24.748	26.718	28.867	31.212	33.770
2	12.849	13.738	14.706	15.759	16.906	18.156	19.517	21.000	22.616	24.377	26.297	28.389	30.669
3	11.938	12.741	13.614	14.563	15.596	16.719	17.942	19.273	20.722	22.300	24.017	25.888	27.925
4	11.118	11.845	12.634	13.492	14.423	15.435	16.536	17.733	19.035	20.451	21.991	23.667	25.491
5	10.380	11.039	11.754	12.530	13.372	14.286	15.279	16.357	17.529	18.802	20.187	21.691	23.327
6	9.712	10.311	10.960	11.664	12.426	13.254	14.151	15.125	16.182	17.329	18.575	19.929	21.399
7	9.108	9.654	10.244	10.883	11.575	12.325	13.138	14.019	14.974	16.010	17.134	18.354	19.677
8	8.559	9.057	9.595	10.177	10.807	11.488	12.225	13.024	13.889	14.826	15.842	16.943	18.137
9	8.061	8.516	9.007	9.538	10.111	10.731	11.402	12.127	12.912	13.761	14.681	15.678	16.757
10	7.606	8.023	8.473	8.958	9.481	10.046	10.657	11.317	12.030	12.802	13.636	14.539	15.516
11	7.191	7.574	7.986	8.430	8.909	9.425	9.982	10.584	11.233	11.935	12.694	13.514	14.400
12	6.811	7.163	7.541	7.949	8.387	8.860	9.369	9.919	10.511	11.151	11.842	12.587	13.393
13	6.462	6.786	7.135	7.509	7.912	8.345	8.812	9.314	9.856	10.440	11.070	11.749	12.483
14	6.142	6.441	6.762	7.107	7.477	7.875	8.303	8.764	9.260	9.794	10.370	10.990	11.659
15	5.847	6.124	6.420	6.738	7.079	7.445	7.839	8.262	8.717	9.206	9.733	10.300	10.911
16	5.575	5.831	6.105	6.399	6.714	7.051	7.413	7.803	8.220	8.670	9.153	9.672	10.231
17	5.324	5.561	5.815	6.087	6.378	6.689	7.024	7.382	7.767	8.180	8.623	9.100	9.612
18	5.092	5.312	5.547	5.799	6.069	6.357	6.665	6.996	7.351	7.731	8.139	8.577	9.048
19	4.876	5.081	5.300	5.533	5.783	6.050	6.336	6.641	6.969	7.320	7.696	8.099	8.532
20	4.675	4.867	5.070	5.288	5.519	5.767	6.032	6.315	6.618	6.942	7.289	7.661	8.059

INFLATION-DISCOUNT FUNCTION FOR N = 20

d, MARKET DISCOUNT RATE (%)	i, ANNUAL INFLATION RATE (%)											
	0	1	2	3	4	5	6	7	8	9	10	11	12
0	20.000	22.019	24.297	26.870	29.778	33.066	36.786	40.995	45.762	51.160	57.275	64.203	72.052
1	18.046	19.802	21.780	24.009	26.524	29.362	32.568	36.190	40.284	44.913	50.150	56.074	62.778
2	16.351	17.885	19.608	21.546	23.728	26.186	28.958	32.084	35.612	39.594	44.093	49.174	54.917
3	14.877	16.221	17.727	19.417	21.317	23.453	25.857	28.564	31.613	35.050	38.926	43.299	48.232
4	13.590	14.771	16.092	17.571	19.231	21.093	23.185	25.536	28.180	31.156	34.506	38.279	42.531
5	12.462	13.503	14.665	15.965	17.419	19.048	20.874	22.922	25.222	27.806	30.710	33.977	37.651
6	11.470	12.391	13.417	14.562	15.840	17.269	18.868	20.659	22.665	24.916	27.442	30.277	33.463
7	10.594	11.411	12.320	13.332	14.459	15.717	17.122	18.692	20.448	22.414	24.617	27.086	29.856
8	9.818	10.546	11.353	12.250	13.247	14.358	15.596	16.977	18.519	20.242	22.169	24.325	26.740
9	9.129	9.779	10.498	11.296	12.181	13.164	14.258	15.476	16.834	18.349	20.039	21.928	24.040
10	8.514	9.096	9.739	10.450	11.238	12.112	13.082	14.160	15.359	16.694	18.182	19.841	21.693
11	7.963	8.487	9.063	9.700	10.403	11.182	12.044	13.001	14.063	15.243	16.556	18.018	19.647
12	7.469	7.941	8.460	9.031	9.661	10.356	11.125	11.977	12.920	13.967	15.129	16.421	17.857
13	7.025	7.451	7.919	8.433	8.998	9.622	10.310	11.070	11.910	12.841	13.872	15.017	16.287
14	6.623	7.009	7.432	7.896	8.406	8.966	9.583	10.263	11.014	11.844	12.762	13.779	14.906
15	6.259	6.610	6.994	7.414	7.874	8.379	8.934	9.545	10.217	10.959	11.779	12.685	13.687
16	5.929	6.249	6.597	6.978	7.395	7.851	8.352	8.902	9.506	10.172	10.905	11.714	12.608
17	5.628	5.920	6.238	6.585	6.963	7.376	7.829	8.325	8.870	9.468	10.126	10.851	11.650
18	5.353	5.620	5.911	6.227	6.572	6.947	7.358	7.807	8.298	8.838	9.430	10.081	10.798
19	5.101	5.347	5.613	5.902	6.216	6.558	6.932	7.339	7.784	8.272	8.806	9.392	10.036
20	4.870	5.096	5.340	5.605	5.893	6.205	6.545	6.916	7.320	7.762	8.245	8.774	9.355

Table 7.9. Residential Example. Form reprinted with permission from Beckman, William A., Klein, Sanford A., and Duffie, John A., Solar Heating Design by the F-Chart Method, New York: John Wiley & Sons, Inc., 1977.

Residential Example

F-Chart Worksheet 7[1]
Economic Analysis

R1.	Collector Area (Worksheet 3)	$0m^2$	$10m^2$	$25m^2$	$40m^2$
R2.	Fraction by Solar (Worksheet 3)	0	0.28	0.50	0.62
R3.	Investment in Solar (K.)(R1.)+(L.)	1500	4200	8250	12300
R4.	1st Year Fuel Expense (Total,E3)(1-R2.)(M.)/(0.)$\div 10^9$	960	691	480	365
R5.	Fuel Savings (X.-R4.)(AA.)	0	6467	11539	14304
R6.	Expenses (Residential) (00.)(R3.)	1836	5141	10098	15055
R7.	Expenses (Commercial) (PP.)(R3.)	--	--	--	--
R8.	Savings (Residential) (R5.)-(R6.)	(1836)	1326	1441	(751)
R9.	Savings (Commercial) (R5.)(1-R.)-(R7.)	--	--	--	--

breakdown of the savings (or losses) occurring over the life of the system. As shown in items R23 and R24 of Table 7.10, positive cash flows do not occur until the seventh year. The savings became greater each successive year because the cost of the alternate fuel used (e.g., electricity, fuel oil, and natural gas) increases.

To summarize the previous residential economic analysis, one should note that in Table 7.9 the 25 m^2 collector area yielded the greatest savings of the three collector sizes examined but is not necessarily the optimum collector size. In other words, a collector size of say 28 m^2 might yield the greatest savings possible and would therefore be the "best" collector size. There are computer programs available which in effect run through this same type of economic analysis for numerous collector sizes until an optimum collector size has been chosen. Finally, even when an optimized collector size has been chosen, this does not automatically mean that a solar installation is economical using the optimized collector size, but only means that the optimized collector size yields the greatest savings or the least economic losses for the solar system installation.

Table 7.10. Residential Example. Form reprinted with permission from Beckman, William A., Klein, Sanford A., and Duffie, John A., Solar Heating Design by the F-Chart Method, New York: John Wiley & Sons, Inc., 1977.

Yearly Savings for Collector Area = $25m^2$

R10. Year (n) (first year=1)	1	2	3	4	5	6	7	8
†R11. Current Mortgage [(R11.)-(R14.)+R18.)]	7425	7379	7329	7274	7213	7146	7073	6992
R12. Fuel Savings (X.-R4.)$(1+U.)^{n-1}$	480	538	602	674	755	846	947	1061
R13. Down Payment (1st year only) (R3.)(J.)	825	--	--	--	--	--	--	--
R14. Mortgage Payment (1-J.)(R3.)/(EE.)	788	788	788	788	788	788	788	788
R15. Extra Insurance & Maintenance (S.)(R3.)$(1+T.)^{n-1}$	82	91	100	110	121	133	146	161
R16. Extra Property Tax (R3.)(Q.)$(1+T.)^{n-1}$	165	182	200	220	242	266	292	322
R17. Sum (R13.+R14.+$15.+R16.)	1860	1061	1088	1118	1151	1187	1226	1271
R18. Interest on Mortgage (R11)(H.)	742	738	733	727	721	715	707	699
R19. Tax Savings (R.)(R16.+R18.)	272	276	280	284	289	294	300	306
*R20. Depreciation (st. line) (R3.)(1-Z.)/(Y.)	--	--	--	--	--	--	--	--
*R21. Business Tax Savings (R.)(R20.+R15.-R12.)	--	--	--	--	--	--	--	--
R22. Salvage Value (R2.)(Z.) (Last Year Only)	--	--	--	--	--	--	--	--
R23. Solar Savings (R12.-R17.+R19.+R21.+R22.)	(1108)	(247)	(206)	(160)	(107)	(47)	21	96
**R24. Discounted Savings (R23.)/$(1+V.)^n$	(1016)	(208)	(159)	(113)	(70)	(28)	12	48

†For the first year use [(R3.)(1-J.)]; for subsequent years use equation with previous years values.
*Income producing property only.
**The down payment should not be discounted.

7.6.2. Economic Analysis for Commercial Buildings

In calculating the economics of using solar energy in a commercial building, the same basic procedures for the previous residential economic analysis are used. Table 7.11 illustrates the calculation procedure for a proposed commercial solar installation. For the commercial building economic example, the solar system can be depreciated as shown in item Y of Table 7.11. In this example a simple straight line depreciation method is used with no equipment salvage value. For other depreciation methods, Tables 7.12 and 7.13 may be used when determining the numerical value of item DD in Table 7.10. Table 7.14 represents the second worksheet table to be examined in the commercial building economic example. In Table 7.14, item R9 shows that the 10 m² collector area size gives the greatest savings over the other two collector sizes examined. Notice that when computing the commercial savings (R9) as opposed to computing the residential savings (R8), the fuel savings shown in item R5 must be reduced in this case by 70% (1.0 - effective income tax bracket) because the fuel savings are taxable since the cost of fuel has already been deducted as an operating expense for the business.

Table 7.15 is presented to help illustrate the cash flows involved in the

Table 7.11.   Commercial Example. Form reprinted with permission from Beckman, William A., Klein, Sanford A., and Duffie, John A., Solar Heating Design by the F-Chart Method, New York: John Wiley & Sons, Inc., 1977.

Economic Parameters

H.	Annual mortgage interest rate	0.10%/100
I.	Term of mortgage	30 Yrs.
J.	Down payment (as fraction of investment)	0.10%/100
K.	Collector area dependent costs	270 $/m^2
L.	Area independent costs	1500 $
M.	Present cost of solar backup system fuel	11.11 $/GJ
N.	Present cost of conventional system fuel	11.11 $/GJ
O.	Efficiency of solar backup furnace	1.0%/100
P.	Efficiency of conventional system furnace	1.0%/100
Q.	Property tax rate (as fraction of investment)	0.02%/100
R.	Effective income tax bracket (state+federal-state x federal)	0.30%/100
S.	Extra ins. & maint. costs (as fraction of investment)	0.01%/100
T.	General inflation rate per year	0.10%/100
U.	Fuel inflation rate per year	0.12%/100
V.	Discount rate (after tax return on best alternative investment)	0.09%/100
W.	Term of economic analysis	20 Yrs.
X.	First year non-solar fuel expense (total, E3)(N.)/(P.)÷10^9	960 $
*Y.	Depreciation lifetime	20 Yrs.
Z.	Salvage value (as fraction of investment)	0 %/100
AA.	Table 7.8 with Yr=(W.), Column=(U.) and Row=(V.)	24.040
BB.	" (W.) " (T.) " (V.)	20.039
CC.	" MIN(I.,W.) " (H.) " (V.)	20.039
+*DD.	" MIN(W.,Y.) " (Zero) " (V.)	9.129
EE.	" (I.) " (Zero) " (H.)	9.427
FF.	" MIN(I.,W.) " (Zero) " (V.)	9.129
GG.	(FF.)/(EE.), Loan payment	0.968
HH.	(GG.)+(CC.)[(H.)-1/(EE.)], Loan Interest	0.846
II.	(J.)+(1-J)[(GG.)-(HH.)(R.)], Capital cost	0.743
JJ.	(S.)(BB.), I&M cost	0.200
KK.	(Q.)(BB)(1-R.), Property tax	0.281
LL.	(Z.)/(1+V.)$^{(W.)}$, Salvage value	0.0178
*MM.	(R.)(DD.)(1-Z)/(Y.), Depreciation	0.137
NN.	Other costs	--
OO.	(II.)+(JJ.)+(KK.)-(LL.+NN.), Residential costs	--
PP.	(II.)+(JJ.)(1-R)+(KK.)-(LL.)-(MM.)+(NN.)(1-R.), Commercial Costs	1.023

* Commercial only.
+ Straight line only. Use Tables 7.12 and 7.13 for other depreciation methods

Table 7.12.   Sum of Digits Depreciation Factors. Reprinted with permission from Beckman, William A., Klein, Sanford A., and Duffie, John A., Solar Heating Design by the F-Chart Method, New York: John Wiley & Sons, Inc., 1977.

MARKET DISCOUNT RATE %	YEARS OF DEPRECIATION					
	5	10	15	20	25	30
0	5.0000	10.0000	15.0000	20.0000	25.0000	30.0000
1	4.8856	9.6126	14.1868	18.6138	22.8988	27.0470
2	4.7757	9.2492	13.4421	17.3741	21.0636	24.5276
3	4.6699	8.9079	12.7586	16.2620	19.4535	22.3646
4	4.5681	8.5868	12.1300	15.2611	18.0345	20.4967
5	4.4702	8.2846	11.5509	14.3577	16.7785	18.8743
6	4.3758	7.9997	11.0161	13.5398	15.6624	17.4572
7	4.2848	7.7310	10.5216	12.7973	14.6664	16.2129
8	4.1970	7.4771	10.0633	12.1213	13.7743	15.1147
9	4.1124	7.2371	9.6379	11.5042	12.9722	14.1407
10	4.0307	7.0099	9.2424	10.9395	12.2484	13.2730
11	3.9518	6.7947	8.8740	10.4214	11.5932	12.4963
12	3.8756	6.5906	8.5303	9.9449	10.9980	11.7983
13	3.8020	6.3969	8.2093	9.5057	10.4556	11.1684
14	3.7308	6.2128	7.9088	9.0999	9.9599	10.5978
15	3.6619	6.0379	7.6272	8.7242	9.5056	10.0791
16	3.5952	5.8713	7.3629	8.3757	9.0879	9.6060
17	3.5307	5.7127	7.1146	8.0517	8.7031	9.1729
18	3.4682	5.5615	6.8808	7.7499	8.3475	8.7753
19	3.4077	5.4173	6.6606	7.4682	8.0182	8.4093
20	3.3490	5.2796	6.4528	7.2050	7.7125	8.0713

Table 7.13.   Double Declining Balance Depreciation Factors. Reprinted with permission from Beckman, William A., Klein, Sanford A., and Duffie, John A., Solar Heating Design by the F-Chart Method, New York: John Wiley & Sons, Inc., 1977.

MARKET DISCOUNT RATE %	5	10	15	20	25	30
0	5.0000	10.0000	15.0000	20.0000	25.0000	30.0000
1	4.8871	9.5701	14.0589	18.3630	22.4916	26.4534
2	4.7787	9.1710	13.2134	16.9394	20.3790	23.5593
3	4.6746	8.7999	12.4516	15.6953	18.5868	21.1738
4	4.5745	8.4542	11.7631	14.6028	17.0554	19.1878
5	4.4783	8.1318	11.1391	13.6388	15.7377	17.5183
6	4.3857	7.8307	10.5719	12.7843	14.5962	16.1017
7	4.2966	7.5489	10.0548	12.0234	13.6007	14.8889
8	4.2108	7.2850	9.5823	11.3430	12.7270	13.8416
9	4.1280	7.0373	9.1492	10.7318	11.9556	12.9299
10	4.0483	6.8047	8.7513	10.1807	11.2705	12.1302
11	3.9713	6.5858	8.3848	9.6818	10.6588	11.4237
12	3.8971	6.3796	8.0465	9.2284	10.1097	10.7955
13	3.8253	6.1852	7.7333	8.8149	9.6145	10.2335
14	3.7561	6.0016	7.4429	8.4365	9.1657	9.7278
15	3.6891	5.8280	7.1729	8.0891	8.7572	9.2704
16	3.6243	5.6637	6.9214	7.7692	8.3840	8.8547
17	3.5617	5.5080	6.6866	7.4737	8.0417	8.4752
18	3.5010	5.3603	6.4671	7.1999	7.7267	8.1274
19	3.4423	5.2201	6.2614	6.9457	7.4357	7.8074
20	3.3854	5.0867	6.0683	6.7089	7.1661	7.5120

Table 7.14.   Commercial Example. Form reprinted with permission from Beckman, William A., Klein, Sanford A., and Duffie, John A., Solar Heating Design by the F-Chart Method, New York: John Wiley & Sons, Inc., 1977.

Economic Analysis

		$0m^2$	$10m^2$	$25m^2$	$40m^2$
R1.	Collector Area (Worksheet 3)				
R2.	Fraction by Solar (Worksheet 3)	0	0.28	0.50	0.62
R3.	Investment in Solar (K.)(R1.)+(L.)	1500	4200	8250	12300
R4.	1st Year Fuel Expense (Total,E3)(1−R2.)(M.)/0.)÷10^9	960	691	480	365
R5.	Fuel Savings (X.−R4.)(AA.)	0	6467	11539	14304
R6.	Expenses (Residential) (OO.)(R3.)	--	--	--	--
R7.	Expenses (Commercial) (PP.)(R3.)	1534	4297	8440	12583
R8.	Savings (Residential) (R5.)−(R6.)	--	--	--	--
R9.	Savings (Commercial) (R5.)(1−R.)−(R7.)	(1534)	230	(363)	(2570)

Table 7.15. Commercial Worksheet. Form reprinted with permission from Beckman, William A., Klein, Sanford A., and Duffie, John A., Solar Heating Design by the F-Chart Method, New York: John Wiley & Sons, Inc., 1977.

Yearly Savings for Collector Area = 10m²

Row	1	2	3	4	5	6	7	8	9	10	11	12	13	14	15
R10. Year (n) (first year=1)	1	2	3	4	5	6	7	8	9	10	11	12	13	14	15
†R11. Current Mortgage [(R11.)−(R14.)+(R18.)]	7425	7379	7329	7274	7213	7146	7073	6992	6903	6805	6697	6579	6449	6306	6149
R12. Fuel Savings $(X.-R4.)(1+U.)^{n-1}$	480	538	602	674	755	846	947	1061	666	746	835	936	1048	1174	1315
R13. Down Payment (1st year only) (R3.)(J.)	825	--	--	--	--	--	--	--	--	--	--	--	--	--	--
R14. Mortgage Payment $(1-J)(R3.)/(EE.)$	788	788	788	788	788	788	788	788	788	788	788	788	788	788	788
R15. Extra Insur. & Maint. $(S.)(R3.)(1+T.)^{n-1}$	82	91	100	110	121	133	146	161	90	99	109	120	132	145	160
R16. Extra Property Tax $(R3.)(Q.)(1+T.)^{n-1}$	165	182	200	220	242	266	292	322	180	198	218	240	264	290	319
R17. Sum [R13.+R14.+R15.+R16]	1860	1061	1088	1118	1151	1187	1226	1271	1058	1085	1115	1148	1184	1223	1267
R18. Interest on Mortgage (R11.)(H)	742	738	733	727	721	715	707	699	690	680	670	658	645	631	615
R19. Tax Savings (R.)(R16.+R18.)	272	276	280	284	289	294	300	306	261	263	266	269	273	276	280
*R20. Depreciation (st. line) $(R3.)(1-Z)/(Y.)$	210	210	210	210	210	210	210	210	210	210	210	210	210	210	210
*R21. Business Tax Savings (R.)(R20.+R15.−R12.)	(56)	(71)	(88)	(106)	(127)	(151)	(177)	(207)	(110)	(131)	(155)	(182)	(212)	(246)	(284)
R22. Salvage Value (R2.)(Z.)(Last year only)	--	--	--	--	--	--	--	--	--	--	--	--	--	--	--
R23. Solar Savings [R12.−R17.+R19.+R21.+R22.]	(1164)	(318)	(294)	(266)	(234)	(198)	(156)	(111)	(241)	(207)	(169)	(125)	(75)	(19)	44
**R24. Discounted Savings $(R23.)/(1+V.)^n$	(1068)	(268)	(227)	(188)	(152)	(118)	(85)	(56)	(111)	(87)	(65)	(44)	(24)	(6)	12

For the first year use [(R3.)(1-J.)]; for subsequent years use equation with previous years values

* income producing property only.

** The down payment should not be discounted.

commercial solar economic example. Notice that in Table 7.15, positive solar savings (item R23) do not occur until the 15th year.

7.7 REFERENCES AND SUGGESTED READING

1. Anderson, Bruce, (ed.), "More on the Solar Tax Incentives," *Solar Age*, **4**, 1:14, (1979).
2. Barach, Arnold B., ed., "Solar Heating Your House — Would It Pay?" *Changing Times*, **32**, 4:6–9, (1978).
3. Barrett, David, Epstein, Peter, and Haar, Charles M., *Home Mortgage Lending and Solar Energy*, Washington, D.C.: U. S. Government Printing Office, February, 1977.
4. Beckman, William A., Klein, Sanford A., and Duffie, John A., *Solar Heating Design by the F-Chart Method*, New York: John Wiley & Sons, Inc., 1977.
5. Brandemuehl, M. J., Beckman, W. A., "Economic Evaluation in Optimization of Solar Heating Systems", *Solar Energy*, Vol. 23, New York: Pergamon Press, 1979, pp. 1–10.
6. Energy Research and Development Administration, *An Economic Analysis of Solar Water and Space Heating*, Washington, D.C.: U. S. Government Printing Office, November, 1976.
7. Federal Energy Administration, *Buying Solar*, Washington, D.C.: U. S. Government Printing Office, June, 1976.
8. Federal Energy Administration, *Economic Thickness for Industrial Insulation*, Washington, D.C.: U. S. Government Printing Office, August, 1976.
9. Florida Solar Energy Center, *A Guide to Solar Water Heating in Florida*, Cape Canaveral, Florida: State University System of Florida, October, 1978, pp. 18–21.
10. McGarity, Arthur E., *Solar Heating and Cooling: An Economic Assessment*, Washington, D.C.: U. S. Government Printing Office, 1978.
11. Oviatt, A. E., *Optimum Insulation Thickness in Wood-Framed Homes*, (USDA Forest Service General Technical Report PNW-32), Washington, D.C.: U. S. Government Printing Office, 1975.
12. Peurifoy, R. L., *Construction Planning, Equipment, and Methods*, New York: McGraw-Hill Book Co., 1970.
13. Reid, R. L., Lumsdaine, R., and Albrecht, L., "Economics of Solar Heating with Home-owner-Type Financing," Vol. 19, New York: Pergamon Press, 1977, pp. 513–517.
14. Roose, Robert W., ed., *Handbook of Energy Conservation for Mechanical Systems and Buildings*, New York: Van Nostrand Reinhold Co., 1978.
15. Ruegg, Rosalie T., *Solar Heating and Cooling in Buildings: Methods of Economic Analysis*, Springfield, Virginia: National Technical Information Service, July, 1975.
16. Saif-Ul-Rehamn, M., "Economic Competitiveness of Solar Energy with Conventional Fuels and Electricity," *Solar Energy*, Vol. 18, New York: Pergamon Press, 1976, pp. 577–579.
17. Schulze, William D., Ben-David, Saul, Balcomb, J. Douglas, Katson, Roberta, Noll, Scott, Roach, Fred, and Thayer, Mark, *The Economics of Solar Home Heating*, Washington, D.C.: U. S. Government Printing Office, January, 1977.
18. Selvon, M. Robert, *Life-Cycle Costing: A Better Method of Government Procurement*, Boulder, Colorado: Westview Press, 1979.
19. Tryon, William J., "Assessing the Economics of Solar Energy," *Heating/Piping/Air-Conditioning*, **50**, 7:49–54 (1978).
20. Upton, J. P., "Determining Solar Energy System Costs," *Plant Engineering*, October 12, 1978, 167–169.
21. U. S. Department of Commerce, *Solar Heating and Cooling of Residential Buildings – Design of Systems*, Washington, D.C.: U. S. Government Printing Office, October, 1977.
22. U. S. Department of Commerce, *Solar Heating and Cooling of Residential Buildings – Sizing, Installation and Operation of Systems*, Washington, D.C.: U. S. Government Printing Office, October, 1977.
23. U. S. Department of Commerce, *Life-Cycle Costing – A Guide for Selecting Energy Conservation Projects for Public Buildings*, Washington, D.C.: U. S. Government Printing Office, September, 1978.

24. U. S. Department of Commerce, *Retrofitting Existing Housing for Energy Conservation: An Economic Analysis,* Washington, D.C.: U. S. Government Printing Office, December, 1974.

25. U. S. Department of Energy, *An Analysis of the Current Economic Feasibility of Solar Water and Space Heating,* Washington, D.C.: U. S. Government Printing Office, January, 1978.

26. U. S. Department of Energy, *DOE Facilities Solar Design Handbook,* Washington, D.C.: U. S. Government Printing Office, January, 1978.

27. U. S. Department of Energy, *Introduction to Solar Heating and Cooling Design and Sizing,* Washington, D.C.: U. S. Government Printing Office, August, 1978.

28. U. S. Department of Housing and Urban Development, *HUD Intermediate Minimum Property Standards Supplement,* Washington, D.C.: U. S. Government Printing Office, 1977.

29. U. S. Department of Housing and Urban Development, *Selling the Solar Home,* Washington, D.C.: U. S. Government Printing Office, April, 1978.

30. U. S. Department of Housing and Urban Development, *State Solar Legislation,* Rockville, Maryland: National Solar Heating and Cooling Information Center.

31. U. S. Department of the Treasury — Internal Revenue Service, Publication 903, "Energy Credits for Individuals."

32. U. S. Department of the Treasury — Internal Revenue Service, Publication 572, "Tax Information on Investment Credit," 1979.

33. U. S. Department of the Treasury — Internal Revenue Service, Publication 17, "Your Federal Income Tax," 1979, pp. 149–150.

34. Ward, Dan S., "Solar Absorption Cooling Feasibility," *Solar Energy,* Vol. 22, New York: Pergamon Press, 1979, pp. 259–268,

35. Weston, J. Fred, and Brigham, Eugene S., *Essentials of Managerial Finance,* Hinsdale, Illinois: The Dryden Press, 1974.

CHAPTER 8
COMPUTER APPLICATIONS

8.1 OVERVIEW

The individual using the computer to solve a problem must first supply the computer with a set of instructions telling how and when to perform various computational operations. This set of instructions represents the computer program. The computer program is responsible for the computer making the right decisions. If a computer program has errors in it, then the computer will follow these instructions and yield errors. This explains the phrase "garbage in—garbage out (gigo)." In essence, a faulty computer program or bad input data will generally yield unsatisfactory output. After a computer program has been written, the next step is debugging. This basically involves running the computer program and comparing the output with data (i.e., answers) known to be correct. If the computer yields incorrect answers then the programmer must make some adjustments to the computer program and run the program again. This testing continues until the program yields consistently correct answers on the computer. Once the program has been fully debugged, the end-user can feel fairly confident that the program will run correctly. Although the accuracy of results from a computer program are very important, an individual using a computer program should also be concerned with the program efficiency. Program efficiency is related to how judiciously the computer program uses computer memory and also how fast a program runs on a given computer. In most computer installations, an individual using the computer pays for both the actual time the computer is running the program [i.e., central processing unit time (CPU time)] and the amount of main memory (or "core") used in running the program. Therefore, an efficient program may require less computer main memory and processing time to run the program than may be the case for an inefficient computer program. Hence, the efficiently constructed computer program will cost less to run on the computer.

The individual purchasing a computer program may not be able to tell before purchase whether one computer program is more efficient than another because the individual firms selling computer programs probably will not allow the prospective purchaser a "sneak peek" at the program source listing. Furthermore, unless the program source listing is well documented and the prospective purchaser well versed in computer programming, viewing the program source listing might not yield much information. Therefore, the purchaser may have to settle for *assuming* that the computer program is efficient. The individual deciding to write a computer program rather than purchase an already available computer program should strive to write it as efficiently as possible to keep computer charges to a minimum when the computer program is run.

8.2 IN-HOUSE VERSUS EXTERNAL COMPUTER PROGRAM

In deciding to use a computer program for solving solar design problems, one has the choice of purchasing, renting, or producing computer programs.

The advantages of purchasing a "canned" program are:

1. One can make modifications to the program if necessary.
2. It may be cheaper than renting or writing an in-house computer program.

The disadvantages are:

1. An individual may purchase a program that does not do exactly what it was wanted for and may be stuck with it. (Some software programs involve signing a contract specifying that the buyer may not resell or give the software program to someone else.)
2. The computer program may not justify paying the initially high purchase price.

In renting a computer program, the user may be forced to pay for both the rental of the program *and* costs of running the program on the rental agency's computer. The advantages of renting are:

1. One may only have to pay for what one uses. In other words, an individual will not be stuck owning a computer program that may become obsolete, or owning a computer program that may be used only a few times.
2. If a better program comes along, the individual can easily switch over to the new program.

The disadvantages are:

1. An individual may not be able to modify the program to fit his exact needs.
2. One may end up spending more renting the program than buying it outright.

The final alternative involves writing one's own computer program.

The advantages are:

1. An individual can tailor the program to fit his *exact* needs without compromise.

2. It *may* be cheaper in the long run depending upon the expertise of the programmer.

The disadvantages are:

1. An individual may beat himself to death trying to come up with a working program when, in fact, a program already exists that will fit his needs. In this case the manhours spent developing the program may be more expensive than just buying or renting an existing program.
2. The individual firm may not have the in-house expertise required to write a workable program and the effort spent might prove futile.

It is hard to say whether to buy, rent, or write one's own computer program because of the many variables involved. The company deciding upon which way to go must make a judgment based upon *their own* needs and programming skills.

Purchased "canned" computer programs are generally sent to the purchaser on either magnetic ("mag") tape or on punched, 80 column cards. Figure 8.1 illustrates some computer program storage mediums.

For the individual desiring to learn more about computer programming, there are numerous books on the subject which help to explain what is involved. Furthermore, most community colleges offer a variety of computer

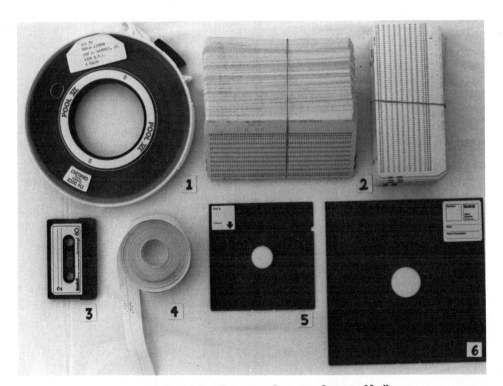

Figure 8.1. Computer Program Storage Mediums.

programming courses for a modest fee. The courses taught generally involve both classroom teaching and actual programming experience using the school's computer.

There are a number of computer "languages" which can be used in writing a computer program. To some extent it can be said that one language may be "better" than another depending upon the programming requirements. For example, one programming language might be considered more powerful than another language because it may allow the programmer to "say" the same thing in fewer programming steps and/or to do more data manipulations. Furthermore, one language may be more efficient (faster) at handling large amounts of input data whereas another language might be better at handling a small amount of input data that requires much number crunching (i.e., data manipulation).

Some of the popular high-level programming computer languages available are:

1. BASIC (Beginners All-Purpose Symbolic Instruction Code)
2. COBOL (Common Business-Oriented Language)
3. FORTRAN (Formula Translator)
4. PL/I (Programming Language One)

8.3 AVAILABLE COMPUTER PROGRAMS

There are numerous computer programs available for use in solar system design. Some are general design programs which can be used in a solar system design analysis. For example, the National Bureau of Standards Heating and Cooling Load Determination Program (NBSLD) performs heat loss–heat gain building load calculations, while such computer programs as SOLCOST perform a complete solar system performance and economic analysis.

Computer programs are available for data processing equipment ranging in size from hand-held programmable calculators to large mainframe computer systems and can be obtained from numerous sources including trade magazines and government software (i.e., computer program) depositories. In a particular trade journal an article may appear describing a calculation procedure for computing the heat gain–heat loss of a building and may typically include a computer program that can be used in a small hand-held programmable calculator. One such program can be found in Appendix IV of this book.

Sometimes government publications may contain listings of computer programs. For example, the National Bureau of Standards (NBS) publication, "Life-Cycle Costing" describes procedures for determining and evaluating investments in solar energy and energy conservation projects and provides a source listing for a small computer program that performs life-cycle cost calculations. The program supplements the discussion on life-cycle costing, whereas another NBS publication entitled "NBSLD, The Computer Program for Heating and Cooling Loads in Buildings" both lists a computer program (written in the UNIVAC version of Fortran V) and explains in great detail the methods and procedures for using the program. In these two instances the cost of the computer program represents the costs of each publication (i.e., $2.75 and $4.60, respectively). Various branches of the Federal Government are sources for programs which may be purchased in magnetic tape or punched card form. The Computer Software Management and Information Center (COSMIC) is one such source. COSMIC is basically a software depository operated by the University of Georgia for the purpose of distributing (selling) computer programs supplied mainly by NASA, but which also has some programs supplied by the Department of Defense and other government agencies. NASA publishes an indexed quarterly journal called "Computer Program Abstracts" which describe recent NASA computer programs available to the public through COSMIC. This quarterly journal can be obtained from the Government Printing Office in Washington, D.C., for an annual subscription fee of $3.30. The Federal Software Exchange Center (FSEC), sponsored by the General Services Administration and the National Technical Information Service, represents another source for obtaining government supplied computer programs. The FSEC publishes a quarterly catalog listing available programs. This catalog can be obtained for a yearly subscription price of $75.00. Both the FSEC catalog subscription and the FSEC computer programs for sale can be purchased through the National Technical Information Service, 5285 Port Royal Road, Springfield, Virginia 22161.

Universities may also be a source for computer programs applicable to solar energy and energy related fields. The University of Wisconsin in Madison is one such example that offers for sale two popular computer programs: the Transient Simulation Program (TRNSYS) and the popular F-Chart computer program. Other universities may have energy related computer programs for sale either from student work, dissertation requirements, or from university-funded energy laboratories located on the campus. Finally, computer programs may be obtained through companies whose sole function is to write computer programs to sell or from companies in energy related fields who may develop a computer program incident to research being done at the time.

Some solar system and collector manufacturers may offer computer design analysis for solar installations incorporating their own collectors or systems. The manufacturer may or may not charge for the computer or technical services if their product is used in the solar design.

Table 8.1 lists some of the available solar computer programs as compiled by the National Solar Heating and Cooling Information Center. ASHRAE also publishes a useful bibliography of available computer programs in the general area of heating, refrigerating, air conditioning, and ventilating which have applications in the field of solar heating and cooling. In Table 8.1, the asterisk (*) is used to signify programs that are particularly appropriate for use by homeowners.

One program that is easily utilized by the homeowner or the design firm that does not want to bother with running their own solar energy computer program, is the U. S. Department of Energy's SOLCOST computer program. SOLCOST will perform both a space heating and service (domestic) hot water load analysis, a life-cycle cost analysis, and a solar collector and system

Table 8.1. Solar Computer Programs. (Source: National Solar Heating and Cooling Information Center).

PROGRAM/CONTACT	OUTPUT/APPLICATION	AVAILABILITY
ACCESS		
Edison Electric Institute 1119 19th St. Washington, D.C. 20002 (202) 862-3800	Output: monthly energy usage and demand levels for each meter in the building (the option of sample calculations for selected days, hours and zones is also available). Applications: SHW, SH, SC, PH, AS.	Service: computer. Cost: $10,000.
BASE TEMPERATURE		
Peter J. Lunde Solar Engineering Consultant 4 Daniel Lane West Simsburg, CT 06092	Output: heating load of active air or liquid heating system, economic analysis. Applications: SHW, SH, AS.	Service: hand held calculator. Materials: prerecorded magnetic program cards, instruction booklet (includes F-Chart); program operates on TI-59 hand held calculator. Cost: $120 to $200.
BLAST		
U. S. Army Construction Attn: Douglas C. Hittle P. O. Box 4005 Champaign, IL 61820 (217) 352-6511 (217) 352-6511	Output: hourly, daily and monthly heating, hot water and cooling loads and energy consumption. Applications: SHW, SH, SC, AS	Service: computer. Materials: source deck (cards), magnetic transmittal tape, user manual; program operates on CDC computer (time-sharing is available). Cost: $300.
CDA SOLAR PROGRAM		
Copper Development Association, Inc. Attn: Raymond A. Weisner 1011 High Ridge Rd. Stamford, CT 06904 (203) 322-7639	Output: hourly heat load. Applications: SHW, SH, AS.	Service: output data. Materials: printouts; program operates on remote terminal. Cost: no charge.
DEROB		
University of Texas Attn: Francisco Arumi 2604 Parkview Drive Austin, TX 78757	Output: daily, monthly and peak day load profiles. Applications: SH, SC, PS	Service: computer. Materials: program source code, user manual; program operates on UNIVAC 1108 computer. Cost: nominal price.
DOE-1		
University of California Lawrence Livermore Lab. Attn: F. C. Winkelmann P. O. Box 803 Livermore, CA 94550	Output: hourly heating and cooling loads, economic analysis Applications: SH, SC, AS.	Service: computer. Materials: source deck (cards), magnetic tape transmittal; program operates on IBM 370/370 computer. Cost: nominal price.
DOE-2		
Los Alamos Scientific Lab Attn: March A. Roschke Mail Stop 985 Los Alamos, NM 87545 (504) 667-3348	Output: heating, hot water and cooling loads and system and component performance, energy consumption, economic analysis Applications: SHW, SH, SC, AS.	Service: output data. Materials: magnetic transmittal tape, source deck (cards), user manuals. Cost: $400.
ECONOMIC ANALYSIS		
Daniel Enterprises, Inc. P. O. Box 2370 La Habra, CA 50631 (312) 943-8883	Output: economic analysis. Applications: SH, SC, AS.	Service: output data Materials: printouts, reference sheets. Cost: $12 (initial printout).
E CUBE III		
American Gas Association Attn: David S. Wood 1515 Wilson Blvd. Arlington, VA 22209 (203) 841-8400	Output: energy consumption of buildings. Applications: SH, AS.	Service: output data. Materials: printout, user manual. Cost: $143.92 (weekend turnaround system printout).

Table 8.1 (Continued)

EMPSS

Arthur D. Little
Attn: Dan Nathanson
20 Acron Park
Cambridge, MA 02140
(617) 864-5770

Output: hourly heating,
hot water and cooling loads,
economic analysis.
Applications: SHW, SH, SC,
AS

Service: computer.
Materials: magnetic transmittal tape, sample probler
worksheets; program
operates on IBM computer.
Cost: $500

ENERGY 1

Gibson-Yackey-Trindale
Associates
311 Fulton Ave.
Sacramento, CA 45811
(916) 483-4369

Output: heating and cooling
loads.
Applications: BL, HVAC, AS,
SH, SC.

Service: output data.
Materials: printouts;
program operates on
minicomputer.
Cost: $2,500.

EP

Energy Management Services
Attn: Robert M. Helm
0436 SW Iowa
Portland, OR 97201
(503) 244-3613

Output: heating and cooling
loads.
Applications: BL, HVAC, AS,
SH, SC.

Service: computer.

ESAS

Ross F. Meriwether &
Associates
1600 NE Loop 410
San Antonio, TX 78209
(512) 824-5302

Output: hour by hour annual
energy consumption, economic
analysis.
Applications: BL, HVAC, AS,
PS, SH, SC

Service: output data.
Materials: printouts,
user manual.
Cost: $350 to $400
(one set of runs).

F-CHART

Design Tool Manager
Market Development Branch
Solar Energy Research
1536 Cole Blvd.

Output: monthly heating and
hot water loads, solar
fraction, economic analysis.
Applications: SHW, SH, AS.

Service: hand held
calculator, computer.
Materials: magnetic transmittal tape, source deck
(cards), prerecorded mag-
netic program cards, user
manual; program operates on
CDC, IBM or UNIVAC computer;
hand held calculators can
also be used.
Cost: $100 (tape),
      $150 (cards),
      $150 (hand held
            calculator
            cards).

FREHEAT

Colorado State University
Mechanical Engineering Dept.
Attn: Dr. C. Byron Winn
Fort Collins CO 80523
(303) 491-6783

Output: detailed daily
passive energy flow
analysis and solar fraction.
Applications: SH, PS

Service: computer.
Materials: magnetic tape
transmittal; program
operates on CDC computer.

GLASIM

GK Associates
157 Stanton Ave.
Auburndale, MA 02166

Output: hourly, daily or
monthly temperature loads.
Applications: SH, PS

Service: computer.
Materials: magnetic tape
transmittal; program
operates on IBM 370/158
computer.

HISPER

National Aeronautics and
Space
Administration
Attn: Robert Elkin
George C. Marshall Space
Flight Center
Huntsville, AL 35812
(205) 453-1757

Output: heating and domestic
hot water loads.
Applications: SHW, SH, SC, AS

Service: computer.
Materials: source deck
(cards), user manual.
Cost: no charge, available
upon request.

NECAP

Computer Software Manage-
ment & Information Center
Attn: Stephen J. Horton
112 Barrow Hall
University of Georgia
Athens, GA 30602
(404) 452-3265

Output: heating loads
Applications: BL, HVAC, AS.

Service: computer.
Materials: 9 track card
image magnetic tape, user
manual; program operates on
CDC 6600 computer.
Cost: $970.
      $46.50 (documentation)

Table 8.1 (Continued)

NRGPBP*

Sun Spot Energy Consultants Attn: Dr. Brian Crissey/ Mr. Mark Chaddon P. O. Box 463 Bloomington, IL 61701 (309) 829-5195 or 828-0154	Output: economic analysis for alternative energy systems for the home and business. Applications: SH, AS.	Service: computer Materials: source deck (cards), user manual; program operates on IBM computer. Cost: $20/hour (computer time).

PASCALC

Total Environmental Action, Inc. 24 Church Hill Harrisville, NH (603) 827-3374	Output: monthly and annual energy consumption of passive buildings. Applications: PS	Service: hand held calculator. Materials: prerecorded magnetic program cards; program operates on TI-59 hand held calculator. Cost: $50.

PEGFIX

Princeton Energy Group 729 Alexander Road Princeton, NJ 08540 (609) 924-7639	Output: heat demand and excess heat, max. fan rate required to remove excess heat, max. hourly auxiliary load. Applications: SH, SC, PC.	Service: hand held calculator. Materials: prerecorded magnetic program cards, user worksheets; program versions can be made for HP-67, HP-97, or TI-59 hand held calculator. Cost: $75, combined with PEGFLOAT.

PEGFLOAT

Princeton Energy Group 729 Alexander Road Princeton, NJ 08540 (609) 924-7639	Output: hourly temperature of space and surface of storage mass and the maximum and minimum of both during 24-hr day. Applications: SH, SC, PS.	Service: hand held calculator. Materials: (see PEGFIX). Cost: $75, combined with PEGFIX.

RSVP

National Solar Heating & Cooling Information Center P. O. Box 1607 Rockville, MD 20850 (202) 223-8105	Output: heating, hot water and cooling loads, economic analysis. Applications: SHW, SH, AS, PS, BL.	Service: computer Materials: computer tape (containing copy of program and weather data), user manual; program operates on CDC and UNIVAC computers. Cost: $175.

SCOTCH PROGRAMS

Attn: Bob McClintock 8325 S.W. 72nd Avenue P. O. Box 430734 Miami, FL 33143 (305) 665-1251	Output: monthly heating load, solar fraction (based on f-chart). Applications: SH, CS, AS.	Service: hand held calculator. Materials: prerecorded magnetic program cards, instruction and diagram sheets; program operates on SR-52, TI-59 (with or with- out printer PC-100A), HP-67 and HP-97 hand held calculators. Cost: $220 (group)

SCOUT

Guard, Incorporated Attn: Robert Henninger 7440 N. Natchez Ave. Niles, IL 60648 (312) 647-9000	Output: heating and cooling loads. Applications: BL, HVAC, PS, HS, SC.	Service: computer. Materials: see contact.

SEE

The Singer Company Climate Control Division Attn: Philip Parkman 62 Columbus Street Auburn, NY 13201 (315) 253-2771, X391	Output: heating and cooling loads. Applications: BL, HVAC, AS, PS, SH, SC	Service: output data. Materials: printouts, worksheets, user manual. Cost: $180. (base price) x $474.60 (per system) x 1 to 4 (number of zones) x $0.80 fee).

Table 8.1 (Continued)

SEEC IV

Solar Environmental Engineering 2524 E. Vine Dr. Ft. Collins, CO 80524	Output: annual heating load, economic analysis Applications: SH, PS.	Service: hand held calculator. Materials: prerecorded magnetic program cards, user manual; program operates on HP-67, HP-97 or TI-59 hand held calculator. Cost: $125.

SESOP

Computer Software & Management & Information Center Attn: Stephen J. Horton 112 Barrow Hall University of Georgia Athens, GA 30602 (404) 542-3265	Output: energy consumption for buildings. Applications: BL, AS.	Service: computer. Materials: 9 track UNIVAC FURPUR formatted tape, manual; program operates on UNIVAC 1100 Series computer. Cost: $530 $10.50 (documentation)

SHASP

University of Maryland Dept. Mechanical Engineering Attn: Dr. D. K. Amand College Park, MD 20742 (301) 454-2411	Output: heating and cooling loads. Applications: SHW, SH, SC, AS.	Service: computer. Materials: program source code, user manual; program operates on UNIVAC computer. Cost: no charge, available upon request.

SIMSHAC

Colorado State University Mechanical Engineering Attn: Dr. C. Byron Winn Fort Collins, CO 80523 (303) 491-6783	Output: daily, monthly, and annual heating and hot water loads, energy consumption, economic analysis. Applications: SHW, SH, AS	Service: computer

SOLAR AVAILABILITY PROGRAM

Daniel Enterprises, Inc. P.O. Box 2370 La Habra, CA 90631 (213) 943-8883	Output: daily and monthly heating and hot water loads and system performance. Applications: SHW, SH, AS, CS, ID.	Service: output data. Materials: printouts, explanation sheets. Cost: $10 (see contact for additional costs).

SOLAR ENERGY SOFTWARE*

Sunshine Power Company Attn: Gary Shramek 1018 Lancer Drive San Jose, CA 95129 (408) 446-2446	Output: sun position, solar radiation levels, system performance. Applications: ID, SP, CS.	Service: hand held calculator. Materials: preprogrammed cards, instruction sheets, description sheets; programs operate on HP-97 and TI-59 hand held calculators. Cost: $30 to $60.

SOLARCON PROGRAM GROUP

Solarcon, Incorporated Attn: Roderick W. Groff Ann Arbor, MI 48103 (313) 769-6588	Output: systems performance. Applications: ID, CS, WD.	Service: hand held calculator. Materials: prerecorded magnetic program cards, instruction sheets; programs operate on SR-52 and TI-59 (with PC-100A printer) HP-67 and HP-97 (with printer). Cost: $85 to $400.

SOLATECH*

Scott Getty 10970 Wayzata Blvd. Minneapolis, MN 55401	Output: systems efficiencies with or without storages. Applications: AS	Service: output data. Materials: printouts. Cost: $10 to $25

SOLCOST*

International Business Services, Inc. 1010 Vermont Ave., NW Suite 1010 Washington, D.C. 20005 or Martin Marietta Aerospace Attn: Rojer Giellis Mail Stop 50484 P.O. Box 179 Denver, CO 80201 (303) 973-3853	Output: heating, hot water and cooling loads, systems performance and economic analysis. Applications: SHW, SH, AS.	Service: output data, computer. Materials: program source sheet, user manual; program operates on CDC, IBM, UNIVAC computers (time sharing is available). Cost: $300

Table 8.1 (Continued)

SOLMET

National Climatic Center Federal Bldg. Ashesville, NC 28801 (704) 258-2850, X319	Output: insolation and weather data for 27 U. S. stations. Applications: ID, WD.	Service: output data. Materials: monthly publi- cation, weather data sheet. Cost: $8.80 (yearly subscription).

SOLOPT

Texas A&M University Dept. Agriculture Attn: Larry O. Degelman College Station, TX 77843 (713) 845-1015	Output: daily, monthly, and annual energy consumption, economic analysis, system performance. Applications: SHW, SH, AS, CS, HVAC.	Service: computer. Materials: program source sheet; program operates on AMDAHL computer. Cost: $20.

SOLSYS

Argonne National Laboratory Argonne Code Center Building 221 9700 S. Cass Ave. Argonne, IL 60439 (312) 972-7250	Output: system and component performance. Applications: AS.	Service: computer. Materials: magnetic tape transmittal, source deck (cards), sample problem worksheets; program operates on CDC 6600. Cost: No fee for DOE contractors.

SOLTES

Sandia Laboratory Attn: Norman Grandjean Kirtland East Division East Albuquerque, NM 87185 (505) 264-8819	Output: daily, monthly and annual heating loads, system performance. Applications: PH, AS, SH, SC.	Service: computer. Materials: program source sheet, user manual.

ST-33 and ST-34*

Solarcon, Inc. Attn: Roderick W. Groff 607 Church St. Ann Arbor, MI 48103	Output: hourly heating loads with various vents and walls. Applications: SH, PS.	Service: hand held calculator Materials: prerecorded mag- netic program cards; program operates on TI-59 hand held calculator with printer. Cost: $135/each.

SUN

Berkeley Solar Groups Attn: Bruce Wilcox 1215 Francisco St. Berkeley, CA 94703 (415) 843-7600	Output: daily, monthly and annual heating load. Applications: SH, PS.	Service: output data. Materials: program source code; programs operate on CDC 6400 computer. Cost: $50 plus $25 hourly.

SUNCAT

National Center for Appropriate Technology Attn: Larry Palmiter or Terry Wheeling P.O. Box 3838 Butte, MT 59701 (406) 723-5475	Output: heating loads in buildings, solar fraction. Applications: BL, SH, PS.	Service: computer. Materials: program source code, user manual; program operates on Data General Eclipse. Cost: nominal charge.

SUNDEX*

Sun Spot Energy Consultants Attn: Dr. Brian Crissey or Mark Chaddon P.O. Box 463 Bloomington, IL 61701 (309) 829-5195 or 828-0154	Output: heating loads, economic analysis. Applications: SH, AS.	Service: computer. Materials: source deck (cards), user manual; program operates on IBM computer. Cost: $20/hour.

SUNSPOT*

Sun Spot Energy Consultants Attn: Dr. Brian Crissey or Mark Chaddon P. O. Box 463 Bloomington, IL 61701 (309) 829-0154	Output: heating loads and system performance. Applications: SH, AS.	Service: computer. Materials: source deck (cards), user manual; program operates on IBM computer. Cost: $20/hour.

Table 8.1 (Continued)

SUN-PULSE II

Sun-Pulse II Joint Venture
138 Mt. Auburn St.
Cambridge, MA 02138
(617) 491-0961

Output: daily and monthly heating loads.
Applications: SH, PS.

Service: hand held calculator.
Materials: prerecorded magnetic program cards; program operates on TI-59 hand held calculator, user manual, weather tables (based on TMY-SOLMET).
Cost: $100.

SYMSYM

Sunworks Computer Service
Attn: Phillip Fine
P.O. Box 1004
New Haven, CT 06508
(203) 934-6301

Output: systems performance.
Applications: SHW, SH, SC, AS, CS.

Service: output data.
Materials: printouts, user manual; program operates on IBM computer.
Cost: $25 (monthly storage cost).

SYRSOL

Syracuse University
Dept. Mechanical Engineering
Attn: Dr. Manas Ucar
Syracuse, NY 13210
(315) 423-3038

Output: daily, monthly, and annual heating loads and energy consumptions, economic analysis.
Applications: SHW, SH, SC, PS.

Service: computer.

TEANET

Total Environmental, Inc.
Church Hill
Harrisville, NH 03450
(603) 827-3374

Output: hourly heating load in building with Trombe wall.
Applications: SH, PS.

Service: hand held calculator.
Materials: prerecorded magnetic program cards; program operates on TI-59 hand held calculator with PC-100A printer.
Cost: $95.

THERMAL SYSTEMS DYNAMICS

GK Associates
157 Stanton Ave.
Auburndale, MA

Output: systems performance.
Applications: PS

Service: computer.
Materials: source deck (cards), tape or disc; program operates on IBM 360/75, IBM 370/158 computer.

TRACE

The Trane Company
Attn: Neil R. Patterson
3600 Pammel Creek Rd.
La Crosse, WI 54601
(608) 782-8000

Output: heating, hot water and cooling loads, monthly fuel and electrical consumption, economic analysis.
Applications: SHW, SH, SC, PH, AS, PS.

Service: computer.
Materials: program will be available on a time-sharing basis as of the fall of 1979; program operates on an IBM 370/168 system.
Cost: $500 to $3,000 (program is not presently available on a purchase or lease arrangement).

TRNSYS

University of Wisconsin
Solar Energy Lab.
Attn: John C. Mitchell
1500 Johnson Drive
Madison, WI 53706
(608) 263-1589

Output: hourly heating, hot water and cooling loads and system and component performance.
Applications: SHW, SH, SC, AS, PS.

Service: hand held calculator computer.
Materials: prerecorded magnetic program cards, source deck (cards), user manual, program operates on CDC, IBM, UNIVAC computers or uses hand held calculator, time-sharing is available.

TWO-ZONE

University of California
Attn: Arthur H. Rosenfeld
Lawrence Berkeley Lab.
Bldg. 90
Berkeley, CA 94720
(415) 843-2740, X 5711
or 486-4000

Output: monthly heating and cooling loads for the residence as well as breakdown by contribution (windows, walls, infiltrations, etc.)
Applications: SH, SC, PS.

Service: computer
Materials: see contact.

Table 8.1 (Continued)

UWENSOL

University of Washington   Output: heating and cooling   Service: computer.
Dept. Mechanical   loads.   Materials: 7 track tape,
Engineering, FU-10   Applications: SH, PS.   source deck (cards), user
Attn: Dr. A. F. Emery   manual; program operates
Seattle, WA 58195   on CDC 6000 computer.
(206) 683-5338   Cost: $200 (tape),
                                                                             $250 (cards),
                                                                             $20 (user manual).

APPLICATIONS: AS - ACTIVE SOLAR SYSTEM   PH - PROCESS HEAT (INDUSTRIAL)
(abbreviation code)
                           BL - BUILDING LOAD   SC - SPACE COOLING

                           CS - COLLECTOR SIZING   SG - SOLAR GREENHOUSE

                         HVAC - HEATING, VENTILATING AND   SH - SPACE HEATING
                                AIR CONDITIONING SYSTEM
                                                          SHW - SERVICE OR DOMESTIC
                           ID - INSOLATION DATA   HOT WATER

                           PS - PASSIVE SOLAR SYSTEM   SP - SUN POSITION

Most of the programs listed above are for a technical audience (i.e. research engineers,
building professionals, commercial businesses, public and private institutions). The
symbol (*) indicates when the program can be utilized by homeowners.

Computer programs for specific collectors are not included in this list. Check with
the solar manufacturers for computer programs designed for their own collectors.

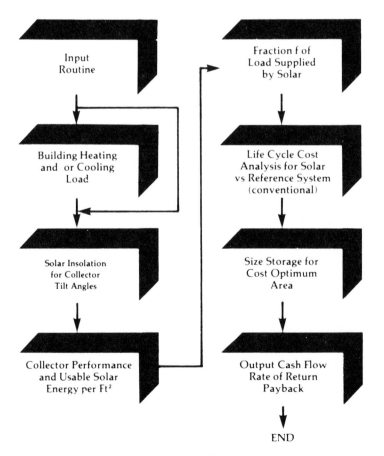

Figure 8.2. Flow Chart for the Solcost Analysis. (Source: Energy Research and Development Administration[11]).

performance analysis. Figure 8.2 shows a flow chart for the SOLCOST analysis.

Basically all the individual has to do is fill out a form, shown in Table 8.2, and mail it in with the required remittance. After the program has been run, the results are returned to the individual. This type of service is great for the homeowner and design firm that does not do much solar system analysis. For the design firm that is involved in analyzing a number of solar systems for clients, the firm should determine the economics of buying the SOLCOST computer program. Table 8.3 shows the data input for a hypothetical case using the SOLCOST program and Table 8.4 shows the output that would be generated by the program based on the input from Table 8.3.

To summarize, there are three basic ways an individual can use the SOL-COST program:

1. The individual can purchase the complete program for $200 from SOL-COST Service Center, 2524 East Vine Drive, Ft. Collins, Colorado 80524.
2. The individual can fill out a form such as shown in Table 8.2 and mail it, with the required remittance, for processing. (The basic SOLCOST analysis presently costs $35.00.)
3. An individual having access to a data terminal and acoustical coupler can use the SOLCOST program in a timeshare mode on National Timeshar-ing Networks. For more information on the SOLCOST timesharing service, the individual should contact:

   a. *Boeing Computer Services Company (BCS)*
      Dr. Kenneth L. Johnson, P.E.
      Boeing Computer Services Company
      565 Andover Park West
      Mail Stop 9C-02
      Tuckwila, Washington 98188 (206) 575-7047

   b. *Control Data Corporation (CYBERNET)*
      Stephen A. Lewis
      Control Data Corporation
      8100 34th Avenue South
      Mailing Address/Box 0
      Minneapolis, MN 55440 (612) 853-6313

   c. *General Electric Information Services Company (GEIS)*
      John D. McCloskey
      General Electric Information Services Company
      600 South Cherry Street, Suite 225
      Denver, CO 80222 (303) 320-3186

   d. *United Computing Systems, Inc. (UCS)*
      Laura Huffman
      United Computing Systems, Inc.
      2525 Washington Street
      Kansas City, MO 64108 (816) 221-9700

Table 8.2. Solcost Data Processing Form. (Source: Solcost Service Center).

SOLCOST
Space Heating and Service Hot Water Form

- Refer to the SOLCOST Space Heating and Service Hot Water Handbook for instructions.
- If you do not have all the information necessary for completing this form, consult your contractor, engineer, solar systems manufacturer, supplier, or utility company for assistance.
- The completed form should be mailed to:

  SOLCOST Service Center
  P.O. Box 1914
  Fort Collins, Colorado 80522   (303) 221-4370

- Service charges and SOLCOST sales information may be obtained from the SOLCOST Service Center listed above.
- Items on this form marked * have values that will automatically be used if the user does not make an entry. Refer to Default Table on reverse side of this form for default values.

** Important — Refer to Handbook for Guidance **

Part 1 — Solar System Sizing and Cost Comparison
A. SOLCOST Analysis Description
1. This SOLCOST analysis is for a _____ owner occupied residential building
   _____ business related, rental building, or light commercial building
   _____ nonprofit organization owner (i.e., public buildings, schools, etc.)
2. SOLCOST determines the optimum solar collector area which maximizes the net present worth of the savings from the solar investment
   Do you want SOLCOST to optimize your collector size? _____ yes _____ no
3. If the previous answer is *no*, you must enter collector area in square feet _____
4. SOLCOST also determines the size for heat exchangers, pumps or blowers, and pipes or ducting.

B. Solar and Conventional System Description
1. Building location _____
   City/State
2. Building type _____ one or two-family residential _____ multifamily residential
   _____ commercial
3. Application _____ retrofit   _____ space heating
   _____ new construction   _____ service hot water
4. Energy sources for conventional space heating and/or service hot water systems which would be installed if solar were not feasible:
   Space Heating/Service Hot Water
   _____ natural gas _____ electricity _____ fuel oil _____ coal _____ propane
5. Auxiliary fuels for space heating and/or service hot water systems (fuel type to be used in addition to solar):
   Space Heating/Service Hot Water
   _____ natural gas _____ electricity _____ fuel oil _____ coal _____ propane
6. If more than one fuel type is to be considered, enter a (2) above for the alternate fuel. A complete SOLCOST analysis will be made for each fuel. Remember to enter separate cost schedules in Subsection I for each fuel designated.

C. Solar Collector Subsystem Description

1. Collector Orientation

a) Azimuth angle _____ degrees
(0 = South; East = (+) degrees, West = (−) degrees)
The azimuth angle represents the direction the solar collector faces, usually due South.

b) Tilt angle _____ degrees
(0 is horizontal, 90 is vertical)

2. Collector Fluid _____ air _____ liquid

3. Collector Type _____ flat plate _____ evacuated tubular

4. This section must be completed with data obtained from the solar systems manufacturer, supplier, or contractor under consideration. Fill in one of the following:

a) FRPRIME-UL product _____ Btu/Hr-ft.^{2}-F° and FRPRIME Tau Alpha product _____

b) Efficienct at $(T_{in} - T_{amb})/1 = 0.0$; _____ (efficiency) and
Efficiency at $(T_{in} - T_{amb})/1 = 0.5$; _____ (efficiency)

c) For evacuated tubular collectors, attach efficiency data vs. hour angle.

d) Manufacturer _____
Address_____
Model Number _____

5. Expected life of solar collector* _____ years.

D. Solar & Conventional System Configuration

The following information is required regardless of application.

LIQUID

1. Solar Storage Tank Size* _____ gallons/sq. ft. of collector

2. Solar Space Heating and/or Service Hot Water Piping Description

Length of pipe required between collector & solar storage tank _____ (ft., total length of supply & return)

3. Freeze Protection (check one)

_____ Anti-Freeze
_____ Drain-Down
_____ None (Tropical Climates only)

4. Heat Exchanger (between collector & storage tank) _____yes _____ no

AIR

1. Solar Storage (Rock Bed)* _____ lbs./sq. ft. of collector

2. Space Heating and/or Service Hot Water Ducting Description

Length of ducting required between collector and solar storage (rock bed) _____ (ft., total length of supply & return)

E. Optional Sizing Information

Liquid Systems

1. Do you want costs computed for the pump, pipe and heat exchanger?
   _____ yes _____ no
   Do you want the costs added into the system costs? _____ yes _____ no

2. Collector flow rate _____ gallons/minute per sq. ft. collector

3. Specific heat of collector fluid _____ Btu/lb/°F

4. Specific heat of storage fluid _____ Btu/lb./°F

5. Heat transfer coefficient of heat exchanger _____ Btu/°F per sq. ft. HX

6. Heat exchanger pressure loss _____ ft. of water

7. Collector pressure drop _____ ft. of water

8. No. valves _____ at $ _____ /ea.

9. No. elbows _____ at $ _____ /ea.

10. No. tees _____ at $ _____ /ea.

11. Additional velocity head loss _____ ft. of water

12. Additional pressure head loss _____ ft. of water

13. Additional elevation head loss _____ ft. of water

14. Pipe cost $ _____ / ft.

15. Costs

Heat Exchanger		Pump		
Area, Sq. Ft.	$	Head, Ft. of water	GPM	$

Table 8.2 (Continued)

Name _____

Address _____

_____

_____ Zip_____

Phone No. _____ Date _____

Contractor/Estimator/Designer Name_____

_____

Phone No. _____

16. Load heat exchanger (water to air cross flow)
    Effectiveness _____ %   Blower flow rate _____ ft³/min.
    Costs: $_____, _____, _____, _____ UA Btu/°F/hr.

Air Systems

1. Do you want cost computed for the blower,
   duct and service hot water heat exchanger?
   _____ yes _____ no
   Do you want the costs added into the
   system costs? _____ yes _____ no

2. Collector flow rate _____ ft³/min. per ft²

3. Total duct length _____ ft.

4. Collector pressure drop _____ ft.

5. Storage pressure drop _____ ft.

6. Service hot water heat exchanger (If present)
   a) pressure drop _____ ft.
   b) effectiveness _____ %
   c) liquid side flow rate _____ gallons/min.

7. Costs

Service Hot Water Heat Exchanger		Fan		
Area, Sq. Ft.	$	Head, Ft. of water	CFM	$

Duct costs $ _____ / ft.

F. Reference System Cost

Difference between conventional and solar system auxiliary initial installed cost* $ _____
Annual maintenance, expressed as percent of initial installed cost* _____ %
(Do not use for retrofit situations)

G. Solar System Cost

1. Size dependent initial costs (enter installed costs)
   a) Collector $ _____ / sq. ft.
   b) Storage
      tank (liquid systems); or $ _____ / gal.
      rock bed (air systems) $ _____ / ton.

2. Fixed Initial Costs
   a) Controls $ _____

   b) Design/engineering $ _____
   c) Building modification $ _____
   d) Installation labor $ _____
   e) Other $ _____
      Total Fixed Cost $ _____

3. Annual maintenance cost for the solar system, as a percent of the initial cost*   _____ %

Table 8.2 (Continued)

H. Finance and Tax Data

Period of economic analysis (usually life of system) _____ years

1. **Residential**
   a) Loan interest rate* _____ %
   b) Loan term* _____ years
   c) Loan down payment* _____ %
   d) Property insurance rate* _____ %
   e) Net property tax rate _____ %
      (percent of actual value)
   f) Personal income tax rate _____ %
      (State and federal)
   g) Inflation rate for maintenance, _____ %
      insurance and property tax

2. **Commercial**
   a) Loan interest rate* _____ %
   b) Loan term* _____ years
   c) Loan down payment* _____ %
   d) Property insurance rate* _____ %
   e) Net property tax rate _____ %
   f) Corporate tax rate or owner _____ %
      income tax rate
   g) Inflation for maintenance, _____ %
      insurance and property tax
   h) Investment income tax credit _____ %
      for equipment in first year
   i) Additional first year* _____ %
      depreciation

3. **If business application (commercial or residential rental, for example) fill out the following:**
   a) Depreciation method (options are straight line or declining balance)
      ___ straight line ___ declining balance
   b) Multiplier used in declining balance depreciation* _____
      (limited to 1.25 for commercial buildings and 2.0 for new residential property)
   c) System useful life for depreciation purposes* _____ years
      (currently 10 years is allowed for building heating, air conditioning, and service hot water systems, except storage tanks, which are allowed 22 years)
   d) Salvage value of the solar system at the end of its useful life* _____ % of total cost

4. **Tax Incentives**
   a) State income tax credit.
      % for the first $ _____ (installed cost)
      % for the next $ _____ (installed cost)
   b) Property tax rate
      _____ % for the first _____ years
      _____ % for the next _____ years
   c) If a federal income tax credit is enacted, attach an explanation of the differences between federal and state.

I. Fuel and Electricity Cost and Price Escalation Data

There are no default values for fuel costs schedules.
Enter current cost for fuel, including local sales tax and fuel cost adjustment factors.

1. **Natural Gas Cost Schedule**
   Escalation rate* _____
   Btu content per cubic foot* _____

Fuel Cost ($/100 cubic feet)	Quantity
Step 1 _____	for first 100 cubic feet
Step 2 _____	for next 100 cubic feet
Step 3 _____	for next 100 cubic feet
Step 4 _____	for next 100 cubic feet
Step 5 _____	for next 100 cubic feet

   Furnace or hot water heater efficiency* _____ %

2. **Electricity Cost Schedule**
   Escalation rate* _____

Fuel Cost ($/kwh)	Quantity
Step 1 _____	for first _____ kwh's
Step 2 _____	for next _____ kwh's
Step 3 _____	for next _____ kwh's
Step 4 _____	for next _____ kwh's
Step 5 _____	for next _____ kwh's
Step 6 _____	for next _____ kwh's
Step 7 _____	for next _____ kwh's

   Furnace or hot water heater efficiency* _____ %

4. **Propane Cost Schedule**
   Escalation rate* _____

Fuel Cost ($/gal.)	Quantity
Step 1 _____	for first _____ gallons
Step 2 _____	for next _____ gallons
Step 3 _____	for next _____ gallons
Step 4 _____	for next _____ gallons
Step 5 _____	for next _____ gallons

   Furnace or hot water heater efficiency* _____ %

5. **Coal Cost Schedule**
   Escalation rate* _____
   BTU content/ton _____

Fuel Cost ($/ton)	Quantity
Step 1 _____	for first _____ tons
Step 2 _____	for next _____ tons
Step 3 _____	for next _____ tons
Step 4 _____	for next _____ tons
Step 5 _____	for next _____ tons

   Furnace or hot water heater efficiency* _____ %

6. **If summer rates apply, attach adjustment factor, and applicable months.**

Table 8.2 (Continued)

3. **Fuel Oil Cost Schedule**

Escalation rate* _____ Fuel oil grade* _____

Fuel Cost ($/gal.)	Quantity
Step 1 _____	for first _____ gallons
Step 2 _____	for next _____ gallons
Step 3 _____	for next _____ gallons
Step 4 _____	for next _____ gallons
Step 5 _____	for next _____ gallons

Furnace or hot water heater
efficiency*   _____ %

J. Hot Water Loads

Fill in *one* of the following:

1. Btus per day required for hot water _____. (Include losses for circulation loop, if applicable)

2. Gallons of hot water used per day _____ (gallons/day)

   Water main temperature_____ (°F)
   Hot water set temperature _____ (°F)

3. Estimated consumption — using fuel bills for summer water heating

   Hot water delivery temperature (if known) _____ (°F)

   Quantity and type of fuel consumed _____ (gallons, 100 ft., etc.)

   Total fuel bill for period of consumption $ _____
   Number of days in consumption period _____ (period when furnace was turned on.)

4. Monthly fuel bill for hot water only for the past 12 months

   $ _____ $ _____ $ _____ $ _____
   $ _____ $ _____ $ _____ $ _____
   $ _____ $ _____ $ _____ $ _____

   Fuel type _____

   Fuel price per unit (actual) $_____

K. Space Heating Loads (Complete *one* of the following)

- **Heat Loss Method**
    1. Building Heat Loss Coefficient _____ Btu/sq. ft. per degree day (°F)
    2. Living Space Floor Area _____ sq. ft.
  or 3. Design Heat Loss Value _____ Btu/hr.
       Indoor Design Temperature _____ °F
       Outdoor Design Temperature _____ °F
    4. Additional Information _____
       _____
       _____

- **ASHRAE Standard 90—75 (applies only to new construction, residential, commercial)**
    1. Building type   _____ One or two-family residence
                           _____ Residential, 3 stories or fewer
                           _____ Light commercial (single zone)
    2. Wall area   _____ Sq. ft.
    3. Ceiling area   _____ Sq. ft.
    4. Window perimeter   _____ Ft.
    5. Sliding glass door area _____ Sq. ft.
    6. Door area   _____ Sq. ft.
    7. Door perimeter   _____ Ft.
    8. Floor area   _____ Sq. ft.
    9. Slab perimeter   _____ Ft.

Table 8.2 (Continued)

- **Fuel Usage Method** (Retrofit only)
  1. Space heating fuel type _____
  2. Quantity of fuel consumed during heating season _____
     (Do not include hot water heating fuel use)
  3. Degree days in heating season _____ (available from local weather stations or fuel suppliers)
  4. Heated floor space _____ Sq. ft.

Part 2 — Energy Conservation/Heat Loads Calculation (Optional)

Space Heating Loads Calculation (Extra cost option)

- **Detailed Building Description**
  1. Exterior Wall Area _____ Sq. ft. (does not include doors or windows)
  2. Wall insulation R—Value _____
     or wall insulation material _____ Thickness _____ Inches
     or wall construction description _____
  3. Window & Sliding Glass Door Area (Fill in as many as are appropriate)
     a) Single pane _____ sq. ft.
     b) Double pane _____ sq. ft.
     c) Triple pane _____ sq. ft.
     d) Single pane w/storm windows _____ sq. ft.
     e) Sash material (check *one*)

     _____ Wood   _____ Steel
     _____ Aluminum   _____ Other
  4. Exterior Door Areas
     a) Number of square ft. _____
     b) Storm doors ____ Yes ____ No If *yes*, area _____ Sq. ft.
  5. Ceiling & Attic Description
     a) Ceiling area _____ Sq. ft.
     b) Ceiling insulation thickness _____ Inches   or R—Value _____
     c) Cathedral ceiling ____ Yes ____ No   If *yes*, area _____ Sq. ft.
        Insulation type_____ Thickness_____ Inches
     d) Attic ____ Yes ____ No
        If *no*, enter roof description as follows:
        Roof area _____ Sq. ft. (Enter only if different from ceiling area)
        Insulation type _____ Thickness _____ Inches
        Roof construction details _____
  6. Floor Description
     a) Floor over unheated basement area _____ Sq. ft.   Carpet ____ Yes ____ No
        Insulation ____ Yes ____ No If *yes*, thickness _____ Inches
     b) Floor over crawl space area _____ Sq. ft.   Carpet ____ Yes ____ No
        Insulation ____ Yes ____ No If *yes*, thickness _____ Inches
     c) Floor on slab area __ _____ Sq. ft.   Slab perimeter ____ Ft.
        Perimeter insulation ____ Yes ____ No   Vapor barrier ____ Yes ____ No
        If *yes*, thickness _____ Inches or R-Value _____
     d) Heated basement floor area _____ Sq. ft.
        Buried wall material _____ (specify)   Area _____ Sq. ft.
  7. Infiltration
     _____ Air changes/hr. for _____ cu. ft. of heated living space
  8. Thermostat settings   Daytime ____ (°F)   Nighttime ____ (°F)
  9. Solar gain components

	W	SW	S	SE	E
a) Glass area					
b) Window type (1, 2, 3 pane)					

Table 8.2 (Continued)

Default Values

Whenever data in this form is asterisked (*) users may either 1) accept the data already in SOLCOST, i.e., the Default Value, by not entering any information, or 2) enter another value of their choosing. Listed below, by section and line number, are the SOLCOST Default Values.

Section C — Solar Collection Subsystems Description

1). a) Azimuth angle 0 (due south) ⎫ Optimum collector size & cash
   b) Tilt angle (lat. +7.5 degrees,/lat. +15 degrees) ⎭ flow sheet provided for both.

4. Expected life for solar collector — 20 years

Solar System Types and Default Values

Solar System Type	Solar System Efficiency*	Collector Inlet Temperature (°F)
Space heating and service hot water heating with liquid collectors, collector/storage HX, fan coils or air duct HX	.95	115.
Space heating and service hot water heating with liquid collectors, fan coils or air duct HX, (no collector/storage HX)	.95	105.
Space heating and service hot water heating with air collectors and rock bed storage	.95	70.

*Efficiency is defined as the ratio of energy delivered to the load divided by the energy delivered to the storage tank from the collector. Does not include the efficiency of solar collectors.

Collector Types and Efficiency Data

Collector Type	Efficiency	Efficiency Intercept	Efficiency at $\Delta T/q = .5$
Liquid, flat plate, 1 cover, paint	Defaulted	.82	.24
Liquid, flat plate, 2 covers, paint	Defaulted	.67	.30
Liquid, flat plate, 1 cover, selective	Defaulted	.70	.30
Liquid, flat plate, 2 covers, selective	Defaulted	.67	.50
Liquid, concentrating	Defaulted	.71	.65
Liquid, tubular array, selective	Defaulted	.62	.49
Air, flat plate, 1 cover, paint	Defaulted	.60	.17
Air, flat plate, 2 covers, paint	Defaulted	.52	.21
Liquid, trickle type, 1 cover	Defaulted	.72	.01

Section D — Solar & Conventional System Configuration

Liquid

1. Solar storage tank size = 1.5 gallons of water/square foot of collector.

Air

1. Solar storage (rock bed) = 50 lbs. of rock/square foot of collector

Section F — Reference System Cost

1. Initial installed cost $0.00 (assumes retrofit, with cost already absorbed)
2. Maintenance cost .01 (1% of initial installed cost per year)

Section G — Solar System Cost

1. Maintenance cost .01 (1% of initial installed cost per year)

Table 8.2 (Continued)

Section H — Finance and Tax Data

1. Residential	New Construction	Retrofit
a) Loan interest rate	.09 (9%)	.11 (11%)
b) Loan term	20 years	7 years
c) Loan down payment	.10 (10% of total loan)	.20 (20% of total loan)
d) Property insurance rate	.005 (.5% per year)	.005 (.5% per year)
g) Inflation rate	0.0	0.0

2. Commercial	New Construction	Retrofit
a) Loan interest rate	.09 (9%)	.11 (11%)
b) Loan term	20 years	7 years
c) Loan down payment	.10 (10% of total loan)	.20 (20% of total loan)
d) Property insurance rate	.005 (.5% per year)	.005 (.5% per year)
g) Inflation rate	0.0	0.0
h) Investment tax credit	.10 (10%)	.10 (10%)
i) Additional first year depreciation	0.0	0.0

3. Business only
   a) Depreciation method is straight line
   b) Multiplier used in declining balance 1.25
   c) System useful life for depreciation purposes 10 years
   d) Salvage value of system at end of useful life .10 (10% of total cost)

Section I — Fuel and Electricity Cost and Price Escalation Data

1. Natural Gas — Heat content per unit 100,000 Btus/100 cubic feet

3. Fuel Oil   — Heat content per unit 139,600 Btus/gallon (no. 2 fuel oil)

4. Propane   — Heat content per unit 91,500 Btus/gallon

5. Coal   — Heat content per unit 30×10^6 Btus/ton

6. Average Annual Escalation Rate

The following estimated rates, through the
year 2000, are provided from DOE research
data.

Natural Gas	8%
Electricity	6%
Fuel Oil	7%
Propane	8%
Coal	6%

Combustion/Usage Efficiency for Fuel Types
for Reference or Auxiliary Systems

Fuel/Utility	Space Heating Efficiency	Service Hot Water Heating Efficiency*
Natural Gas	.70	.63
Electricity	1.00	.90
Fuel Oil	.65	.59
LP Gas	.70	.63
Coal	.65	.59

Table 8.3. Solcost Input for Boston Residents Example. (Source: Energy Research & Development Administration[11]).

```
Name    Sam Sunray         Phone Number  (617) 222 5858
Address 1000 Sunshine Blvd. Date  12 Apr 77
        Boston             Contractor/Estimator/Designer Name
        MA       Zip  02151   J. Smith, Hot Water, Inc.
                            Phone Number  (617) 220 6111
```

A. SOLCOST Analysis Description

1. This SOLCOST analysis is for a
   __√__ owner occupied residential building
   _____ business related, rental building or commercial building
   _____ non-profit organization owner (i.e., public buildings,
          schools, etc.)

2. SOLCOST determines the optimum solar collector area which
   maximizes the rate of return (or present worth) of the
   solar investments. Do you want SOLCOST to optimize your
   collector size? __√_ yes _____no

3. If the previous answer is no, you must enter collector
   area in square feet _____.

4. SOLCOST also determines the optimum size for heat exchangers,
   pumps or blowers, and pipes or ducting.

B. Solar and Conventional System Description

1. Building location _____Boston, MA_____
                              city,state

2. Building type __√__ one or two family residential
                 _____ multi-family residential
                 _____ commercial

3. Application __√_ retrofit
               _____ new construction

4. Energy source for a conventional service hot water system
   which would be installed, if solar were not feasible:

   ___ natural gas   ___ fuel oil   ___ propane
   _√_ electricity   ___ coal

5. Auxiliary fuel for service hot water system (fuel type to
   be used in addition to solar).

   ___ natural gas   ___ fuel oil   ___ propane
   ___ electricity   ___ coal

6. If more than one fuel type is to be considered, enter
   a (2) for the alternate fuel. A complete SOLCOST analysis
   will be made for each fuel.

C. Solar Collector Subsystem Description

1. Collector Orientation

a) Azimuth Angle* ____ degrees.
   (0 is South; East = (+)
   degrees, West = (-) degrees)
   The azimuth angle represents
   the direction the solar collector
   faces, usually due South.

b) Tilt Angle* ____ degrees
   (0 is horizontal, 90 is
   vertical)

2. Collector Type

   ___ air   _√_ liquid

3. Collector Efficiency

   This section must be completed with data obtained from the
   solar systems manufacturer, supplier, or contractor under
   consideration. Fill in one of the following:

   ·FRPRIME-UL product_____Btu/Hr-Ft^2-°F and FRPRIME
    Tau-Alpha product _____
   or ·Efficiency at $(T_{in}-T_{amb})/I=0.0$; _.67_ (efficiency) and
      Efficiency at $(T_{in}-T_{amb})/I=0.5$; _.30_ (efficiency)
   or ·Manufacturer_____/_____
      Address_____
      Model Number_____

4. Expected life of solar collector* _20_ years.

Table 8.3 (Continued)

D. Solar and Conventional System Configuration

In the Handbook are three system configurations that are
commonly used in residential and commercial solar hot water
applications.
SOLCOST uses these systems as a model when calculating
optimum sizes and costs.
   For one and two family residential applications,
   calculations will be based on the configuration A.
   For multi-family and commercial applications,
   SOLCOST will automatically select configuration B
   or C, depending on optimal perforamcne and cost.

The following information is required regardless of
application.

1. Solar storage tank

What is the size of the hot water storage tank, if
known _____(gallons).
(If solar storage tank size is not know, SOLCOST will
provide a reasonable size estimate.)

2. Solar hot water piping (ducting) description
Length of pipe (or ducting) required between collector
and solar storage tank __75__ (feet, total length of supply
and return).

3. Solar hot water freeze protection (liquid collectors
only - check one)

Anti-freeze___ Drain-down_√_ None___(Tropical climate only)

4. Solar System Heat Exchanger?

E. Hot Water Loads

1. Residential Application
·   Fill in one of the following:

      ·Btu's per day required for hot water ____
   or ·Gallons of hot water used per day __80__
                                    (gallons/day)
       Number of occupants __4__
       Hot Water set temperature __140__ (°F)

   or ·Estimated Consumption - using fuel bills for summer
      water heating.
      Hot water delivery temperature (if known) _____°F
      Quantity of fuel consumed ____ (kwh, gallons, tons,
      100 cu.ft., etc.)
      Total fuel bill for period of consumption $_____
      Number of days in consumption period _____
      (period when furnace was turned off)

2. Commercial Application
    Fill in one of the following:

      ·Btu's per day required for hot water _____
       (Include losses for circulation loop, if applicable)
   or ·Monthly fuel bill for hot water only for the past 12
      months

      $_____ $_____ $_____ $_____
      $_____ $_____ $_____ $_____
      $_____ $_____ $_____ $_____

      Fuel type _____
      Fuel price per unit (actual) $_____
   or ·Gallons of hot water used per day _____
                                    (gallons/day)
      Hot water heater set temperature _____(°F)
      Water main temperature _____(°F)
      If circulation loop is involved, estimate losses in
       Btu's per day _____

F. Reference System Cost

1. Reference (conventional) system, initial installed
    cost* $_____.

2. Reference (conventional) system, annual maintenance
    cost $_____ (average annual maintenance costs)
    Annual maintenance may be expressed as percent of
    initial installed cost* _____% (Do not use for
    retrofit situations)

Table 8.3 (Continued)

G. Solar System Cost

1. Solar Component Cost Information

    a) Collector cost/square foot $ 12.50 /sq.ft.
    b) Solar storage tank cost/gallon $ 1.30 /gallon
    c) Controls $ -
    d) Other $ -

2. Installation Costs of Solar System

    a) Design/engineering costs $ 50.00
    b) Modifications to building and
        existing system - labor costs $100.00
    c) Installation labor costs $450.00
    d) Misc. materials cost (paint,
        wallboard, shingles, etc.) $ 80.00

        Total installation $680.00

3. Solar System Maintenance Cost

    (Calculate average annual maintenance cost
    over life of system)* $ _____
    Annual maintenance may be entered as
    percent of solar system costs*   _____ %

H. Finance and Tax Data

1. Residential

    a) Loan interest rate*   _____ %
    b) Loan term*   6 years
    c) Loan down payment*   _____
    d) Property insurance rate* $_____
    e) Property tax rate   2-1/2%
    f) Personal income tax rate   33 %

2. Commercial

    a) Loan interest rate*   _____ %
    b) Loan term*   _____ years
    c) Loan Down payment * $_____
    d) Property insurance rate*   _____ %
    e) Property tax rate   _____ %
    f) Corporate tax rate or owner income
        tax rate   _____ %

3. If business application (commercial or residential rental,
    for example) check ___ and fill out the following:

    a) Depreciation method (options are straight
        line or declining balance)   _____
    b) Multiplier used in declining balance
        depreciation.* (limited to 1.5 for
        commercial buildings and 2.0 for new
        residential property)   _____
    c) System usefule life for depreciation
        purposes.* (currently 10 years is
        allowed for building heating, air
        and service hot water systems, except
        storage tanks, which are allowed 22
        years)   _____ years
    d) Salvage value of the solar system at
        the end of its useful life*   _____ % of total
                                          cost

4. Tax Incentives

    a) State income tax credit.
        % for the first $_____ (installed cost)
        % for the next $_____ (installed cost)
    b) Property tax rate
        ___ % for the first ____ years
        ___ % for the next ____ years
    c) If a federal income tax credit is enacted, attach an
        explanation of the differences between federal and
        state.

I. Fuel and Electricity Cost and Price Escalation Data

Table 8.3 (Continued)

1. Natural Gas Cost Schedule
   Btu content per cubic foot* _____

Fuel Cost ($/cubic ft.)		Quantity (no. of units)	
Step 1	_____	for first	_____
Step 2	_____	for next	_____
Step 3	_____	for next	_____
Step 4	_____	for next	_____
Step 5	_____	for next	_____

2. Electricity Cost Schedule

Fuel Cost ($/kwh)		Quantity (no. of units)	
Step 1	.147	for first	15
Step 2	.074	for next	35
Step 3	.061	for next	50
Step 4	.052	for next	50
Step 5	.038	for next	234
Step 6	.043	for next	616
Step 7	.043	for next	-

3. Fuel Oil Cost Schedule
   Fuel oil grade* _____.

Fuel Cost ($/gallon)		Quantity (no. of units)	
Step 1	_____	for first	_____
Step 2	_____	for next	_____
Step 3	_____	for next	_____
Step 4	_____	for next	_____
Step 5	_____	for next	_____

Table 8.4. Solcost Outputs. (Source: Energy Research & Development Administration[11]).

Energy Balance by Month for 56.3 Sq. Ft. Collector

Month	Fraction by Solar	Average Useful Solar Per Day (Btu/day-Sq.Ft.)	Total Useful Solary Energy (Mil Btu/Mo)	Auxiliary Energy (Mil Btu/Mo)	Conventional System Energy (Mil Btu/Mo)
1	.228	357.3	.62	2.35	3.04
2	.318	551.7	.87	2.08	3.04
3	.373	584.8	1.02	1.91	3.04
4	.371	600.8	1.01	1.91	3.04
5	.395	619.7	1.08	1.84	3.04
6	.426	690.0	1.17	1.75	3.04
7	.455	713.9	1.25	1.66	3.04
8	.445	698.2	1.22	1.69	3.04
9	.423	685.2	1.16	1.76	3.04
10	.344	539.6	.94	2.00	3.04
11	.247	400.0	.68	2.29	3.04
12	.217	341.0	.60	2.38	3.04
	Totals		11.62	23.62	36.48

Note 1: Conventional energy and solar auxiliary energy are gross values
(i.e., includes tank insulation and/or combustion losses).

Solar Components Sizing

Solar storage tank size for this collector is 84 gallons
Heat Exchanger: None
Pump Characteristics: 1/20 horsepower, 1.40 gallons per minute.
Pipe Diameter (Flow Area): 3/4 inch

Collector Size Optimization by SOLCOST

Collector Type - Flat Plate 1 Glass Selective
Best Solar Collector Size for Tilt Angle of 43 Degrees is 56 Sq. Ft.
Solar System Costs (dollars) - 400.00 fixed
                               700.00 Collector
                               105.00 Storage
Installation Costs (dollars) - 680.00

Table 8.4 (Continued)

Yr.	(A) Fuel/Utility Savings	(B) Maint. + Insur.	(C) Property Tax	(D) Annual Interest	(E) Tax Savings	(F) Loan Payment	(G) Net Cash Flow
							−384. (Down Payment)
1	163.	29.	48.	169.	72.	363.	−205.
2	174.	30.	50.	148.	65.	363.	−203.
3	186.	31.	52.	124.	58.	363.	−201.
4	199.	32.	54.	98.	50.	363.	−200.
5	213.	34.	56.	68.	41.	363.	−198.
6	228.	35.	58.	36.	31.	363.	−197.
7	244.	36.	61.	0.	20.	0.	167
8	261.	38.	63.	0.	21.	0.	181
9	279.	39.	66.	0.	22.	0.	196.
10	299.	41.	68.	0.	23.	0.	212.
11	320	43.	71.	0.	23.	0.	229.
12	342.	44.	74.	0.	24.	0.	248.
13	366.	46.	77.	0.	25.	0.	268.
14	392.	48.	80.	0.	26.	0.	290.
15	419.	50.	83.	0.	27.	0.	313.
16	448.	52.	87.	0.	29.	0.	339.
17	480.	54.	90.	0.	30.	0.	366.
18	513.	56.	94.	0.	31.	0.	395.
19	549.	58.	97.	0.	32.	0.	426
20	588.	61.	101.	0.	33.	0.	459.
Totals	6663.	857.	1430.	643.	683.	2178.	2885.

```
Payback time for fuel savings to equal total investment - 8.9 years
Payback time for net cash flow to offset down payment - 13.3 years
Rate of return on net cash flow - 8.5 percent
Annual portion of load provided by solar - 35.3 percent
Annual energy savings with solar system - 11.6 million Btus
```

```
Tax Savings = Income Tax Rate x (C + D)
Net Cash Flow = A - B - C + E - F
```

8.4 REFERENCES AND SUGGESTED READING

1. American Society of Heating, Refrigerating, and Air-Conditioning Engineers, Inc., New York, *Bibliography on Available Computer Programs in the General Area of Heating, Refrigerating, Air-Conditioning and Ventilating,* 1975.

2. American Society of Heating, Refrigerating, and Air-Conditioning Engineers, Inc., New York, *Procedure for Determining Heating and Cooling Loads for Computerizing Energy Calculations – Algorithms for Building Heat Transfer Subroutines,* 1976.

3. American Society of Heating, Refrigerating, and Air-Conditioning Engineers, Inc., New York, *Procedures for Simulating the Performance of Components and Systems for Energy Calculations,* 1976.

4. Boillot, Michel H., Gleason, Gary M., and Horn, L. Wayne, *Essentials of Flowcharting,* Dubuque, Iowa: Wm. C. Brown Co., 1975.

5. Burroughs Corporation, El Monte, California, *Burroughs B6700 Basic Language Information Manual,* 1975.

6. Coan, James S., *Advanced Basic,* Rochelle, New Jersey, Hayden Book Co., Inc., 1976.

7. Coan, James S., *Basic Basic,* Rochelle, New Jersey, Hayden Book Co., 1978.

8. Conway, Richard, and Gries, David, *An Introduction to Programming,* Cambridge, Massachusetts: Winthrop Publishers, Inc., 1979.

9. COSMIC, The University of Georgia, 112 Barrow Hall, Athens, Georgia 30602.

10. Deitel, Harvey M., *Introduction to Computer Programming with the Basic Language,* Englewood Cliffs, New Jersey: Prentice-Hall, Inc., 1977.

11. Energy Research and Development Administration, *Solcost Solar Hot Water Handbook,* Washington, D.C.: U. S. Government Printing Office, September, 1977.

12. Ford, Donald H., *Basic Fortran IV Programming,* Homewood, Illinois: Richard D. Irwin, Inc., 1974.

13. Gilbert, Jack, *Advanced Applications for Pocket Calculators*, Blue Ridge Summit, Pennsylvania: Tab Books, 1975.

14. Hamming, Richard W., and Feigenbaum, Edward A., *Programming Systems and Languages*, New York: McGraw-Hill Book Co., 1967.

15. Hewlett-Packard, Santa Clara, California, *HP 2000/Access Basic*, 1976.

16. IBM Corp., Kingston, New York, *System/370 VS Basic Language*, 1974.

17. Kusuda, Tamami, *NBSLD, The Computer Program for Heating and Cooling Loads in Buildings*, Washington, D.C.: U. S. Government Printing Office, 1976.

18. Lott, Richard W., *Basic With Business Applications*, New York: John Wiley & Sons, Inc., 1977.

19. Marateck, Samuel L., *Basic*, New York: Academic Press, 1975.

20. Murrill, Paul W., Smith, Cecil L., *An Introduction to Fortran IV Programming - A General Approach*, New York: Intext Educational Publishers, 1975.

21. National Aeronautics and Space Administration, *COSMIC - A Catalog of Selected Computer Programs*, National Aeronautics and Space Administration Technology Utilization Office, Washington, D.C., 1977.

22. Pollack, Seymour V., *A Guide to Fortran IV*, New York: Columbia University Press, 1965.

23. Ruegg, Rosalie T., McConnaughey, John S., Sav, G. Thomas, and Hockenbery, Kimberly, A., *Life-Cycle Costing – A Guide for Selecting Energy Conservation Projects for Public Buildings*, Washington, D.C.: U. S. Government Printing Office, 1978.

24. Sammet, Jean E., *Programming Languages: History and Fundamentals*, Englewood Cliffs, New Jersey: Prentice-Hall, Inc., 1969.

25. Solar Energy Research Institute, *Analysis Methods for Solar Heating and Cooling Applications - Passive and Active Systems*, Golden, Colorado, January, 1980.

CHAPTER 9
COMPUTER SYSTEMS

9.1 DATA PROCESSING CHOICES

Once an individual has obtained a computer program, the next step involves determining the method of running the program. The individual has a number of choices available as to the method used in running the program.

9.1.1. Timeshare Operation

The first choice involves timesharing or "on-line" computer usage via telephone lines. This generally involves accessing someone else's computer using a data terminal such as the one shown in Fig. 9.1.

While an individual is using a computer service bureau's timeshare computer, other individuals may also be using the service bureau's computer at the same time. Figure 9.2 is shown to illustrate how one central computer can handle a number of users using (accessing) the computer simultaneously.

Figure 9.1. Data Terminal. (Courtesy of Texas Instruments, Inc.).

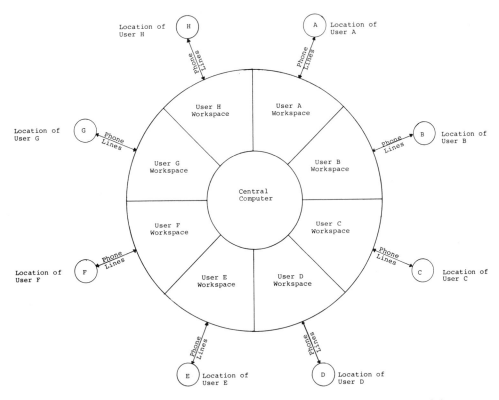

Figure 9.2. Computer User Workspace. (Workspace Refers to Certain Amount of Computer Memory Allocated to Each User. The Partitioned Workspace is Located in the Computer's Memory.)

Suppose there are eight people at different geographical locations communicating with a central timeshare computer using telephones and data terminals (such as the one shown in Fig. 9.1). Each individual using the computer would have their own special job code. For each job code there would be a certain amount of computer memory allocated. The computer would handle all eight users by processing a little of each user's program at a time. This could involve spending a fraction of a second doing user A's calculations, and then spending a fraction of a second doing calculations for user B, and then user C, and so forth until the computer returns to user A where the process is repeated. In reference to Fig. 9.2, one can think figuratively of the computer as traveling in a clockwise direction doing some quick calculations for a particular user and then moving on to the next user. The computer does this so quickly that each of the eight users shown in Fig. 9.2 may think that they are the only one using the computer at the time. Each of the eight users would in effect be using the computer in a "timeshare" mode since each user would be sharing the computer with other users.

The individual deciding to use a timeshare computer should first look in the yellow pages of the telephone book under "Data Processing Service" and consult the various local firms listed to find out whether they offer timeshare services via telephone lines. Not all firms offer timeshare services and may require the individual to physically bring the computer program to their computer center whereby they (the computer center staff) run the program. In some cases, the individual wanting to use a timeshare computer may not find a local computer firm that offers such services. An out-of-town computer may be the answer but with the added expense of long distance telephone dialing charges.

One big advantage of using a timeshare computer is the ease and quickness of operation. All the individual has to do is to dial the computer center's telephone number, place the telephone headset in the cradle located on the data terminal (such as shown in Fig. 9.1), type in the correct user identification and password, wait for the computer to acknowledge the identification and password, and begin entering a new program or data.

A possible disadvantage of using timesharing may be cost. The individual may be better off either owning or leasing a computer. Another possible disadvantage is the risk of theft of data the individual may have stored on the central computer. There have been cases where one timeshare user has gained illegal access to another timeshare user's data and programs stored on a central timeshare computer. The individual deciding to use a central timeshare computer should therefore question the computer center's director about safeguards used to prevent unauthorized access and what the center's policy is if the safeguards are breached.

9.1.2. Batch Mode Operation

Running a computer program in batch mode generally refers to loading both the program and data into the computer at the same time. In the timeshare mode described earlier, the individual user could run a computer program which would prompt the user for appropriate input. For example, the user may have a computer program such as the one shown in Appendix VI that computes both the annual fraction of thermal energy supplied to a building by solar collectors and the most economical size solar collector. Notice in the computer program shown in Appendix VI that the first question asked is: "What is the latitude (in degrees)?" The user would then input a number into the data terminal for transmittance to the computer via telephone. The computer would then ask the second question: "What is the ground reflectance (between 0.2 and 0.7)?" The user running this program is prompted by questions and required to answer each one before the computer elicits the next question. This type of interaction between the individual user and the computer is one feature generally available with timeshare. To run a computer program in batch mode, the user would submit both the computer program and data statements to the computer operator who would input them. The computer program and data statements submitted to the computer operator for running in batch mode are generally in the form of either computer cards or magnetic tape as illustrated in Fig. 9.3.

One advantage of submitting both the computer program and input data to the computer operator for running in batch mode is that the computer-use charges may be less than using other methods such as timeshare. A possible disadvantage is the time spent physically taking the computer programs and/or

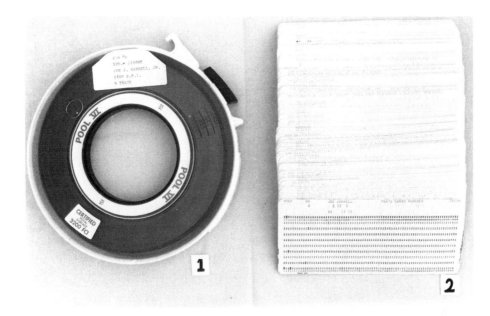

Figure 9.3. Two Popular Computer Program Storage Mediums.

data cards to the computer center and the time spent waiting to receive the computer printout of the results. Tables 9.1 and 9.2 show actual charges for computer time and other supplies and services provided by the Auburn University Computer Center. These charges should give the reader a ballpark idea of the relative charges for various computer services.

9.1.3. Computer Leasing

Computer leasing involves leasing (or renting) a computer that is generally placed on the premises of the lessee. With reference to the architect or engineer, the computer leased would generally be on the order of a small minicomputer or microcomputer. The firm leasing the computer, such as an architect/engineering firm for example, usually receives some training by the computer leasing firm as to the procedures used in operating the computer. Generally, there is also a service and maintenance contract whereby the computer vendor periodically checks the computer for proper operation and will be on call to repair any breakdowns.

   An advantage of leasing a computer may be a reduction of computing costs when compared to other alternatives such as outright ownership of a computer. One big advantage to leasing, as opposed to owning a computer, is that the individual is not stuck owning an outdated computer when a new generation computer comes on the market. The firm leasing the computer generally has the option to renegotiate the lease to obtain a newer model computer. Depending upon the computing needs and cash flow of a particular firm, cost may be a negative factor to leasing.

9.1.4. Ownership of a Computer

Numerous firms owning a computer have used minicomputer systems such as the Hewlett-Packard system shown in Fig. 9.4. In the past, smaller companies unable to afford purchasing such a system, generally had to either lease a computer system or timeshare to fulfill the necessary data processing requirements in their business.

Table 9.1. Auburn University Request for Services from the Computer Center. (Auburn University Computer Center).

A.U.C.C. 001
Rev. May 15, 1976

AUBURN UNIVERSITY
REQUEST FOR SERVICES FROM
THE COMPUTER CENTER
1977 - 78

Blue—Computer Center
Yellow—Applicant-Project
Green—Head of Department

INSTRUCTIONS: See Other Side

Job Code No ☐☐☐☐

1. DEPARTMENT_____ PURCHASE ORDER NO. _____

   APPLICANT_____ TELEPHONE NUMBER _____

2. Project Title ☐☐☐☐☐☐☐☐☐☐☐☐☐☐☐☐☐☐☐☐☐☐☐☐

3. Has a Job Code previously been assigned this project? Yes ☐ No ☐. If yes, what was it?_____

Project Type	Check	Project Type	Check
Instruction in Course No.		Extension	
Masters Thesis		Administration	
Doctoral Dissertation		Computer Center	
Research		External Use (Higher Rates)	

5. Computing, Supplies, Services

Equipment	Est. Quantity	*Rate	Estimated Cost
Computer Time (CPU)	.... hrs.	$250/hr.	$....................
Printer	.... lines	21¢/1000 lines	$....................
Printer (Labels)	.... labels	21¢/1000 labels	$....................
Card Reader	.... cards	12¢/1000 cards	$....................
Card Punch	.... cards	23¢/1000 cards	$....................
Subtotal, Computing			$....................
Supplies			
Paper, 14⅞" x 8½" 1 part	.... pgs.	$ 7/1000 pages	$....................
Paper, 14⅞" x 11" 1 part	.... pgs.	$10/1000 pages	$....................
2 part	.... pgs.	$20/1000 pages	$....................
3 part	.... pgs.	$30/1000 pages	$....................
4 part	.... pgs.	$40/1000 pages	$....................
Paper, 9½" 1 part	.... pgs.	$ 6/1000 pages	$....................
2 part	.... pgs.	$15/1000 pages	$....................
3 part	.... pgs.	$25/1000 pages	$....................
4 part	.... pgs.	$35/1000 pages	$....................
Cards (5081)	.... boxes (2000)	$ 5/box of 2000 cards	$....................
Labels	.... labels	$ 4/1000 labels	$....................
Plotter paper, 12"	.... feet	10¢/foot	$....................
30"	.... feet	40¢/foot	$....................
Tape Rental	.... reels	50¢/reel—month	$....................
Disk Storage (on-line)	.... tracks	15¢/track—month	$....................
Disk Storage (off-line)	.... packs	$75/pack—month	$....................
Services			
Consulting	.... hrs.	$15/hour	$....................
Programming and Testing	.... hrs.	$10/hour	$....................
Keypunching/Verifying	.... cards	$ 1/50 cards	$....................
Interpreting	.... cards	$ 2/1000 cards	$....................
Forms Mount/Carriage Tape Mount	.... mounts	$ 1/each	$....................
Subtotal, Supplies and Services			$....................

   Total of Estimated Costs . . . . $....................

*Subject to change

6. Recommend Approval:

   a. Major Prof., Researcher, or Instructor   b. Computer Center Consultant

7. SOURCE OF FUNDS

   a. UNSUPPORTED
   I certify no funds are available for these services from SPONSORED RESEARCH OR OTHER DEPARTMENTAL FUNDS. However, funds are available in the above Purchase Order Account for supplies and personnel services, and I authorize the Computer Center to bill the account accordingly.

   b. SUPPORTED
   This project is supported by funding external to Auburn University. I certify that funds are available and authorize the Computer Center to begin work immediately and bill the account accordingly. In the event funds are depleted in the account indicated, other funds will be made available to pay for the work completed by the Computer Center.

   Head of Department   Head of Department

8. Approved _____ Date _____
   Director, Computer Center

Table 9.2. Computer Usage Form Instructions (For Table 9.1). (Auburn University Computer Center).

INSTRUCTIONS

The Job Code Number will be assigned by the Computer Center. The Job Code Number is not transferable, but should be used only for the project for which it was established and only by those persons who are approved by the signers of items 6 and 7. A purchase order can be used for several job code numbers; a job code number can have only one purchase order.

To request Computer Center services follow all numbered instructions below.

1. Applicant fills in Department name and standard abbreviation, his own name, phone number, and purchase order number if appropriate (see item 7). The latter may not be available until after item 7 is completed. The applicant should be that person who will be the primary contact with the Computer Center regarding this work.

2. The project title must fit in the 23 spaces provided.

3. Give the previous number for a continuation, even from one fiscal year to another. For a continuation involving any disk data sets, write "DISK" after the job code number.

4. A check mark must be made indicating the Project Type. If a thesis or dissertation is a part of a larger research effort, separate job code numbers must be requested and a reasonable allocation made between thesis computing and non-thesis computing. If the research is supported, then both categories of computing should be supported. If more than one thesis is involved, single or multiple job code numbers may be used as desired. A single purchase order may be used for all job code numbers requested except external use. External use refers to computing services rendered directly to non-AU organizations. Please give full details on any Request for Services where it might be necessary to rule that external rates apply; the Computer Center is subject to audit on this matter.

5. Estimates should be based on past experience as well as analysis of the current problem. For assistance, call the Director (4285) for an appointment with him or a Computer Center Consultant. Experienced researchers may make estimates without consultation. Item 5 must be completed even for unsupported projects. If data or programs are to be stored on magnetic tape or disk, be sure to include these items in estimated cost. This action constitutes a request for the user's job code to become a valid index in the catalog. Attempts to create datasets without the indexing will result in datasets which are not catalogued, hence not accessible for future processing. Please estimate the amount of disk tracks needed under item 5.

6a. The signer agrees to monitor the computer usage made as a result of this request.

6b. This may be omitted. See instruction 5.

7. Choose the appropriate paragraph for the department head's signature. Statements and invoices will be sent to the department head who signs. The Computer Center does not send invoices to students. A departmental purchase order is required with payment by Auburn University voucher.

8. Send all three copies of this form to the Computer Center for the Director's approval. If the request is approved, a job code number will be assigned and two copies of the form returned. If a new purchase order is required, the purchase order form should be sent independently to the purchasing office. It will be forwarded to the Computer Center.

This form can be used for estimating computer costs for contract and grant budget proposals to be submitted to the Vice President for Research as well as for requesting Computer Center services. To use this form as a computer services cost estimate, write COMPUTER COST ESTIMATE in large letters across the top of the form. Follow all numbered instructions except that No. 7 should be omitted, the purchase order should be omitted, and no job code number will be assigned. After approval, the Director will retain the white copy and send the yellow and green copies to the applicant. The green one should be forwarded to the Vice President for Research with the proposal. If the proposal becomes a project, a normal Request For Services form will be required. Of course, the Computer Cost Estimate can be used as a basis.

Figure 9.4. Minicomputer System. (Photograph courtesy of Hewlett-Packard Company).

With the introduction of fairly sophisticated and low-cost microcomputers, the whole ball game of computer ownership has changed. The computer systems that many firms were once unable to afford are now affordable. Firms such as Radio Shack (a division of Tandy Corporation) offer microcomputer systems for both home use and more powerful systems for business use. One such system suitable for business use is shown in Fig. 9.5.

Other companies, that have traditionally been thought of as manufacturers of large computer systems, are also entering the market to fill the need for the user desiring a "small" computer system. The IBM 5110, shown in Fig. 9.6, is such a system.

Another microcomputer which is comparatively small in size and easily portable is the Hewlett-Packard HP-85 system shown in Fig. 9.7. This system like most other microcomputer systems, uses a version of the BASIC computer programming language.

The following article, originally appearing in *Computerworld,* is included to show how one individual was able to use his microcomputer for both business and pleasure.

Avocation Turns to Vocation
MICRO WORKING FOR BUSINESS AS WELL AS PLEASURE

Simi Valley, Calif.—Contract bids that used to take Jemco Plumbing, Inc., as long as four hours to estimate now require 15 minutes or less, thanks to a microcomputer business system that company head Jay E. Moss recently installed in his home.

Moss, a confirmed hobbyist with interests in various avocations, originally acquired the system a little more than a year ago as a personal computing "toy," which he soon programmed to outwit him in four-dimensional tic-tac-toe.

Whenever Moss misplays a game, the system taunts its owner with CRT-displayed invectives like, "That was really a stupid move, Jay. I'm going to finish you off in three moves."

About six months after he installed the system, however, Moss "realized its potential for my plumbing contracting business" and had soon adapted the hardware for more serious applications than simple home amusement. Today, in addition to estimating contract bids worth about $2.5 million each year, the computer oversees Jemco's payroll and bookkeeping and prints both invoices and form letters.

The system uses the same information source as Jemco's plumbing equipment suppliers to check the company's supplier invoices.

Accountant Fees Down

Since Moss automated his payroll, Jemco's accountant has lowered his monthly fee by $75/mo. the company head reported. "I've spent about $9,000 on the system in all, and I figure it has paid for itself several times over in this first year alone," he added.

To estimate a contract bid for a residential subdivision or other large-scale construction project, Moss programmed his Altair 8800A system to request answers to a long list of business questions like:

—How many houses of a particular plan are in a tract?
—How many fixtures are in the plan?
—How much does each fixture cost?—How far apart are the fixtures?

Figure 9.5. Microcomputer System. (Source: Radio Shack).

Figure 9.6. Computer System. (Source: IBM).

Figure 9.7. Portable Microcomputer System. (Illustration courtesy of the Hewlett-Packard Company).

With no human assistance, the computer can also calculate the amount of labor Jemco will need to emplace a home plumbing system because Moss has supplied the machine with years of records showing the typical turnaround time for each of the company's 15 installation steps.

"To check whether the computer's estimates on labor were correct, I took old housing plans from four to five years ago and had the computer estimate them," Moss recalled. "Then I checked those figures with the actual labor and costs, and they were fantastically accurate."

Using all the information Moss has fed it, the system produces an itemized cost estimate for each job, including profit and overhead statements. For a tract with four house plans, a bid estimate takes about 45 minutes to prepare.

"Most small businessmen would probably need only about $5,500 worth of (computer) equipment," he concluded. "But the big thing is that they shouldn't be afraid of using a computer. They don't have to know how a computer works, but how their own business works."

Moss' own experience provides a convenient case in point. When he first became involved with hobby computing about a year ago, "I knew nothing about computers beyond how to spell the word."

His interest in DP began after he read a news magazine article about hobbyists buying computer kits for as little as $400 and assembling small systems that provided as much power as giant computers of a decade ago.

One day after reading the article, Moss and his son went to a computer store and spent $2,700 for the Altair 8800A microcomputer kit and peripherals, both of which are manufactured and marketed by Pertect Computer Corp.'s Microsystems Division.[23]

There are numerous microcomputer systems on the market being manufactured by both well established leaders in the computing field and by firms relatively new to the field. The surge in manufacturers building and selling microcomputers is due in great part to the advances in the field of electronics whereby fairly sophisticated computers can be built at a relatively low cost. This has opened the door to the now available low-cost microcomputer systems for both home and business use.

Individuals deciding to purchase a microcomputer should analyze their individual needs. For example, some microcomputer systems can do little more than process relatively simple computer programs while other more powerful microcomputer systems can do everything from run payrolls to processing fairly sophisticated computer programs. Additionally, the individual should also be concerned with determining the impact of a computer break down. For instance, the homeowner buying a relatively inexpensive microcomputer may be able to part with a microcomputer for a number of weeks while it is being repaired at the factory, whereas the businessman may suffer great hardships if the system were to be inoperative for even a day. Therefore the businessman would be advised to ascertain whether the computer vendor has a service and maintenance program and what the average expected "down-time" will be should the computer system need service. The individual may find that, for example the XYZ microcomputer company, located 500 miles away with no regional offices, may take eight weeks to fix the computer upon receipt of the computer through the mail. On the other hand, a larger and more well-established computer firm may have branch offices in almost every large city across the country and have technicians on call around the clock who will come to the individual's office to repair his computer. In such cases the computer "down-time" may be very short. Naturally, the computer owner who cannot afford to have the computer out of service for very long will generally want to own a computer system whose vendor has a good maintenance and service record. Additionally the professional should steer clear of unknown computer vendors who are so new that they have not established a reputation. In the case of any new growth market, there are a few bad apples whose only aim is to get rich quick and to go out of business almost as fast as they started. Therefore prudence in selecting a microcomputer system and vendor is imperative for the professional wanting to avoid the headaches associated with buying a poor system and/or a system sold by an unreliable manufacturer.

The next step down from the microcomputer is the hand-held programmable calculator. The popular models are the Hewlett-Packard HP-67 and the Texas Instruments TI-58 shown, respectively, in Figs. 9.8 and 9.9. For some professionals the programmable calculator may be all the computing power necessary and for others may act as a supplement to a more powerful computer system. Most programmable hand-held calculators generally use either the algebraic operating system (AOS) such as for the TI-58 or the reverse polish notation (RPN) type programming logic as used in the HP-67 calculator. Basically, the algebraic system allows the user to enter the arguments and functions for an equation in the same order as they are found in the equation. This involves pressing the parenthesis button as needed to separate order of execution (See Fig. 9.9). The reverse polish notation system involves placing the operator immediately after the operands in the equation. When the operator is pressed,

Figure 9.8. Hewlett-Packard Calculator. (Illustration courtesy of the Hewlett-Packard Company).

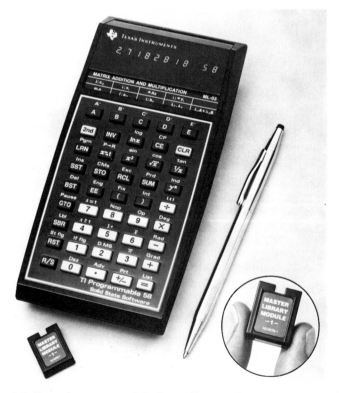

Figure 9.9. Texas Instruments Calculator. (Source: Texas Instruments, Inc.).

the step is immediately executed. (Hewlett-Packard publishes a booklet that contrasts the two systems which the reader may find useful.)

The programmable hand-held calculator does offer a substantial amount of computing power that is sometimes taken for granted. In retrospect, it was not too many years ago that such compact computer power was not available at any price. The advent of large scale integration (LSI), whereby hundreds of electronic components could be put on a "chip" the size of a pinhead, opened the door for the present breed of calculators. With further advances in electronic technology, the computing power of hand-held calculators will continue to increase.

9.2 SUMMARY

The individual must determine whether more computing power is needed than a hand-held type calculator offers. If more computing power is deemed necessary, the individual must decide whether to timeshare, pay a computer center to run the program in batch mode, lease, or purchase a computer. The individual must analyze data processing requirements, cash flow, tax situation, etc., before determining which of the four choices is most appropriate. There is no pat answer for determining which is the best option. It depends on the computer requirements and financial position of each firm. Basic accounting and finance methods can be used in determining which particular choice is "best" for a specific firm. This is why earlier, when the advantages and disadvantages were discussed for each of the four alternatives, cost was generally listed as being a possible advantage or disadvantage for each choice. For one firm it may be more advantageous to buy a computer whereas another firm may find it more economical to lease, etc. (For the individual desiring more information on data processing equipment and system, some popular books on the subject are listed in Section 9.3.)

One final clarification involves the words *timeshare* and *batch mode* which were used earlier in reference to an individual using a computer service bureau's computer for a fee. An individual owning a computer with several data terminals located at the computer location may have the capabilities to run in either batch or a timesharing mode. For example, some engineering firms may have their own in-house computer and several data terminals located in the same room which are directly connected to the computer. The computer may allow for timesharing (i.e., multi-processing) so that an engineer might be running a technical program while a secretary, using another terminal, might be using the computer to run a business payroll program. This is still operating the computer in a timeshare mode even though both users and the computer are in the same room as opposed to the earlier discussion of timesharing whereby the users were geographically located far from the central computer and communicating via phone lines.

9.3 REFERENCES AND SUGGESTED READING

1. Barden, William Jr., *How to Buy and Use Minicomputers and Microcomputers*, Indianapolis, Indiana: Howard W. Sams & Co., Inc., 1976.

2. Bernard, Dan, Emery, James C., Nolan, Richard L., and Scott, Robert H., *Charging for Computer Services: Principles and Guidelines*, New York: Petrocelli Books, 1977.

3. Brown, John A., and Workman, Robert S., *How a Computer System Works*, New York: Arco Publishing Co., Inc., 1975.

4. Bueschel, Richard T., Stephenson, Andrew G., and Whitney, Douglas C., *Commercial Time-Sharing Services and Utilities*, New York: American Management Association, Inc., 1969.

5. Bunnell, David, *Personal Computing*, New York: Hawthorn Books, Inc., 1978.

6. Farr, M. A. L., Chadwick, B., and Wong, K., *Security for Computer Systems*, London, England: NCC Publications, 1972.

7. Gruenberger, Fred, and Babcock, David, *Computing with Minicomputers*, Los Angeles, California: Melville Publishing Co., 1973.

8. Hamilton, Peter, *Computer Security*, London, England: Associated Business Programmes, Ltd., 1972.

9. Hoffman, Lance J., *Security and Privacy in Computer Systems*, Los Angeles, California: Melville Publishing Co., 1973.

10. Joslin, Edward O., *Computer Selection*, Fairfax Station, Virginia: The Technology Press, Inc., 1977.

11. Korn, Granino A., *Microprocessors and Small Digital Computer Systems for Engineers and Scientists*, New York: McGraw-Hill Book Co., 1977.

12. Lewis, T. G., and Doerr, J. W., *Minicomputers Structure and Programming*, Rochelle Park, New Jersey: Hayden Book Co., Inc., 1976.

13. Parker, Donn B., *Crime By Computer*, New York: Charles Scribner's Sons, 1976.

14. Pritchard, J. A. T., *Quantitative Methods in On-Line Systems*, Rochelle Park, New Jersey: Hayden Book Co., Inc., 1976.

15. Rosenthal, Paul H. and Mish, Russell K., eds., *Multi-Access Computing – Modern Research and Requirement*, Rochelle Park, New Jersey: Hayden Book Co., Inc., 1974.

16. Sanders, Donald H., *Computers in Business - An Introduction*, New York: McGraw-Hill Book Co., 1972.

17. Scherr, Allan Lee, *An Analysis of Time-Shared Computer Systems*, Cambridge, Massachusetts: The M.I.T. Press, 1967.

18. Sippl, Charles J., *Microcomputer Handbook*, New York: Petrocelli/Charter, 1977.

19. White, James, *Your Home Computer*, Menlo Park, California: Dymax, 1977.

20. Wilkes, M. V., *Time-Sharing Computer Systems*, New York: American Elsevier Publishing Co., Inc., 1975.

21. Wooldridge, Susan, Corder, Colin R., and Johnson, Claude, *Security Standards for Data Processing*, New York: Halsted Press, 1973.

22. Ziegler, James R., *Time-Sharing Data Processing Systems*, Englewood Cliffs, New Jersey: Prentice-Hall, Inc., 1967.

23. Lundell, S. E. Drake, Jr. (ed.), *Computerworld*, **12**, 29:67, (1978).

CHAPTER 10
PURCHASING AND INSTALLING SOLAR EQUIPMENT

10.1 BUYING TIPS

The following twelve tips may prove useful to someone contemplating the purchase of a solar system.[1,3]

1. Deal only with licensed contractors and insist on a written contract. Any questions or complaints regarding contractors should be referred to your State Contractors' License Board.
2. Ask for proof that the product will perform as advertised. The proof could come from an independent laboratory or a university. You should have the report itself, not what the manufacturer claims the report states. Have an engineering consultant go over the report if you don't understand it.
3. If there is a warranty, examine it carefully. Remember that, according to the law, the manufacturer must state that the warranty is full or limited. If it is limited, know what the limitations are. How long does the warranty last? Are parts, service and labor covered? Who will provide the service? Does the equipment have to be sent back to the manufacturer for repair? Ask the seller what financial arrangements, such as an escrow account, have been made to back the warranty.
4. Solar components are like stereo components—some work well together, others don't. If the system you are purchasing is not sold as a single package by one manufacturer, then you should obtain assurance that the seller has had the professional expertise for choosing properly.
5. Ask for the person who owns one. Ask the seller for a list of previous purchasers and their addresses, and then ask the owners about their experiences.
6. Be careful of sellers who use P. O. Box numbers. Though many legitimate businesses use these outlets as a convenient way to receive bills and orders, a common tactic of the fly-by-night artist is to use a Post Office Box number, operate a territory until the law starts closing in, then move and take a new name in a new territory. Find out from the seller where his place of business is, how long he has been there, and ask for his financial references.
7. Be sure that you know specifically who will service the solar system if something goes wrong. Don't settle for a response that any plumber or handyman will do.
8. Don't try a do-it-yourself kit unless you really have a solid background as a handyman. One or two mistakes could make a system inoperable and you will have no one to blame but yourself.
9. Don't change your use habits simply because you are getting plenty of free energy. Conservation of energy still counts if you want to bring your monthly bills down. Don't blame the seller of a solar heating system if you keep your doors open during the middle of winter.

10. Don't forget your local consumer office or your Better Business Bureau. Both may be able to help you determine whether a seller is reputable or not. Check, too, to see whether there is a local volunteer citizens solar organization who can advise you.
11. If the seller makes verbal claims that are not reflected in the literature handed out, ask him to write these claims down, and to sign his name to it. Compare what he said with what he wrote. Save that statement.
12. If you have what appears to be a legitimate complaint, notify the local district attorney's office immediately, the Better Business Bureau, the local consumer protection agency, or the Contractors' State License Board. Be as specific in your complaint as possible, and give as much documentation as you can.

10.2 INSTALLATION CHECKLIST

The following checklist[2] is presented as a guide for the installation of solar domestic hot water (DHW) systems. Where the solar system manufacturer's instructions differ from any of these guidelines, the manufacturer's recommendations should be followed.

1. SITING AND ORIENTATION

1-1. Are collectors oriented in a proper southerly direction?
DISCUSSION: Most manufacturers recommend true (not magnetic) south plus or minus 15°. Variations outside these limits reduce the efficiency of the system.

1-2. Do solar collectors have an unobstructed view in a southerly direction between 9:00 A.M. and 3:00 P.M.?
DISCUSSION: While shading problems are possible all year, low winter sun angles may cast shadows of distant obstructions across the collectors that would not be a problem in summer. Remember that trees which shed leaves in the fall are less prone to create such problems than evergreens.

1-3. Are collectors tilted within acceptable limits?
DISCUSSION: Most manufacturers recommend latitude plus up to 10° for domestic hot water installations. Variations outside these limits may reduce the efficiency of the system. Increasing the tilt favors winter energy collection, decreasing it favors summer collection.

1-4. Are system components located in such a manner as to harmonize with surroundings, to minimize vandalism and obstruction to pedestrian or vehicular traffic, and to facilitate emergency access?
DISCUSSION: Solar energy system components may include elements which are large and visually dominant when viewed from off-site. If

not carefully designed and located, such elements can produce a detrimental effect on the overall appearance of a residential area. The potential hazard and nuisance of collector reflections should not be ignored when planning system locations. Also, solar hot water systems should not block exits, roads, or walkways and, since they might be said to constitute an "attractive nuisance" (like an unfenced swimming pool), should be fenced off to prevent unauthorized access.

1-5. Are system components located in such a manner as to allow easy access for cleaning, adjusting, servicing, examination, replacement or repair, especially without trespassing on adjoining property?

DISCUSSION: Components should not be located in places which are difficult to reach. Storage tanks may need periodic inspection, maintenance, and possible replacement. Care must be taken that installers do not damage existing roofing or flashing.

1-6. Is there safe and easy access to gutters, downspouts, flashing, and caulked joints to allow minor repairs and preventative maintenance?

1-7. Are collectors located to minimize heat losses?

DISCUSSION: In order to avoid heat losses, the location of the solar collector should be planned to avoid low spots where freezing ground fog can collect or unprotected ridges where winds can be more extreme.

1-8. If ground-mounted, are collectors located to minimize interference from drifting snow, leaves, and debris?

DISCUSSION: Collector surfaces are often smooth and slippery, warmer than their surroundings, and located in elevated positions at steep angles. Adequate space should be provided for melting snow to slide off collectors. Poorly placed ground-mounted collectors and components may encounter snow cover and drifting problems beyond the capabilities of collectors to clean themselves. Collectors should be a minimum of 18 inches above ground level at their lowest point to reduce drift coverage and mud splashing.

1-9. If roof-mounted, are existing roof structures capable of supporting the additional load imposed by the collectors?

2. COLLECTOR MOUNTING

2-1. Is the framework constructed to support collectors under anticipated extreme weather conditions (wind loading up to 100 MPH, ice, rain, etc.)?

DISCUSSION: A tilted collector array can markedly increase the wind loading (both positive and negative) imposed on a roof.

2-2. Has the framework been treated to resist corrosion?

2-3. Are joints between the framework and the rest of the building caulked and/or flashed to prevent water leakage and are collectors installed so as not to contribute to moisture build-up, rotting, or deterioration of the roof or wall of the building?

DISCUSSION: All holes in the roof should be flashed and sealed according to the National Roofing Contractors Association Manual of Roofing Practices, 1970.

2-4. Are collectors installed so that water flowing off warm collector surfaces cannot freeze in cold weather and damage roof or wall surfaces?

DISCUSSION: Keep collectors several feet from eaves to prevent ice dams from forming and backing water up under shingles. In the case of ground-mounted systems, a gutter and downspout might be necessary to prevent excessive erosion around the footings.

2-5. Have collectors been mounted with weep holes (if provided) at the lowest end of the collector?

DISCUSSION: Proper provision for runoff of condensation within the collectors minimizes the problem of fogging of the inside of the collector glazing. Holes should be blocked with glass fiber to prevent entry of dirt.

2-6. In areas that have snow loads over 20 pounds per square foot or greater, have provisions been made to deflect snow or ice that may slide off roof-mounted components and endanger vehicles or pedestrians?

3. PIPING AND VALVES

3-1. Have the required building, plumbing, and electrical permits (if necessary) been obtained prior to the start of installation?

DISCUSSION: In some localities it may be necessary to supply background information on the operation of solar domestic hot water systems to the local building inspector.

3-2. Have solar components been ordered well in advance of the scheduled date to begin installation?

DISCUSSION: Installers frequently underestimate the time required to complete their first few installations. Special order equipment should be received before installation begins.

3-3.   Are connections to potable water lines being made by a licensed plumber where required by local codes?

3-4.   Is all piping properly insulated to maintain system efficiency?

DISCUSSION: All piping should be insulated to at least R-4. The first 5-10 feet of pipe coming from the conventional water heater tank also should be insulated.

3-5.   Is all exposed insulation protected from weather damage?

DISCUSSION: Fiberglass, rigid pipe insulation should be wrapped with weather-protective material. Neoprene insulation must be painted with a flexible finish or wrapped to prevent ultraviolet degradation. It is important to seal all joints, because water seepage will damage insulation and reduce system efficiency.

3-6.   Are sufficient pipe hangers, supports, and expansion devices provided to compensate for thermal effects?

DISCUSSION: Insulation should not be interrupted for pipe hangers or supports. Pipes should be insulated as completely as possible. Straight pipe runs of 100 feet or more call for an expansion joint and anchors to avoid pipe or equipment damage.

3-7.   In ground-mounted systems, are insulated pipes to and from collectors buried below the frost lines?

DISCUSSION: Not only do unburied pipes present a poor appearance, but inadequately buried pipes may be subjected to abnormal stresses due to frost heaving (insulation can absorb some stress). Because buried pipe insulation is subject to absorption of ground water, use only thoroughly waterproofed closed-cell foam insulation. Local codes may prohibit normal 95-5 or 50-50 soldered joints underground. Use 45 percent silver solder, brazing alloys, or whatever local codes require. Take care when backfilling the ditch that rocks and construction debris do not touch pipes and create stress points.

3-8.   Is piping for draindown systems properly pitched to facilitate draining of fluid from the collectors?

DISCUSSION: Draindown systems with piping not buried below the frost line are vulnerable to frost heaving, changes in the piping pitch, air lock, and subsequent incomplete draindown resulting in frozen pipes, and collector damage. Closed loop systems should also be pitched to facilitate draining, filling, and venting.

3-9. If ground-mounted collectors are used, is the run of pipe to storage and back reduced to the absolute minimum?

DISCUSSION: Long pipe runs between collectors and storage tanks reduce the efficiency of the system by increasing heat losses and pressure drops. If the run from collectors to storage and back is greater than 100 feet, use thicker insulation and consult standard plumbing tables for pipe size required. Provisions must be made for expansion.

3-10. Have isolation valves been provided so that major components of the system (pumps, heat exchangers, storage tank) can be serviced without system draindown?

3-11. Have air bleed valves been provided at high points in the system so that air can be removed from the liquid circulation loop during both filling and normal operation?

3-12. Have suitable connections been supplied for filling, flushing, and draining both the collector loop and the potable water piping of the system?

3-13. Has piping been leak tested to 1½ times system design pressure for at least 1 hour at constant temperature (with collectors covered) prior to backfilling and insulating?

3-14. Has corrosion between dissimilar metals been avoided by the use of suitable inhibitors in the system as well as dielectric washers in the mounting?

3-15. Has care been taken not to short out the insulating effect of dielectric washers between dissimilar metals by pipe hangers, control systems connections, etc.?

3-16. Will heat transfer fluids be safe and stable at both stagnation temperature and normal running temperatures?

DISCUSSION: Glycol heat transfer fluids should not be subjected to more than 250°F, because this will shorten service life. The flash point of any oil should be compared with the maximum collector stagnation temperature, particularly if the collector manifold is inside the building envelope.

3-17. If a system using antifreeze is used, have a fill valve and a drain (for sampling) been provided in the collector loop?

DISCUSSION: An extra gate valve and drain may have to be installed to blow out the system with compressed air if it is not pitched to drain properly. Make sure that the fill port is upstream of the check valve to prevent air being trapped in the system when filling the collector loop with antifreeze.

3-18.   Has a tempering valve or other temperature limiting device been installed to limit exit temperature of the hot water to a safe level?

DISCUSSION: Reducing the output water temperature prevents scalding, saves energy, and may be required by law for unregulated DHW water heaters.

3-19.   If a system containing antifreeze is used, have threaded joints been taped with tightly drawn Teflon[R] tape?

DISCUSSION: Antifreeze solutions will often leak through joints that normally will contain water. Therefore, special attention should be paid to joint leakage.

3-20.   Are all systems, subsystems, and components clearly labeled with appropriate flow direction, fill weight, pressure, temperature, and other information useful for servicing or routine maintenance?

3-21.   Are there vacuum relief valves in the system to prevent the collapse of storage or expansion tanks?

3-22.   Has care been taken to install the circulator pumps so that fluid is flowing in the proper direction?

DISCUSSION: Improper pump installation is a common problem. Check to see that the small paper gasket in a Bell and Gossett[R] pump (if used) is still intact after rotating the pump for proper flow direction. Pumps manufactured by other firms will indicate flow direction by another method. Isolate pump vibration from structural members.

3-23.   Has the expansion tank been located on the suction side of the pump?

3-24.   Has a check valve been installed in the collector loop to prevent reverse circulation by thermosiphoning at night?

DISCUSSION: Such reverse circulation causes system inefficiency and heat losses.

3-25.   Are vacuum relief valves protected from freezing?

4. STORAGE TANK

4-1. Is the storage tank insulated to at least R-11?

4-2. Are the piping connections to the tank located to promote thermal stratification?

DISCUSSION: The cold city water inlet should be at the bottom of the storage tank, as should the pickup for the cold water to be supplied to the heat exchanger or collector. Hot fluid returning from the heat exchanger or collector should be introduced at the top of the tank. Do not block the top of a gas heater or the entrance of combustion air at its base.

4-3. If a storage tank is installed on a roof or in an attic, is it provided with a drip pan and an outlet to an adequate drain?

4-4. Is the storage tank properly connected to the conventional water heater?

DISCUSSION: To maximize system efficiency, the collector loop must be permitted to operate independently of the hot water demand. A separate tank, in addition to the conventional heater tank, will provide greatest system efficiency. The higher end of the solar storage tank, through which the collector loop circulates, may be connected to the low end of the conventional hot water tank so that cold city water goes into the solar tank and forces the preheated water from the top of the solar tank into the conventional tank.

4-5. Are buried storage tanks anchored to prevent flotation in case of high groundwater levels?

5. SYSTEM SAFETY

5-1. Are all surfaces with running temperatures at 120°F. or higher isolated from pedestrian traffic in order to prevent burns?

5-2. Are temperature and/or pressure relief valves installed so that pedestrians or equipment are not exposed to effects of venting valves?

5-3. Are temperature and/or pressure relief valves installed so as to prevent system pressures from rising above working pressure and temperatures?

5-4. When toxic or flammable fluids are used in the system, will fluids overflow or discharge into sewers or storage in a manner acceptable to the local applicable codes?

DISCUSSION: As an added safety precaution, the end of the pipe draining the toxic or flammable heat transfer fluid should NOT BE THREADED to prevent any type of hose hookups for any accidental use.

5-5. If supplied water pressure is in excess of 80 pounds per square inch or the working pressure rating of any system components, has an approved pressure regulator preceded by an adequate strainer been installed?

DISCUSSION: Pressure should be reduced below 80 psig or system working pressure. Each regulator and strainer should be located and isolated by valves so that the strainer is readily accessible for cleaning without removing the regulator or strainer body or disconnecting the supply piping.

5-6. Has the system been designed so that any direct connection between wastes from the system and potable water is impossible?

5-7. Is there an approved backflow preventer at the cold water supply inlet if required?

DISCUSSION: When a backflow preventer is installed, an expansion tank may be required for the hot water system.

5-8. Is there a double-walled heat exchanger in the system or another approved method of separating nonpotable heat transfer fluids from potable water?

DISCUSSION: HUD's Intermediate Minimum Property Standards notes that single-walled heat exchanger designs that rely solely on potable water pressure or extra thick walls to prevent contamination are not acceptable.

5-9. Have all outlets and faucets on the nonpotable water lines of the system that might be used by mistake for drinking or domestic uses been marked "DANGER— WATER NOT DRINKABLE"?

DISCUSSION: It is suggested that valve handles be removed or tool-operated valves be added if the system is accessible to children too young to read.

5-10. If hazardous fluids are used in the system, have proper procedures for their use, including first-aid, handling, and safe disposal been supplied to the owner?

DISCUSSION: Nonpotable heat transfer fluids should be colored as a safety precaution.

5-11.   Is adequate drainage available in the collector piping array for leaks in collectors and discharges from pressure relief valves?

DISCUSSION: Suitable high-temperature weather-resistant piping should be utilized. Because some oils are corrosive to asphalt shingles, take special care when oils are used in the collector loop and collectors are roof-mounted.

6. ELECTRICAL SYSTEM

6-1.   Does field electrical wiring comply with all applicable local codes and equipment manufacturer's recommendations?

6-2.   Is there a properly grounded and protected power outlet for the system controls?

6-3.   Has control circuit wiring been color-coded or otherwise labeled so that wires are readily traceable?

6-4.   Are the sensors for collectors and storage tank attached tightly for the best possible thermal transfer and located per equipment manufacturer's instructions?

6-5.   Is the collector temperature sensor located in a collector or near the exit from the collector array?

7. CHECKOUT AND START-UP OF SYSTEM

7-1.   Has a person qualified in both solar and conventional hot water systems put the system through at least one start-up and shutdown cycle, including putting the system through all modes of operation?

DISCUSSION: Installation labor should include time allotted to balance the system for proper flow. Temperature differences under known conditions can be used in lieu of a flow meter.

7-2.   Has the owner been instructed in the proper start-up and shutdown procedures, including the operation of emergency shutdown devices, and fully instructed in the importance of routine maintenance of the system, including cleaning collector glazing and other components, draining and refilling the system, air venting, corrosion control, and other procedures?

DISCUSSION: A clear understanding between installer and owner as to what task could or should be undertaken by each party is a valuable tool for increasing solar business. Some owners may be interested in doing some or all of their system's routine maintenance. Others will gladly enter a service contract arrangement when the full extent of the routine maintenance tasks is clearly understood. While you might be more interested in an installation—rather than maintenance—based business, remember that regular service calls give the owner an opportunity to have a system "tuned" to greatest efficiency.

7-3.   Do operating instructions include provisions for the system if the owner leaves for a vacation and hot water use is nil?

7-4.   Has the system been designed so that both solar and conventional systems can operate independently of each other?

10.3 REFERENCES AND SUGGESTED READING

1. Dunn, Robert J. (ed.), *Hudson Home Guide Home Plans and Projects #93*, May 1977, Vol. XXiii, No. 5, p. 51.
2. Franklin Research Center, *Installation Guidelines for Solar DHW Systems in One- and Two-Family Dwellings*, U. S. Government Printing Office; Washington, D.C., May 1980.
3. Interactive Resources, Point Richmond, California, *Buying Tips*, 1977.
4. U. S. Department of Energy, *Active Solar Energy System Design Practice Manual*, National Technical Information Service; Springfield, Va., October, 1979.

CHAPTER 11
SOLAR ENERGY BUSINESS OPPORTUNITIES
FOR THE PROFESSIONAL

11.1 OVERVIEW

With the present energy crisis, alternate energy sources such as solar power are gaining popularity as a feasible supplement to fossil fuels. Problems in the nuclear industry highlighted by the Brown's Ferry and Three Mile Island accidents have served to further focus interest on solar energy as an acceptable and as yet largely untapped reservoir of limitless energy.

The future for solar energy, as a means of supplementing the nation's requirements, is very promising. This translates into an excellent opportunity for architects, engineers, and contractors. For the professional desiring additional exposure to solar energy design methods and installation techniques, numerous workshops sponsored by the Department of Energy (DOE) and other organizations are held periodically all over the country. These workshops enable the professional to learn from others in the field and to hear speakers and lecturers presenting workshops. Numerous colleges across the country are now offering at least one solar related course, while others offer complete solar curriculum. The *National Solar Energy Education Directory* gives a state by state listing of the colleges offering solar related courses. The individual desiring to take one of the solar related courses at a university, without credit for the course, can generally register as an auditor.

Although presently in many areas of the nation the local demand for solar designers and contractors may be low, the field is expected to open up as the energy crunch becomes more pervasive and acute. Presently many of the solar designers are architects and engineers first, and *solar* designers second. This is also generally the case for solar contractors who are heating, ventilating, and air-conditioning (HVAC) contractors first and *solar* contractors second. This is simply because in many cases the contractor or design firm cannot generate enough business to operate solely as either a solar contractor or solar designer.

11.2 GOVERNMENT GRANTS

There are options that the solar contractor or design firm can explore to increase solar related business. One specific avenue involves submitting bids for federal grant money or contracts. These calls for bids by the Federal Government can generally be found in the *Federal Register, Commerce Busi-*

ness Daily, and the *Catalog of Federal Domestic Assistance.* The bids can range in scope from research and development type grants to solar heating and cooling demonstration programs. For example, one such grant announced by the Department of Housing and Urban Development (HUD) in July, 1976 called for submissions from individuals wanting to install solar energy equipment in residences. From this particular HUD project, almost $4 million in grants were awarded to individuals for purchasing and installing solar energy equipment in a total of 1,411 residences. For a list of some of the recipients of this particular grant and a diagram with description of the recipient's particular solar system, one can refer to the HUD publication, *Solar Heating and Cooling Demonstration Program–A Descriptive Summary of HUD Cycle 2 Solar Residential Projects–Fall 1976.*[17] HUD has had other grants such as the one announced on May 31, 1977, whereby $6 million was awarded to recipients for buying and installing solar energy equipment in 3,468 residences.

There are a number of other agencies and organizations supplying grants to individuals and businesses. The grants available are of all different types. For example, the two previously described HUD grants were for designing and installing solar systems in residences. Other types of grants include grants for research and development (R & D) programs. The R & D grant programs include such things as developing solar collectors, thermal energy storage systems, passive systems, and systems analysis. In Chapter 8 of this book the SOLCOST computer program was highlighted. This computer program was developed by the Martin Marietta Corporation with a $153,385 grant. The following paragraphs describe the R and D grant to Martin Marietta Corporation.[12] (This particular R & D grant would fall under the "system analysis" grant category.

PROJECT TITLE:
   Solar Heating and Cooling Computer
   Analysis—A Simplified Sizing Design
   Method for Non-Engineers

INSTITUTION:
   Martin Marietta Corporation
   P. O. Box 179
   Denver, CO 80201

PRINCIPAL INVESTIGATOR:
   R. T. Grellis

CONTRACTOR NUMBER: EY-76-C-02-2876

STARTING DATE: 03/08/76

DURATION: 18 months

AMOUNT: $153,383

PROJECT DESCRIPTION

The objective of this project is to develop a validated and documented solar heating and cooling system design tool for use by persons untrained in solar system design and thermal analysis. A version of this tool will also be developed for engineering consultants. It will be a generalized solar energy heating and cooling computer program and a simple form designed to define the building's thermal characteristics. It will provide for the input of data to a widely distributed time-sharing computer system, and define a solar system for a specific building in a specific location according to minimum life cycle cost. With modifications, if necessary, it can be applied to any computer network.

The water heating design procedure is currently available.

Another R and D grant is listed below:

PROJECT TITLE:
A Proposal to Develop Very Low Cost Non-Concentrating Collectors

INSTITUTION:
Payne, Inc.
1910 Forest Drive
Annapolis, MD 21401

PRINCIPAL INVESTIGATOR:
P. R. Payne

CONTRACT NUMBER: EG-77-G-04-4138

STARTING DATE: 09/30/77

DURATION: 12 months

AMOUNT: $129,261

PROJECT DESCRIPTION

This project applies experience with trickle-type lightweight concrete water heaters to low temperature air heating collectors. This is accomplished by optimizing, designing and building three different types of low temperature integral wall air heaters all of which are "building blocks" intended to comprise part or all of a south-facing wall. Two of the intended configurations are "precision blocks," and are bonded together in the wall with epoxy resin—a technique which is already gaining favor with conventional concrete blocks. The third scheme uses conventional blocks and mortar, and may be retrofitted to an existing building. Fairly detailed mathematical models of collector performance will be devised and careful comparisons made between the predictions of these models and actual performance. The project also includes an analysis of thermal stress effects, and the structural load capabilities of the various designs.

This project is divided into the following tasks:

1. Definition of optimum materials and establishment of thermal and structural characteristics
2. Analytical optimization
3. Design
4. Tool construction
5. Preliminary performance measurements
6. Field evaluation
7. Reporting

There are also R & D grants awarded to individuals:

PROJECT TITLE:
Evaluation of Costs and Performance of Four Solar Heating Building Experiments

INSTITUTION:
Norman B. Saunders
Experimental Manor
15 Ellis Road, Sunshine Circle
West, MA 02193

PRINCIPAL INVESTIGATOR:
Norman B. Saunders, P.E.

CONTRACT NUMBER: EG-77-X-04-1078

STARTING DATE: 09/30/77

DURATION: 12 months

AMOUNT: $9,999

PROJECT DESCRIPTION

The objective of this project is to measure and document the performance and costs of four solar heated buildings and to disseminate the results. This will be done by installing instrumentation and data-logging devices in the subject buildings: assembling, analyzing, and evaluating the performance data; and publishing the results in quarterly reports with a recapitulation in the report for the last quarter of the project. Suitable procedures for conducting the research were established under ERDA project WA-75-4947.

PROJECT TITLE:
Skytherm Design Study

INSTITUTION:
Harold R. Hay
2425 Wilshire Boulevard
Los Angeles, CA 90057

PRINCIPAL INVESTIGATOR:
Harold R. Hay
Daniel A. Aiello

CONTRACT NUMBER: EG-77-G-04-4134

STARTING DATE: 09/19/77

DURATION: 12 months

AMOUNT: $39,980

PROJECT DESCRIPTION

This project will study the use of thermoponds for passive heating and cooling.

In the first phase of the study, professionals will assess untested Skytherm roof designs and select two for ground-level accelerated testing. The design will then be installed on a full-scale house and tested.

In Phase Two, the information and design details from Phase One will be combined with what has already been learned about Skytherm and with FHA Minimum Building Standards. The result of this synthesis will be a design manual for low-cost Skytherm applications. Based on this manual, the selected and proved low-cost design will undergo further analysis.

Also separate structural analysis will be conducted, above-ceiling elements will be detailed and the working plans for the low-cost structures will be costed and thermally and structurally analyzed. A revised design manual will be distributed.

The previous information describing the R & D grants may be found in the U. S. Department of Energy (DOE) Publication: *Solar Heating and Cooling Research and Development Project Summaries.* This DOE publication also describes numerous other grants awarded. There are also energy related loan

Table 11.1. Market Summary — Millions of 1976 Dollars. (Source: International Compendium[8]).

	1982		1987		1992	
	A. With Present Programs	B. Plus New Programs	A. With Present Programs	B. Plus New Programs	A. With Present Programs	B. Plus New Programs
Water Heaters						
New Residential	53	120	233	435	680	960
Retrofit Residential	180	683	725	1,730	1,925	2,550
Nonresidential	4.5	8	57	250	290	635
Space Heating						
New Residential	55	90	315	1,200	1,760	1,940
Retrofit Residential	11	30	205	1,100	4,500	8,935
Nonresidential	4	16	143	600	2,400	4,935
Air Conditioning						
Residential	.5	1	9	120	225	675
Nonresidential	2	2	63	265	1,220	3,870
Total	310	950	1,750	5,700	13,000	24,500

Table 11.2.   Market Summary — Percentage Breakdown. (Source: International Compendium[8]).

	1982 A. With Present Programs	1982 B. Plus New Programs	1987 A. With Present Programs	1987 B. Plus New Programs	1992 A. With Present Programs	1992 B. Plus New Programs
Water Heaters						
New Residential	17.1%	12.6%	13.3%	7.6%	5.2%	3.9%
Retrofit Residential	58.1	71.9	41.4	30.4	14.8	10.4
Nonresidential	1.5	.8	3.3	4.4	2.2	2.6
Space Heating						
New Residential	17.7	9.5	18.0	41.1	13.5	7.9
Retrofit Residential	3.5	3.2	11.7	19.3	34.7	36.5
Nonresidential	1.3	1.7	8.2	10.5	18.5	20.1
Air Conditioning						
Residential	.2	.1	.5	2.1	1.7	2.8
Nonresidential	.6	.2	3.6	4.6	9.4	15.8

programs that periodically appear which allow one to borrow money for less than the going market rate. These types of loans provide added incentive for incorporating solar systems and energy saving devices in buildings.

Tables 11.1 and 11.2 present summary figures for projected growth in the solar field. The "new programs" refer to tax incentives offered by the government.

11.3 REFERENCES AND SUGGESTED READING

1. Catalog of Federal Domestic Assistance (CFDA), Washington, D.C.: Superintendent of Documents, U. S. Government Printing Office (published yearly).
2. Code Development Team, *State of Florida Model Energy Efficiency Building Code – Code for Energy Conservation and New Building Construction With Florida Amendments,* Tallahassee, Florida: State Energy Office, November, 1978.
3. *Commerce Business Daily,* Washington, D.C.: Superintendent of Documents, U. S. Government Printing Office.
4. Council on Environmental Quality, *Solar Energy – Progress and Promise,* Washington, D.C.: U. S. Government Printing Office, April, 1978.
5. Energy Research and Development Administration, *Solar Energy and America's Future,* Washington, D.C.: U. S. Government Printing Office, March, 1977.
6. *Federal Register,* Washington, D.C.: Superintendent of Documents, U. S. Government Printing Office.
7. Federal Trade Commission, Bureau of Competition, *The Solar Market: Proceedings of the Symposium on Competition in the Solar Energy Industry,* Washington, D.C.: U. S. Government Printing Office, June, 1978.
8. International Compendium — A Division of Solar Science Industries, Inc., *Solar Energy Business.*
9. Library of Congress, *A Guide to Federal Programs of Possible Assistance to the Solar Energy Community,* Washington, D.C.: U. S. Government Printing Office, 1977.
10. Solar Energy Research Institute, *National Solar Energy Education Directory,* Washington, D.C.: U. S. Government Printing Office, January, 1979.
11. U. S. Department of Energy, *National Program Plan for Research and Development in Solar Heating and Cooling for Building, Agricultural, and Industrial Applications,* Washington, D.C.: U. S. Government Printing Office, August, 1978.

12. U. S. Department of Energy, *Solar Heating and Cooling Research and Development Project Summaries,* Washington, D.C.: U. S. Government Printing Office, May, 1978.

13. U. S. Department of Housing and Urban Development, *Building the Solar Home,* Washington, D.C.: U. S. Government Printing Office, June, 1978.

14. U. S. Department of Housing and Urban Development, *HUD Solar Status – Installing Solar: Training Expands to Meet the Need,* Washington, D.C.: U. S. Government Printing Office, September, 1978.

15. U. S. Department of Housing and Urban Development, *Selling the Solar Home,* Washington, D.C.: U. S. Government Printing Office, April, 1978.

16. U. S. Department of Housing and Urban Development, *Solar Heating and Cooling Demonstration Program - A Descriptive Summary of HUD Solar Residential Demonstration Cycle 1,* Washington, D.C.: U. S. Government Printing Office, July, 1976.

17. U. S. Department of Housing and Urban Development, *Solar Heating and Cooling Demonstration Program - A Descriptive Summary of HUD Cycle 2 Solar Residential Project - Fall 1976,* Washington, D.C.: U. S. Government Printing Office, April, 1977.

18. U. S. Department of Housing and Urban Development, *Solar Heating and Cooling Demonstration Program - A Descriptive Summary of HUD Cycle 3 Solar Residential Projects - Summer 1977,* Washington, D.C.: U. S. Government Printing Office, November, 1977.

CHAPTER 12
SOURCES OF INFORMATION

12.1 SOLAR ENERGY INFORMATION

With constant change and expansion within the solar energy field, keeping abreast of progress in the industry can be a challenge. One excellent source of such current (solar) energy information is the U. S. Government. Within the U. S. Department of Energy there are a number of offices, some of which publish their own newsletters. For example, the Department of Energy's Office of Consumer Affairs publishes *The Energy Consumer,* and the Department of Energy's Office of Public Affairs publishes *Energy Insider.* Both of these periodic publications may be obtained free of charge. Rather than writing to every department office requesting to be put on their mailing list, a "Solar Energy Information Request" form will serve to have the individual's name placed on several mailing lists for free periodic newsletters from various branches of the government. Write to *both* the Solar Energy Research Institute and the U. S. Department of Energy and request their "Solar Energy Information Request" forms. Each of the request forms is basically a questionnaire asking questions about what topics of information one is interested in, occupation or profession, etc.

An excellent source of vast amounts of free solar energy literature and a place where questions concerning solar energy are answered is the National Solar Heating and Cooling Information Center. They have a toll-free number for literature requests or to answer questions. For Pennsylvania residents the number is (800) 462-4983, for Alaska and Hawaii residents the number is (800) 423-4700, and for the rest of the United States the number is (800) 523-2929.

Additional sources of solar energy information include:

1. Department of Energy Regional Offices
2. State Energy/Solar Offices
3. Regional Solar Energy Centers
4. Solar Energy Journals and Newsletters
5. Solar Directories
6. Testing Facilities for Solar Equipment

Tables 12.1 through 12.7 list items falling within each of the previously mentioned six category sources for energy information.

Table 12.1.   U. S. Department of Energy Regions. (Source: National Solar Heating and Cooling Information Center[5]).

REGION I   (BOSTON)
Connecticut
Maine
Massachusetts
New Hampshire
Rhode Island
Vermont

REGION II   (NEW YORK)
New Jersey
New York
Peurto Rico
Virgin Islands

REGION III   (PHILADELPHIA
Delaware
District of Columbia
Maryland
Pennsylvania
Virginia
West Virginia

REGION IV   (ATLANTA)
Alabama
Florida
Georgia
Kentucky
Mississippi
North Carolina
South Carolina
Tennessee

REGION V   (CHICAGO)
Illinois
Indiana
Michigan
Minnesota
Ohio
Wisconsin

REGION VI   (DALLAS)
Arkansas
Louisiana
New Mexico
Oklahoma
Texas

REGION VII   (KANSAS CITY)
Iowa
Kansas
Missouri
Nebraska

REGION VIII   (DENVER)
Colorado
Montana
North Dakota
South Dakota
Utah
Wyoming

REGION IX   (SAN FRANCISCO)
Arizona
California
Hawaii
Nevada
Guam
Pacific Trust Territories

REGION X   (SEATTLE)
Alaska
Idaho
Oregon
Washington

Table 12.2.   DOE Regional Offices. (Source: National Solar Heating and Cooling Information Center[5]).

REGION I   (BOSTON)
Analex Building
Room 700
150 Causeway Street
Boston, Massachusetts   02114

REGION II   (NEW YORK)
26 Federal Plaza
Room 3206
New York, New York   10007

REGION III   (PHILADELPHIA)
1421 Cherry Street
10th Floor
Philadelphia, Pennsylvania 19102

REGION IV   (ATLANTA)
1655 Peachtree St., N.E.
8th Floor
Atlanta, Georgia   30309

REGION V   (CHICAGO)
175 West Jackson Blvd.
Room A-333
Chicago, Illinois   60604

REGION VI   (DALLAS)
P. O. Box 35228
2626 West Mocking Bird Lane
Dallas, Texas   75235

REGION VII   (KANSAS CITY)
Twelve Grand Building
1150 Grand Avenue
Kansas City, Missouri   64106

REGION VIII   (DENVER)
P. O. Box 26247
Belmar Branch
1075 South Yukon Street
Lakewood, Colorado   80226

REGION IX   (SAN FRANCISCO)
111 Pine Street
3rd Floor
San Francisco, California 94111

REGION X   (SEATTLE)
1992 Federal Building
915 Second Avenue
Seattle, Washington   98174

Table 12.3. State Energy/Solar Offices. (Source: National Solar Heating and Cooling Information Center[7]).

AL Alabama Solar Energy Center
    Johnson Environmental &
      Energy Center
    University of Alabama/
      Huntsville
    P. O. Box 1247
    Huntsville, AL   35806
    (205) 895-6361

    Alabama Energy Management
      Board
    Rm. 203, Executive Bldg.
    312 Montgomery St.
    Montgomery, AL   36104
    (205) 832-5010

AK Alaska Dept. of Commerce
    Office of Energy Conversation
    7th Floor
    338 Denali St.
    Anchorage, AK   99501
    (907) 272-0527

AZ Arizona Solar Energy Research
      Commission
    Capital Tower - Rm. 500
    1700 W. Washington
    Phoenix, AZ   85007
    (602) 255-3303

AR Arkansas Energy Office
    960 Plaza West Bldg.
    Little Rock, AR   72205
    (501) 371-1370

CA California Energy Commission
    1111 Howe Ave.
    Sacramento, CA   95825
    *(800) 852-7516
     (916) 920-6430

CO Colorado Office of Energy
      Conservation
    1600 Downing St.
    Denver, CO   80218
    (303) 839-2507

CT Connecticut Office of Policy
      & Management
    Energy Division
    80 Washington St.
    Hartford, CT   06115
    *(800) 842-1648
     (203) 566-2800

DE Delaware Governor's Energy
      Office
    P. O. Box 140
    114 W. Water St.
    Dover, DE   19901
    *(800) 282-8616
     (302) 678-5644

FL Florida Solar Energy Center
    300 State Rd., 401
    Cape Canaveral, FL   32920
    (305) 783-0300

    Florida State Energy Office
    301 Bryant Bldg.
    Tallahassee, FL   32301
    (904) 488-6764

GA Georgia Office of Energy
      Resources
    Rm. 615
    270 Washington St., S.W.
    Atlanta, GA   30334
    (404) 656-5176

HI Hawaii State Energy Office
    Dept. of Planning & Economic
      Development
    P. O. Box 2359
    Honolulu, HI   96804
    (808) 548-4150
    (808) 548-4080 (on Oahu)
    Dial 0 ask for 8016
    (toll-free within Hawaii,
    off Oahu)

ID Idaho State Office of Energy
    State House
    Boise, ID   83720
    (208) 384-3800

IL Institute of Natural
      Resources
    325 West Adams
    Springfield, IL   62706
    (217) 782-1926

IN Indiana Dept. of Commerce
    Energy Group
    7th Floor-Consolidated Bldg.
    115 N. Pennsylvania St.
    Indianapolis, IN   46204
    (317) 633-6753

IA Iowa Solar Office
    Energy Policy Council
    Capitol Complex
    Des Moines, IA   50319
    (515) 281-8071

KS State of Kansas Energy Office
    Room 241
    503 Kansas Ave.
    Topeka, KS   66603
    *(800) 432-3537
     (913) 296-2496

KY Kentucky Department of Energy
    Capitol Plaza Tower
    Frankfort, KY   40601
    *(800) 372-7978
     (802) 564-7416

LA Louisiana Research &
      Development Division
    Dept. of Natural Resources
    P. O. Box 44156
    Baton, Rouge, LA   70804
    (504) 342-4594

ME Maine Office of Energy
      Resources
    55 Capitol St.
    Augusta, ME   04330
    (207) 289-2195

MD Maryland Energy Policy
      Office
    301 W. Preston St.
    Baltimore, MD   21201
    *(800) 494-5903
     (301) 383-6810

MA Massachusetts Office of
      Energy Resources
    73 Tremont St.
    Room 700
    Boston, MA   02108
    (617) 727-4732

MI Michigan Energy Administration
    Dept. of Commerce
    P. O. Box 30228
    6520 Mercantile Way, Suite 1
    Lansing, MI   48909
    *(800) 292-4704
     (517) 374-9090

Table 12.3. (Continued)

MN Minnesota Solar Office
980 American Center Bldg.
150 East Kellog
St. Paul, MN 55101
*(800) 652-9747
(612) 296-5175

MS Mississippi Fuel & Energy
Management Commission
Suite 228, Barfield Complex
455 North Lamar St.
Jackson, MS 39201
(601) 354-7406

MO Missouri Dept. of Natural
Resources
Division of Policy
Development
P. O. Box 1309
Jefferson City, MO 65120
*(800) 392-8269
(314) 751-4000

MT Montana Dept. of Natural
Resources & Conversation
32 South Ewing
Helena, MT 59601
(406) 449-3940

NE Nebraska State Solar Office
W-191 Nebraska Hall
University of Nebraska
Lincoln, NE 68588
(402) 472-3414

Nebraska State Energy Office
State Capitol
Lincoln, NE 68588
(402) 471-2867

NV Nevada Dept. of Energy
1050 E. Williams, Suite 405
Capitol Complex
Carson City, NV 89710

NH New Hampshire Governor's
Council on Energy
26 Plesant
Concord, NH 03301
*(800) 562-1115
(603) 271-2711

NJ New Jersey Office of
Alternate Technology
NJ Dept. of Energy
101 Commerce St.
Newark, NJ 07102
*(800) 492-4242
(201) 648-6293

NM New Mexico Energy &
Minerals Dept.
Energy Conservation &
Management Division
P. O. Box 2270
Santa Fe, NM 87501
*(800) 432-6782
(505) 827-2472
(505) 827-2386

NY New York State Energy Office
Agency Building #2
Empire State Plaza
Albany, NY 12223
(518) 474-8181

NC North Carolina Dept. of
Commerce
Energy Division
430 North Salisbury St.
Raleigh, NC 27611
*(800) 662-7131
(919) 733-2230

ND North Dakota State Solar
Office
1533 N. 12th St.
Bismarck, ND 58501
(701) 224-2250

OH Ohio Solar Office
30 East Broad St.
34th Floor
Columbus, OH 43215
*(800) 282-9284
(614) 466-7915

OK Oklahoma Department of
Energy
4400 North Lincoln Blvd.
Suite 251
Oklahoma City, OK 73105
(405) 521-3941

OR Oregon Dept. of Energy
Rm. 111
Labor & Industries Bldg.
Salem, OR 97310
*(800) 452-7813
(503) 378-4040

PA Pennsylvania Governor's
Energy Council
1625 N. Front St.
Harrisburg, PA 17102
*(800) 882-8400
(717) 783-8610

PR Puerto Rico Office of Energy
Energy Information Program
41089 Minillas Station
Santurce, PR 00940
(809) 727-8877

RI Rhode Island Governor's
Energy Office
80 Dean St.
Providence, RI 02903
(401) 277-3774

SC South Carolina Energy
Management Office
1205 Pendleton St.
Columbia, SC 29201
*(800) 922-1600
(803) 758-2050

SD South Dakota State Solar
Office
Capital Lake Plaza Bldg.
Pierre, SD 57501
*(800) 592-1865
(605) 224-3603

TN Tennessee Energy Authority
707 Capitol Blvd. Bldg.
Nashville, TN 37219
*(800) 642-1340
(615) 7412994

TX Texas Office of Energy
Resource
7703 North Lamar Street
Austin, TX 78752
(512) 475-5491

UT Utah Energy Office
231 East 400 South
Empire Building, Suite 101
Salt Lake City, UT 84111
*(800) 662-3633
(801) 533-5424

VT Vermont State Energy Office
State Office Bldg.
Montpelier, VT 05602
*(800) 642-3281
(802) 828-2393

Table 12.3. (Continued)

VA Virginia Division of Energy
310 Turner Street
Richmond, VA 23235
*(800) 552-3831
 (804) 745-3245

WA Washington State Energy
Office
400 East Union St.
1st Floor
Olympia, WA 98504
(206) 754-1370

WV West Virginia Fuel and
Energy Office
126-1/2 Greenbrier St.
Charleston, WV 25311
*(800) 642-9012
 (304) 348-8860

WI Wisconsin Division of
Energy
1 West Wilson - Rm. 201
Madison, WI 53702
(608) 266-9861

WY Wyoming Energy Conservation
Office
320 West 25th St.
Capital Hill Bldg.
Cheyenne, WY 82001
*(800) 442-6783
 (307) 777-7131

VI Virgin Islands Energy Office
P. O. Box 2996
St. Thomas
U. S. Virgin Islands 00801
(809) 774-6726

*Toll-free number in state only.

Table 12.4. Regional Solar Energy Centers. (Source: National Solar Heating and Cooling Information Center[5]).

MID-AMERICAN SOLAR
ENERGY COMPLEX

8140 26th Avenue, South
Bloomington, MN 55420

Illinois	South Dakota
Indiana	Wisconsin
Iowa	
Kansas	
Michigan	
Minnesota	
Missouri	
Nebraska	
North Dakota	
Ohio	

SOUTHERN SOLAR
ENERGY CENTER

61 Perimeter Park
Atlanta, GA 30341

Alabama	Mississippi
Arkansas	North Carolina
Delaware	Oklahoma
District of	Puerto Rico
Columbia	South Carolina
Florida	Tennessee
Georgia	Texas
Kentucky	Virginia
Louisiana	West Virginia
Maryland	Virgin Island

NORTHEAST SOLAR
ENERGY CENTER

70 Memorial Drive
Cambridge, MA 02142

Connecticut
Maine
Massachusetts
New Hampshire
New Jersey
New York
Pennsylvania
Rhode Island
Vermont

WESTERN SOLAR
UTILIZATION NETWORK

921 S.W. Washington St.
Suite 160
Portland, OR 97205

Alaska	New Mexico
Arizona	Oregon
California	Utah
Colorado	Washington
Hawaii	Wyoming
Idaho	
Montana	
Nevada	

Table 12.5. Solar Energy Journals and Newsletters — A Selected List. (Source: Florida Solar Energy Center[2]).

Advanced Solar Energy Technology Newsletter. Monthly. $60.
per year. 1609 West Windrose, Phoenix, AZ 85029.

AERO Sun-Times. $10 (includes one year membership in the Society)
Alternative Energy Resources, 417 Stapleton Bldg., Billings,
MT 59101.

Alternative Sources of Energy. Bimonthly. $10 per year.
Alternative Sources of Energy, Inc., Rt. 2, Box 90 A,
Milaca, NM 56353

Table 12.5. (Continued)

Applied Solar Energy. (Translation of Geliotekhnika) Bimonthly.
$125 per year. Allerton Press, Inc., 150 Fifth Avenue, New
York, NY 10011.

Biotimes. Bimonthly. $10 individual membership. The
International Biomass Institute, Suite 600, 1522 "K" Street, N.W.,
Washington, D.C. 20005.

California Energy Commission News. Monthly. Free. Office of
Communications, 1111 Howe Avenue, Sacramento, CA 98525.

Department of Energy Information Weekly Announcements. Free.
Department of Energy, Office of Public Affairs, Washington,
D.C. 20585.

Energy. Quarterly. $52 per year. Business Communications
Co., P. O. Box 2070C, Stanford, CT 06906

Energy: A Newsletter of Energy and the Built Environment.
Monthly. At present available only to subscribers of AIA
Energy Notebook ($60 per year for AIA members, $80 per year for
non-members). American Institute of Architects, 1735 New York
Avenue, N.W., Washington, D.C. 20006

Energy: The International Journal. Bimonthly. $165 per year.
Pergamon Press, Maxwell House, Fairview Park, Elmsford, NY 10523.

Energy and Buildings. Quarterly. $89.25 per year. Elsevier
Sequoia S.A., P. O. Box 851, 1001 Lausannel, Switzerland.

Energy Consumer. Monthly. Free. Department of Energy,
Office of Consumer Affairs, Washington, D.C. 20585

Energy Digest-Southeastern Region Newsletter. Irregular.
Free. Southern Rural Action, Inc., 1202 West Marietta St. N.W.,
Atlanta, GA 30318

Energy Insider. Biweekly. Free. Department of Energy,
Washington, D.C. 20585

Energy News. Monthly. Free. Tennessee Energy Authority,
226 Capitol Blvd., Suite 707, Nashville, TN 37219.

Energy Research Digest. Biweekly. $150 per year. Dulles
International Airport, P. O. Box 17162, Washington, D.C. 20041

Energy Review. Bimonthly. Free. Minnesota Energy Agency,
980 American Center Bldg., 150 E. Kellogg Blvd., St. Paul,
MN 55101.

Energy Studies. Bimonthly. Free. Center for Energy Studies,
ENS 143, The University of Texas at Austin, Austin, TX 78712.

Energy Today. Published twice a month. $120 per year. Trends
Publishing, Inc., National Press Building, Washington, D.C.
20045

Energyline. Monthly. Free. Department of Energy. Office of
Information Services, 101 Commerce Street, Newark, NJ 07102

Florida Energy Action Report. Bimonthly. Free. Governor's
Energy Office. The Capitol, Tallahassee, FL 32401.

Florida Solar Coalition Newsletter. Individual membership.
$10 per year. Environmental Information Center, 935 Orange
Ave., Winter Park, FL 32789.

Florida SUN. Quarterly, with intermittent bulletins. $6
(includes one-year membership in the Florida Solar Users
Network. $20 for corporations). Florida SUN, 1086 Coronado
Drive, Rockledge, FL 32955

HUD Solar Status. Irregular. Free. Solar Status, P. O. Box
1607, Rockville, MD 20850

ISES News. Quarterly. Membership rates: Individuals $20;
Students $12; Combined membership with American Section $40
per year. International Solar Energy Society, c/o American
Technical University, P. O. Box 1416, Killeen, TX 76541.

In Review: A SERI Monthly Update. Free. Solar Energy
Research Institute, 1536 Cole Blvd., Golden, CO 80401

International Journal of Solar Energy. Quarterly. $60 per
year (new publication to begin in 1979). Harwood Academic
Publishers, P. O. Box 786, Cooper Station, New York, NY 10003.

Table 12.5. (Continued)

Journal of Energy. Bimonthly. $7 per year for members; others
apply for rates. American Institute of Aeronautics and
Astronautics, Inc., 1290 Avenue of the Americas, New York,
NY 10019.

Journal of Solar Energy Engineering. Quarterly. $50 per year
for non-members; $25 per year for members (new publication to
begin in 1980). American Society of Mechanical Engineers,
345 E. 47th Street, New York, NY 10017.

NCSEA Newsletter. Monthly. Individual membership: $10 per
year. Northern California Solar Energy Association, P. O.
Box 1056, Mountain View, CA 94042.

New England Solar Energy Newsletter. $20 (includes membership
in New England Solar Energy Association). N.E.S.E.A., P. O.
Box 541, Brattleboro, VT 05301.

New Jersey Sun. $5 per year. N.J. Solar Action, 32 W. Lafayette
St., Trenton, NJ 08608.

New Mexico Solar Energy Association Monthly Bulletin.
Individual membership $10 per year. Institutional membership
$50 per year. N.M.S.E.A., P. O. Box 2004, Sante Fe, NM 87501.

People and Energy. Bimonthly. $15 for individuals; $25 for
institutions. Institute for Ecological Policies, 1413 "K"
Street, N.W., 8th Floor, Washington, D.C. 20005.

SEIA News. Monthly. Available to members only. Solar Energy
Industries Association, Inc., 1001 Connecticut Avenue, N.W.
Suite 800, Washington, D.C. 20036.

Solar Age. Monthly. $20 per year (Also included with membership
in American Section, International Solar Energy Society).
SolarVision, Inc., Church Hill, Harrisville, NH 03450.

Solar Cells. Quarterly. $89 per year (new publication
announced to begin in 1979). Elsevier Sequoia S.A., P. O.
Box 851, CH-1001 Lausannel, Switzerland.

Solar Energy. Monthly. $140 per year (Also included with
membership in International Solar Energy Society). Pergamon
Press, Maxwell House, Fairview Park, Elmsford, NY 10523.

Solar Energy Digest. Monthly. $30.50 per year. Solar Energy
Digest, P. O. Box 17776, San Diego, CA 93117.

Solar Energy Intelligence Report. Weekly. $90 per year.
Business Publishers, Inc., P. O. Box 1067, Silver Springs,
MD 20910.

Solar Energy Materials. 4 issues per year. $65 per year.
North-Holland Publishing Co., 52 Vanderbilt Avenue, New York,
NY 10019.

Solar Energy News: A Compendium. Biweekly. $67.50 per year.
Solar Energy News, 178 Miller Avenue, Providence, RI 02905.

Solar Energy Research & Development Report. Irregular. Free.
Department of Energy. Washington, D.C. 20545.

Solar Energy Update. Monthly. $27.50 per year. (Abstracts
of current scientific and technical literature.) National
Technical Information Service, Springfield, VA 22161.

Solar Engineering. Monthly. $15 per year. Solar Engineering
Publishers, Inc., 8435 N. Stemmons Freeway, Suite 880, Dallas,
TX 75247.

Solar Greenhouse Digest. Bimonthly. $10 per year. P. O.
Box 3218, Kingman, AZ 86401.

Solar Heating & Cooling. Bimonthly. (Monthly in 1980). $8
per year. Gordon Publications, P. O. Box 2126-R, Morristown,
NJ 07960.

Solar Law Reporter. Bimonthly. $7.50 per year. Solar Energy
Research Institute, P. O. Box 5400, Denver, CO 80217.

Solar Life. Monthly. Individual membership dues $15 per year.
Solar Energy Institute of America, 1110 6th Street, N.W.,
Washington, D.C. 20001.

Solar News: The Newsletter of the Mid-Atlantic Solar Energy
Assocation. Individual membership: $10 per year. MASEA,
2233 Gray's Ferry Avenue, Philadelphia, PA 19146.

Table 12.5.   (Continued)

Solar News & Views. Monthly (Centerfold of Solar Age Magazine).
$25 per year (includes membership in the American Section of
ISES). American Section of ISES, c/o American Technological
University, P. O. Box 1416, Killeen, TX 76541.

Solar Ocean Energy Liaison. Monthly. $95 per year. Popular
Products, Inc., 1303 South Michigan Avenue, Chicago, IL 60605.

Solar Resources. Bimonthly. Free. Minnesota Solar Resource
Advisory Panel, P. O. Box 9815, Minneapolis, MN 55440.

Solar Science and Technology Magazine. $20 per year (new
publication announced to begin in 1979). 8585 N. Stemmons
Freeway, Suite 912, Dallas, TX 75247.

Solar Times. Monthly $12.50 per year. Solar Times Co.,
901 Bridgeport Avenue, Shelton, CT 06484.

Solar Utilization News. Monthly. $8 per year to individuals;
$15 per year to commercial institutions and libraries. Alternate
Energy Institute, P. O. Bxo 3100, Estes Park, CO 80517.

Sun at Work in Britain: The Magazine of the UK Section of ISES.
$10 per year. Pergamon Press, Maxwell House, Fairview Park,
Elmsford, NY 10523.

Sun Times. Monthly. $15 per year. Solar Lobby. 1028
Connecticut Avenue, N.W., Washington, D.C. 20036.

Sun Up: Energy News Digest. Monthly. $6.50 per year.
55888 Yucca Trail, Suite 2, Yucca Valley, CA 92284.

Sunworld. Quarterly. $12 per year (Also included with
membership in ISES). Pergamon Press, Inc., Maxwell House,
Fairview Park, Elmsford, NY 10523.

Texas Energy and Mineral Resources. Published ten times a year.
Free. Editor, Center for Energy and Mineral Resources,
Texas A&M University, College Station, TX 77843.

The Solar Collector. Quarterly. Free. Editor, Florida Solar
Energy Center, 300 State Road 401, Cape Canaveral, FL 32920.

The Sunspot. Quarterly. Members $5 per year. Student members
$2.50 per year. Alabama Solar Energy Association, c/o The
University of Alabama in Huntsville, Center for Environmental
and Energy Studies, P. O. Box 1247, Huntsville, AL 35807.

Vermont Energy News. Monthly. Free. Virginia Office of
Emergency and Energy Services. 310 Turner Road, Richmond,
VA 23225.

Vermont Energy News. Monthly. Free. Vermont State Energy
Office, State Office Building, Montpelier, VT 05602.

Virginia Energy Report. Monthly. Free. Virginia Office of
Emergency and Energy Services, 310 Turner Road, Richmond,
VA 23225.

Weekly Government Abstracts-Energy (NTIS). $75 per year.
National Technical Information Service, 5285 Port Royal Road,
Springfield, VA 22161.

Wind Energy Report. Monthly. $75 per year. P. O. Box 14,
Rockville Center, NY 11571.

Table 12.6.   Solar Directories. (Source: National Solar Heating and Cooling Information Center[6]).

ARIZONA SOLAR ENERGY DIRECTORY
  (1977; Free to Arizona residents;
   $1.00)
Judi Wright
Solar Energy Research Commission
1700 West Washington St.
Room 502
Phoenix, AR 85007
(602) 271-3682

SOLAR COMPENDIUM
  (1977; $25.00 prepaid)
Environment Information Center
124 E. 39th St.
New York, NY 10016
(212) 949-9494

ASE SPECTRUM
  (1975; $2.00)
Alternative Sources of Energy
Route 2, Box 90A
Milaca, MN 56353

Table 12.6. (Continued)

DESK DIRECTORY OF MANUFACTURERS
(1977; $1.50)
Solar Engineering
8435 N. Stemmons Freeway
Suite 880
Dallas, TX 75247

DIRECTORY OF THE SOLAR INDUSTRY
(1977; $12.50)
Solar Data
13 Evergreen Road
Hampton, NH 03842
(603) 926-8082

INFORMATION DIRECTORY OF THE
ORGANIZATIONS AND PEOPLE
INVOLVED IN THE SOLAR HEATING
OF BUILDINGS
(1977; 3rd edition; $9 prepaid)
W. A. Shurcliff
19 Appleton St.
Cambridge, MA 02138

SOLAR AGE CATALOGUE
(1977; $8.50; June/July issue
of Solar Age)
SolarVision, Inc.
Box Z
Port Jervis, NY 12771
(914) 856-6633

SOLAR COLLECTOR MANUFACTURING
ACTIVITY
(1976; Free)
National Energy Information
Center
Federal Energy Administration
Federal Building, Room 4358
12th & Pennsylvania Ave., NW
Washington, D.C. 20461
(202) 961-8685

SOLAR DIRECTORY
(1975; $20.00 prepaid)
Carolyn Pesko (editor)
Ann Arbor Science Publishers
Box 1425
Ann Arbor, MI 48106

SOLAR DIRECTORY - Vol. 1
(1977; $20.00)
Home Free
4924 Greenville Ave.
Dallas, TX 95206
(214) 368-8850

SOLAR ENERGY DIRECTORY
(1976; $7.50)
Centerline Corporation
401 S. 36th Street
Phoenix, AZ 85034
(602) 267-0014

SOLAR INDUSTRY INDEX
(1977; $8.00 + $2.00 mailing)
Solar Energy Industries
Association (SEIA) Publications
1001 Connecticut Ave. NW
Sutie 632
Washington, D.C. 20036

1977 SUN CATALOGUE
(1977; $2.00)
Solar Usage Now
Box 306
Bascom, OH 44809
(419) 937-2226

DIRECTORY OF SOLAR ENERGY
EQUIPMENT MANUFACTURERS &
SOLAR ENERGY ARCHITECTS &
ENGINEERS
(1976; Free)
Illinois Dept. of Business
& Economic Development,
Division of Energy
222 S. College Street
Springfield, IL 62706
(217) 782-7500

FLORIDA SOLAR ENERGY EQUIPMENT
AND SERVICE
(1976; Free)
Florida Solar Energy Center
300 State Road 401
Cape Canaveral, FL 32920
(305) 783-0300

NEW ENGLAND "YELLOW PAGES" OF
SOLAR ENERGY DEVELOPMENT
(1976; $5.00)
New England Solar Energy
Association
P. O. Box 541
Brattleboro, VT 05301
(802) 254-2386

SYNERGY: DIRECTORY OF ENERGY
ALTERNATIVES
(1976; $8.00/copy; $14.00
update subscription)
SYNERGY
P. O. Box 4790
Grand Central Station
New York, NY 10017

SOLAR ENERGY AND RESEARCH
DIRECTORY
(1977; $22.50)
Ann Arbor Science Publishers
Box 1425
Ann Arbor, MI 48106
(313) 761-5010

SOLAR ENERGY SOURCEBOOK
(1977; Free - Members; $12.00
prepaid - Non-members)
C. W. Martz (editor)
Solar Energy Institute of
America (SEINAM)
P. O. Box 6068
Washington, D.C. 20005
(202) 667-6611

SOLAR REGISTER
(1977; $50.00)
9 Allen Street
Toms River, NJ 08753
(201) 341-4500

SURVEY OF THE EMERGING SOLAR
ENERGY INDUSTRY
(1977; $60.00)
Solar Energy Information
Services
P. O. Box 024
San Mateo, CA 94401
(415) 347-2640

WESTERN REGIONAL SOLAR ENERGY
DIRECTORY
(1976; $2.35)
Southern California Solar
Energy Association
202 C Street - 11B
San Diego, CA 92101
(714) 232-3914

Table 12.7. Testing Facilities for Solar Equipment. (Source: National Solar Heating and Cooling Information Center[8]).

The following testing facilities have indicated that they test collectors according to ASHRAE 93-77 and HUD 4930.2 standards. Some test both flat plate and concentrating collectors.

William Parkyn
APPROVED ENGINEERING TEST LAB.
20988 Golden Triangle Road
Canyon Country, CA 91351
(805) 259-8184

Dr. Stanley A. Mumma, PE
ARIZONA STATE UNIVERSITY
College of Architecture
Tempe, AZ 85281
(602) 965-2764

A. R. Lunde
BOEING KENT FACILITY
P. O. Box 3999, Mail Stop 8601
Seattle, WA 98124
(206) 773-8516

J. O. Bradley or Chuck Miller
DESERT RESEARCH INSTITUTE
Energy Systems Center
1500 Buchanan Blvd.
Boulder City, NV 89005
(702) 293-4217

Gene A. Zerlaut, Pres.
DESERT SUNSHINE EXPOSURE TESTS,
  INC.
Box 185, Black Canyon State
Poenix, AZ 85020
(602) 465-7525

M. C. Falk, PE
PITTSBURGH TESTING LABORATORY
850 Poplar Street
Pittsburgh, PA 15220
(412) 922-4000

Dr. John Minardi
RESEARCH INSTITUTE
University of Dayton
Dayton, OH 15220
(513) 229-4235

Dr. James R. Clinton
SOLAR ENERGY ANALYSIS LABORATORY
4325 Donald Avenue
San Diego, CA 92117
(714) 270-3781

Danny M. Deffenbaugh
SOUTHWEST RESEARCH INSTITUTE
Department of Mechanical Sciences
8500 Culebra Road
P. O. Drawer 28510
San Antonio, TX 78284
(512) 684-5111 Ext. 2384

David Jackson
UNIVERSITY OF CONNECTICUT
Department of Mechanical
  Engineering
Box U-139
Storrs, CT 06268
(203) 486-2189

James Roland
FLORIDA SOLAR ENERGY CENTER
300 State Road 401
Cape Canaveral, FL 32920
(305) 783-0300

David L. Christensen
JOHNSON ENVIRONMENTAL & ENERGY
  CENTER
The University of Alabama in
  Huntsville
P. O. Box 1247
Huntsville, AL 35807
(205) 895-6257

Dr. Matt McCargo
LOCKHEED PALO ALTO RESEARCH
  LAB
3251 Hanover Street
Palo Alto, CA 94304
(415) 493-4411 Ext. 45193

W. C. Stevens
NEW MEXICO STATE UNIVERSITY
Physical Science Laboratory
Box 3-PSL
Las Cruces, NM 88003
(505) 522-4400

Richard W. Dixon
UNIVERSITY OF FLORIDA
SEECL, Room 325 MEB
Mechanical Engineering Dept.
Gainesville, FL 32611

Dr. R. Bannerot of Dr. Stan
  Kleis
UNIVERSITY OF HOUSTON
Department of Mechanical
  Engineering
4800 Calhoun
Houston, TX 77004
(713) 749-4462 or 4464

Dr. Lynn Russell
UNIVERSITY OF TENNESSEE AT
  CHATTANOOGA
School of Engineering
Chattanooga, TN 37402
(615) 755-4121

W. C. Thomas
VIRGINIA POLYTECHNIC INSTITUTE
  & STATE UNIVERSITY
Department of Mechanical
  Engineering
Blacksburg, VA 24061
(703) 961-7464

Robert E. Losey, Solar Program
  Devl.
WYLE LABORATORIES
Eastern Operations
7800 Governor's Drive West
Huntsville, AL 35807
(205) 837-4411

Table 12.7.   (Continued)

The following testing facilities test collectors
according to their own engineering methods, not according to
the methods of ASHRAE 93-77.

Harold A. Blum, Pres.
DEVICES & SERVICES COMPANY
3501-A Milton
Dallas, TX 75205
(214) 368-5749

Richard E. Crane
FIRL, INC.
Engineering Dept.
20th & Race Sts.
Philadelphia, PA 19103
(215) 448-1258

12.2 REFERENCES AND SUGGESTED READING

1. Electric Power Research Institute, *EPRI Guide – A Directory of Technical Reports and Audiovisual Materials, Computer Programs and Data Bases, Licensable Inventions, Research and Development Information System, Other Information Services*, Palo Alto, California, 1980.
2. Florida Solar Energy Center, *Solar Energy Journals and Newsletters – A Selected List*, Cape Canaveral, Florida: August, 1979.
3. National Solar Energy Education Campaign, *Solar Energy Books*, Mamaroneck, New York: Imprints (division of Dimondstein Book Co.), 1977.
4. National Solar Heating and Cooling Information Center, *Solar Data Bank Report - U. S. Department of Energy Regions and DOE Regional Offices*, Rockville, Maryland, May, 1979.
5. National Solar Heating and Cooling Information Center, *Solar Directories*, Rockville, Maryland, January, 1978.
6. National Solar Heating and Cooling Information Center, *Solar Source Sheet - State Energy/ Solar Offices*, Rockville, Maryland, July, 1979.
7. National Solar Heating and Cooling Information Center, *Solar Source Sheet - Testing Facilities for Solar Equipment*, Rockville, Maryland, June, 1979.
8. Solar Energy Information Data Bank, Solar Energy Research Institute, 1617 Cole Boulevard, Golden, Colorado 80401.
9. Solar Energy Research Institute, *Analysis Methods for Solar Heating and Cooling Applications - Passive and Active Systems*, Golden, Colorado, January, 1980, page 1.
10. Solar Energy Research Institute, *Solar Energy Information Locator*, Golden, Colorado.
11. U. S. Department of Energy, Room E-020-GTN, Washington, D.C. 20545.

APPENDIX I
CONVERSION FACTORS

Product	Btu	Unit
Coal:		
Anthracite (Penn.)	25,400,000	Ton
Bituminous	26,200,000	Ton
Blast furnace gas	100	ft³
Briquettes and package fuels	28,000,000	Ton
Coke	24,800,000	Ton
Coke breeze	20,000,000	Ton
Coke-oven gas	550	ft³
Coal tar	150,000	gallon
Coke-oven and manufactured gas products, light oils	5,460,000	barrel
Natural gas (dry)	1,035	ft³
Natural gas liquids (average)	4,011,000	barrel
Butane	4,284,000	barrel
Propane	3,843,000	barrel
Petroleum:		
Asphalt	6,640,000	barrel
Coke	6,024,000	barrel
Crude oil	5,800,000	barrel
Diesel	5,806,000	barrel
Distillate fuel oil	5,825,000	barrel
Gasoline, aviation	5,048,000	barrel
Gasoline, motor fuel	5,253,000	barrel
Jet fuel:		
Commercial	5,670,000	barrel
Military	5,355,000	barrel
Kerosene	5,670,000	barrel
Lubricants	6,060,000	barrel
Miscellaneous oils	5,588,000	barrel
Refinery still gas	5,600,000	barrel
Heavy fuel oil	6,287,000	barrel
Road oils	6,640,000	barrel
Wax	5,570,000	barrel
Electricity	3,412	kWh

	Specific Heat	Density
Pure Water (68°F)	4190 J/Kg°C 1.00 Btu/lb°F	1000 Kg/m³ 62.42 lb/ft³
50/50 Ethylene Glycol/Water (73°F)	3350 J/Kg°C 0.80 Btu/lb°F	1065 Kg/m³ 66.50 lb/ft³
50/50 Propylene Glycol/Water (73°F)	3562 J/Kg°C 0.85 Btu/lb°F	
Rock	838 J/Kg°C 0.2 Btu/lb°F	2400 Kg/m³ 150 lb/ft³
Air (68°F)	1012 J/Kg°C 0.241 Btu/lb-°F	1.204 Kg/m 0.07516 lb/ft³

Solar Constant

1353 W/m^2

1.940 cal /cm^2min

428 Btu/ft^2hr

4871 K J/m^2hr

1.94 l/min

Solar Energy Flux (Power Density)

1 langley/hr = 11.63 W/m^2

1 langley/min = 697.8 W/m^2

1 langley/min = 3.688 Btu/min ft^2

1 Btu/hr-ft^2 = 3.155 W/m^2

1 Btu/min ft^2 = 0.1892 kW/m^2

1 Btu/min ft^2 = 0.2711 l/min

1 Btu/hr -ft^2-f = 5.678 W/m^2-C

1 Btu/hr-ft-f = 1.703 W/m-C

1 cal/cm^2-min = 697.4 W/m^2

Energy

1 kWh = 3412 Btu

1 kWh = 3.6 MJ

1 kWh = 859.8 Kg-cal

1 langley = 41.84 K J/m^2

1 Btu = 1.055 K Joules

1 Btu = 2.931×10^{-4} kWh

1 Btu = 0.252 Kg-cal

1 Therm = 105.506 MJ

1 Kg-Cal = 4.184 KJ

1 KJ = 0.9478 Btu

1 KJ = 1,000 Watt-sec

1 KJ - 2.778×10^{-4} kWh

Power

1 ft-lb/sec = 1.356 W

1 Ton (refrigeration) = 3.517 kW

1 Kg m/sec = 9.807 watts

1 kW = 0.9478 Btu/sec

1 kW = 0.2388 Kg-Cal/sec

1 hp = 0.7457 kW

1 hp = 0.7068 Btu/sec

1 hp = 550 ft pounds/sec

Volumetric Flow Rate

1 gal/min-ft^2 = 0.6791 l/s-m^2

1 gal/min = 0.06309 l/s

1 cfm = 0.47195 l/s

1 cfn/ft^2 = 0.1968 l/s-n^2 (air)

Mass Flow Rate

1 lb/hr-ft^2 = 0.001356 Kg/s-m^2

1 lb/hr = 0.000126 Kg/s

Volume

1 ft^3 = 7.48 gal

1 ft^3 = 28.3168 1

1 gal/ft^2 - 0.02454 l/m^2

1 gal = 3.78544 1

1 yd^3 = 0.7645 m^3

The basic data in the following tables were compiled from the references listed below. References are indicated by number (enclosed in parentheses) on the table.

1. Beck, Field, *Solar Heating of Buildings and Domestic Hot Water,* 1976, Technical Report R835, Naval Facilities Engineering Command.
2. U. S. Department of Commerce Environmental Science Services Administration Environmental Data Service, *Climatic Atlas of the U. S.,* 1968. Reprinted by the National Oceanic and Atmospheric Administration, 1974.
3. U. S. Department of Commerce National Oceanic and Atmospheric Administration Environmental Data Service, *Monthly Normals of Temperature, Precipitation, and Heating and Cooling Degree Days 1941-1970.,* Climatography of the U. S. No. 81 (by State).
4. Data from SOLMET Program—Communication from the National Oceanic and Atmospheric Administration (NOAA), 1978.
5. National Weather Bureau, Washington, D.C.
6. Klein, S. A., Beckman, W. A., and Duffie, J. A., *Monthly Average Solar Radiation on Inclined Surfaces for 261 North American Cities,* Madison, Wisconsin: University of Wisconsin, 1978.

ABILENE   TX   (LAT. 32.4)

		JAN	FEB	MAR	APR	MAY	JUNE	JULY	AUG	SEP	OCT	NOV	DEC
HORIZ. RAD.	(4)	10.48	13.42	17.89	20.92	23.12	25.07	24.28	22.20	18.13	14.93	11.44	9.80
KT	(6)	.53	.55	.59	.58	.58	.61	.60	.60	.56	.57	.55	.54
AVE. TEMP.	(5)	6.0	9.0	12.0	18.0	22.0	27.0	29.0	29.0	24.0	19.0	12.0	8.0
DEGREE-DAYS	(5)	367.	266.	197.	58.	6.	0.	0.	0.	0.	49.	187.	321.

AKRON   OH   (LAT. 40.9)

		JAN	FEB	MAR	APR	MAY	JUNE	JULY	AUG	SEP	OCT	NOV	DEC
HORIZ. RAD.	(4)	4.86	7.37	10.94	15.40	18.93	20.87	20.28	18.11	14.43	10.30	5.73	4.01
KT	(6)	.34	.37	.41	.45	.48	.50	.50	.50	.49	.47	.36	.31
AVE. TEMP.	(3)	-3.0	-2.0	2.0	9.0	15.0	20.0	22.0	21.0	18.0	12.0	5.0	-1.0
DEGREE-DAYS	(3)	667.	580.	496.	275.	128.	18.	5.	9.	56.	205.	405.	613.

ALBANY   NY   (LAT. 42.7)

		JAN	FEB	MAR	APR	MAY	JUNE	JULY	AUG	SEP	OCT	NOV	DEC
HORIZ. RAD.	(4)	5.70	8.70	12.54	16.95	19.99	22.14	22.22	19.29	14.94	10.42	5.96	4.53
KT	(6)	.42	.46	.48	.51	.51	.54	.55	.54	.52	.50	.41	.38
AVE. TEMP.	(3)	-5.0	-4.0	.0	8.0	14.0	19.0	22.0	21.0	17.0	10.0	4.0	-2.0
DEGREE-DAYS	(3)	749.	646.	544.	302.	141.	22.	5.	12.	75.	234.	423.	673.

ALBUQUERQUE   NM   (LAT. 35.0)

		JAN	FEB	MAR	APR	MAY	JUNE	JULY	AUG	SEP	OCT	NOV	DEC
HORIZ. RAD.	(4)	11.54	15.23	20.06	25.29	28.81	30.40	28.24	25.99	22.38	17.55	12.87	10.53
KT	(6)	.64	.66	.68	.71	.73	.74	.70	.70	.71	.71	.67	.63
AVE. TEMP.	(3)	1.0	4.0	7.0	12.0	17.0	22.0	25.0	23.0	20.0	13.0	6.0	1.0
DEGREE-DAYS	(3)	513.	389.	331.	157.	32.	0.	0.	0.	4.	121.	342.	496.

ALLENTOWN   PA   (LAT. 40.6)

		JAN	FEB	MAR	APR	MAY	JUNE	JULY	AUG	SEP	OCT	NOV	DEC
HORIZ. RAD.	(4)	5.99	8.66	12.24	16.00	18.58	20.17	20.03	17.55	14.05	10.51	6.45	4.88
KT		.41	.43	.46	.47	.47	.49	.50	.49	.48	.48	.41	.37
AVE. TEMP.	(3)	-2.0	-1.0	3.0	10.0	16.0	21.0	23.0	22.0	18.0	12.0	6.0	-1.0
DEGREE-DAYS	(3)	641.	554.	463.	252.	106.	12.	0.	3.	47.	191.	378.	591.

AMARILLO TX (LAT. 35.2)

		JAN	FEB	MAR	APR	MAY	JUNE	JULY	AUG	SEP	OCT	NOV	DEC
HORIZ. RAD.	(4)	10.90	14.11	18.51	22.91	25.10	27.16	25.88	23.87	19.98	15.93	11.72	9.89
KT		.61	.61	.63	.65	.63	.66	.64	.65	.64	.64	.61	.60
AVE. TEMP.	(5)	2.0	4.0	8.0	14.0	19.0	24.0	26.0	25.0	21.0	15.0	8.0	4.0
DEGREE-DAYS	(5)	499.	393.	334.	153.	45.	6.	0.	0.	11.	114.	312.	457.

ANNAPOLIS MD (LAT. 39.0)

		JAN	FEB	MAR	APR	MAY	JUNE	JULY	AUG	SEP	OCT	NOV	DEC
HORIZ. RAD.	(1)	7.33	10.17	14.24	17.54	20.43	23.32	22.69	19.64	16.03	12.31	7.91	6.49
KT		.47	.49	.52	.51	.52	.56	.56	.54	.53	.54	.47	.46
AVE. TEMP.	(3)	1.0	2.0	6.0	12.0	17.0	22.0	24.0	23.0	20.0	14.0	8.0	2.0
DEGREE DAYS	(3)	526.	454.	376.	183.	58.	0.	0.	0.	16.	137.	293.	484.

ANNETTE AK (LAT. 55.0)

		JAN	FEB	MAR	APR	MAY	JUNE	JULY	AUG	SEP	OCT	NOV	DEC
HORIZ. RAD.	(2)	2.64	4.81	9.88	15.24	18.30	18.34	18.34	14.28	10.80	5.11	2.47	1.72
KT		.44	.42	.51	.52	.49	.45	.47	.44	.47	.37	.34	.36
AVE. TEMP.	(3)	.0	2.0	3.0	5.0	9.0	11.0	15.0	14.0	12.0	8.0	4.0	2.0
DEGREE-DAYS	(3)	527.	440.	460.	370.	269.	177.	128.	117.	183.	312.	418.	501.

APALACHICOLA FL (LAT. 29.7)

		JAN	FEB	MAR	APR	MAY	JUNE	JULY	AUG	SEP	OCT	NOV	DEC
HORIZ. RAD.	(4)	9.68	12.8	16.73	21.32	23.73	22.68	20.58	19.15	17.42	15.56	11.80	9.28
KT		.46	.49	.53	.58	.60	.56	.51	.51	.53	.57	.53	.47
AVE. TEMP.	(3)	12.0	13.0	15.0	19.0	23.0	26.0	27.0	27.0	26.0	21.0	16.0	13.0
DEGREE-DAYS	(3)	204.	161.	97.	17.	0.	0.	0.	0.	0.	12.	88.	177.

ARCATA CA (LAT. 41.0)

		JAN	FEB	MAR	APR	MAY	JUNE	JULY	AUG	SEP	OCT	NOV	DEC
HORIZ. RAD.	(4)	6.00	9.00	12.86	18.01	20.91	22.26	20.52	17.92	15.23	10.62	6.73	5.33
KT		.42	.46	.48	.53	.53	.54	.51	.50	.52	.49	.43	.41
AVE. TEMP.	(3)	7.0	8.0	8.0	10.0	12.0	14.0	15.0	15.0	15.0	13.0	10.0	7.0
DEGREE-DAYS	(3)	365.	288.	307.	262.	208.	143.	110.	98.	105.	181.	263.	344.

ASHEVILLE NC (LAT. 35.4)

		JAN	FEB	MAR	APR	MAY	JUNE	JULY	AUG	SEP	OCT	NOV	DEC
HORIZ. RAD.	(4)	8.19	11.02	14.82	18.92	20.48	21.05	20.16	18.46	15.44	13.02	9.63	7.46
KT		.46	.48	.51	.53	.52	.51	.50	.50	.49	.53	.51	.45
AVE. TEMP.	(5)	3.0	4.0	8.0	13.0	18.0	21.0	23.0	23.0	19.0	14.0	8.0	4.0
DEGREE-DAYS	(5)	467.	398.	329.	155.	56.	8.	0.	0.	28.	149.	312.	453.

ASTORIA OR (LAT. 46.2)

		JAN	FEB	MAR	APR	MAY	JUNE	JULY	AUG	SEP	OCT	NOV	DEC
HORIZ. RAD.	(4)	3.57	6.19	9.83	14.22	18.25	18.45	19.82	17.01	13.43	8.09	4.39	2.96
KT		.32	.37	.41	.44	.47	.45	.50	.49	.49	.43	.35	.30
AVE. TEMP.	(3)	5.0	6.0	6.0	8.0	11.0	13.0	15.0	15.0	14.0	11.0	8.0	5.0
DEGREE-DAYS	(3)	420.	333.	355.	287.	219.	142.	91.	84.	112.	210.	308.	382.

ATLANTA GA (LAT. 33.6)

		JAN	FEB	MAR	APR	MAY	JUNE	JULY	AUG	SEP	OCT	NOV	DEC
HORIZ. RAD.	(4)	8.14	11.00	14.79	19.14	21.04	21.7	20.57	19.39	16.14	13.62	10.02	7.65
KT		.43	.46	.50	.54	.53	.53	.51	.52	.51	.53	.50	.44
AVE. TEMP.	(3)	6.0	7.0	11.0	16.0	20.0	24.0	25.0	25.0	22.0	17.0	11.0	7.0
DEGREE-DAYS	(3)	389.	311.	246.	80.	15.	0.	0.	0.	4.	76.	227.	371.

AUGUSTA GA (LAT. 33.4)

		JAN	FEB	MAR	APR	MAY	JUNE	JULY	AUG	SEP	OCT	NOV	DEC
HORIZ. RAD.	(4)	8.52	11.52	15.19	19.62	21.17	21.61	20.47	18.92	16.00	13.84	10.40	8.18
KT		.45	.48	.51	.55	.53	.53	.51	.51	.50	.54	.52	.46
AVE. TEMP.	(3)	8.0	9.0	13.0	18.0	22.0	26.0	27.0	26.0	23.0	18.0	12.0	8.0
DEGREE-DAYS	(3)	334.	264.	192.	50.	6.0	0.	0.	0.	0.	58.	191.	321.

AVOCA PA (LAT. 41.3)

		JAN	FEB	MAR	APR	MAY	JUNE	JULY	AUG	SEP	OCT	NOV	DEC
HORIZ. RAD.	(4)	5.16	7.81	11.25	15.20	18.05	19.97	19.81	17.17	13.61	10.17	5.56	4.18
KT		.36	.40	.42	.45	.46	.48	.49	.48	.47	.47	.36	.33
AVE. TEMP.	(3)	-3.0	-3.0	2.0	9.0	15.0	20.0	22.0	21.0	17.0	11.0	5.0	-2.0
DEGREE-DAYS	(3)	672.	587.	499.	275.	122.	16.	4.	10.	64.	217.	403.	618.

BAKERSFIELD   CA   (LAT. 35.4)

	JAN	FEB	MAR	APR	MAY	JUNE	JULY	AUG	SEP	OCT	NOV	DEC
HORIZ. RAD. (4)	8.70	12.50	18.10	23.77	28.48	31.20	30.45	27.47	22.61	16.55	10.69	7.69
KT	.49	.55	.62	.67	.72	.76	.76	.74	.72	.67	.56	.47
AVE. TEMP. (3)	9.0	11.0	14.0	17.0	21.0	25.0	29.0	28.0	25.0	19.0	13.0	9.0
DEGREE-DAYS (3)	302.	196.	148.	78.	12.	0.	0.	0.	0.	31.	153.	294.

BALTIMORE   MD   (LAT. 39.2)

	JAN	FEB	MAR	APR	MAY	JUNE	JULY	AUG	SEP	OCT	NOV	DEC
HORIZ. RAD. (4)	6.66	9.53	13.19	16.89	19.45	21.33	20.69	18.15	15.10	11.32	7.49	5.67
KT	.43	.46	.48	.49	.9	.52	.51	.50	.50	.50	.45	.40
AVE. TEMP. (5)	1.0	2.0	6.0	12.0	18.0	22.0	25.0	24.0	20.0	14.0	8.0	2.0
DEGREE-DAYS (5)	544.	470.	382.	189.	61.	0.	0.	0.	15.	139.	315.	512.

BATON ROUGE   LA   (LAT. 30.5)

	JAN	FEB	MAR	APR	MAY	JUNE	JULY	AUG	SEP	OCT	NOV	DEC
HORIZ. RAD. (4)	8.91	11.96	15.65	19.08	21.24	21.86	19.81	19.03	16.62	14.77	10.44	8.36
KT	.43	.47	.50	.52	.54	.54	.49	.51	.51	.55	.48	.43
AVE. TEMP. (3)	11.0	12.0	15.0	20.0	24.0	27.0	28.0	28.0	25.0	20.0	15.0	12.0
DEGREE-DAYS (3)	251.	186.	116.	18.	0.	0.	0.	0.	0.	30.	116.	212.

BILLINGS   MT   (LAT. 45.8)

	JAN	FEB	MAR	APR	MAY	JUNE	JULY	AUG	SEP	OCT	NOV	DEC
HORIZ. RAD. (4)	5.52	8.66	13.50	17.32	21.71	24.67	27.05	22.95	16.68	1.20	6.37	4.78
KT	.48	.51	.55	.53	.56	.60	.68	.66	.61	.59	.50	.47
AVE. TEMP. (5)	-6.0	-3.0	.0	7.0	12.0	17.0	22.0	21.0	15.0	10.0	2.0	-3.0
DEGREE-DAYS (5)	742.	585.	558.	340.	185.	73.	6.	8.	123.	271.	488.	658.

BINGHAMPTON   NY   (LAT. 42.2)

	JAN	FEB	MAR	APR	MAY	JUNE	JULY	AUG	SEP	OCT	NOV	DEC
HORIZ. RAD. (4)	4.38	6.53	9.77	14.09	16.98	19.08	18.83	16.17	12.84	8.84	4.70	3.37
KT	.32	.34	.37	.42	.43	.46	.47	.45	.45	.42	.31	.28
AVE. TEMP. (5)	-6.0	-5.0	.0	7.0	13.0	18.0	21.0	20.0	16.0	10.0	3.0	-4.0
DEGREE-DAYS (5)	741.	657.	581.	338.	18.	42.	12.	22.	96.	253.	447.	682.

BIRMINGHAM   AL   (LAT. 33.6)

	JAN	FEB	MAR	APR	MAY	JUNE	JULY	AUG	SEP	OCT	NOV	DEC
HORIZ. RAD. (4)	8.02	10.98	14.71	18.99	21.07	21.77	20.54	19.56	16.51	13.74	9.74	7.51
KT	.42	.46	.49	.53	.53	.53	.51	.53	.52	.54	.49	.43
AVE. TEMP. (5)	7.0	8.0	12.0	17.0	21.0	25.0	27.0	26.0	23.0	17.0	11.0	7.0
DEGREE-DAYS (5)	363.	287.	216.	64.	11.	0.	0.	0.	3.	76.	217.	341.

BISMARCK   ND   (LAT. 46.8)

	JAN	FEB	MAR	APR	MAY	JUNE	JULY	AUG	SEP	OCT	NOV	DEC
HORIZ. RAD. (4)	5.30	8.80	13.26	16.56	20.97	23.38	24.78	21.30	15.37	10.30	5.76	4.23
KT	.48	.54	.55	.51	.54	.57	.62	.61	.57	.55	.47	.45
AVE. TEMP. (3)	-13.0	-11.0	-3.0	6.0	12.0	17.0	21.0	20.0	14.0	7.0	-1.0	-9.0
DEGREE-DAYS (3)	978.	801.	687.	367.	188.	68.	10.	19.	140.	313.	602.	851.

BOISE   ID   (LAT. 43.6)

	JAN	FEB	MAR	APR	MAY	JUNE	JULY	AUG	SEP	OCT	NOV	DEC
HORIZ. RAD. (4)	5.51	9.53	14.80	20.73	25.84	27.95	29.65	24.93	19.71	12.91	7.13	4.96
KT	.43	.52	.58	.62	.66	.68	.74	.70	.70	.63	.50	.43
AVE. TEMP. (3)	-1.0	1.0	5.0	9.0	14.0	18.0	23.0	22.0	17.0	11.0	4.0	.0
DEGREE-DAYS (3)	620.	487.	412.	267.	140.	54.	0.	7.	71.	226.	420.	567.

BOSTON   MA   (LAT. 42.4)

	JAN	FEB	MAR	APR	MAY	JUNE	JULY	AUG	SEP	OCT	NOV	DEC
HORIZ. RAD. (4)	5.40	8.05	11.53	15.05	18.39	20.62	19.85	16.87	14.30	10.10	5.71	4.57
KT	.40	.42	.44	.45	.47	.50	.49	.47	.50	.48	.38	.38
AVE. TEMP. (3)	-1.0	.0	3.0	9.0	15.0	20.0	23.0	22.0	18.0	13.0	7.0	1.0
DEGREE-DAYS (3)	617.	538.	463.	273.	121.	15.	0.	5.	42.	167.	330.	551.

BOULDER   CO   (LAT. 40.0)

	JAN	FEB	MAR	APR	MAY	JUNE	JULY	AUG	SEP	OCT	NOV	DEC
HORIZ. RAD. (2)	8.41	11.22	16.79	19.26	19.26	21.98	21.77	18.38	17.25	12.98	9.29	7.62
KT	.56	.55	.62	.58	.49	.53	.54	.51	.58	.58	.57	.56
AVE. TEMP. (3)	.0	1.0	3.0	9.0	14.0	19.0	23.0	22.0	17.0	12.0	5.0	2.0
DEGREE-DAYS (3)	551.	459.	449.	268.	131.	49.	3.	0.	77.	204.	383.	503.

```
                    BROWNSVILLE        TX          (LAT. 25.9)

                    JAN    FEB    MAR    APR    MAY    JUNE   JULY   AUG    SEP    OCT    NOV    DEC
HORIZ. RAD. (4)    10.36  12.89  16.54  19.71  21.87  24.01  25.11  23.01  19.22  16.33  11.97   9.79
KT                   .44    .47    .51    .53    .55    .60    .63    .61    .56    .56    .49    .44
AVE. TEMP.  (3)    15.0   17.0   20.0   23.0   26.0   28.0   28.0   28.0   27.0   24.0   19.0   16.0
DEGREE-DAYS (3)    125.    84.    49.     0.     0.     0.     0.     0.     0.     3.    19.    81.


                    BUFFALO            NY          (LAT. 42.9)

                    JAN    FEB    MAR    APR    MAY    JUNE   JULY   AUG    SEP    OCT    NOV    DEC
HORIZ. RAD. (4)     3.96   6.20  10.08  14.92  18.12  20.47  20.16  17.17  13.07   8.90   4.58   3.22
KT                   .30    .33    .39    .45    .46    .50    .50    .48    .46    .43    .32    .27
AVE. TEMP.  (3)    -5.0   -4.0     .0    7.0   13.0   19.0   21.0   20.0   16.0   11.0    4.0   -2.0
DEGREE-DAYS (3)    711.   632.   567.   335.   178.    32.     7.    18.    77.   233.   420.   639.


                    BURLINGTON         VT          (LAT. 44.5)

                    JAN    FEB    MAR    APR    MAY    JUNE   JULY   AUG    SEP    OCT    NOV    DEC
HORIZ. RAD. (4)     4.83   7.70  11.98  16.45  20.05  22.13  22.14  18.99  14.35   9.47   4.92   3.63
KT                   .39    .43    .48    .50    .52    .54    .55    .54    .51    .48    .36    .33
AVE. TEMP.  (3)    -8.0   -7.0   -2.0    6.0   13.0   18.0   21.0   20.0   15.0    9.0    3.0   -5.0
DEGREE-DAYS (3)    830.   722.   618.   367.   184.    35.    11.    27.   106.   279.   467.   730.


                    BURNS              OR          (LAT. 43.6)

                    JAN    FEB    MAR    APR    MAY    JUNE   JULY   AUG    SEP    OCT    NOV    DEC
HORIZ. RAD. (4)     5.56   8.99  13.47  18.71  23.29  25.87  27.92  23.64  18.39  11.83   6.74   4.89
KT                   .43    .49    .53    .56    .60    .63    .70    .67    .65    .58    .48    .43
AVE. TEMP.  (3)    -4.0   -1.0    2.0    7.0   11.0   15.0   20.0   19.0   15.0    8.0    2.0   -2.0
DEGREE-DAYS (3)    686.   529.   498.   347.   223.   114.    17.    38.   126.   305.   487.   639.


                    CAPE HATTERAS      NC          (LAT. 35.3)

                    JAN    FEB    MAR    APR    MAY    JUNE   JULY   AUG    SEP    OCT    NOV    DEC
HORIZ. RAD. (4)     7.78  10.81  15.05  20.13  22.26  23.11  21.80  19.36  16.69  12.90   9.91   7.48
KT                   .43    .47    .52    .57    .56    .56    .54    .52    .53    .52    .52    .45
AVE. TEMP.  (3)     8.0    8.0   10.0   14.0   19.0   23.0   25.0   25.0   23.0   18.0   13.0    9.0
DEGREE-DAYS (3)    339.   299.   254.   104.    26.     0.     0.     0.     0.    42.   154.   298.


                    CARIBOU            ME          (LAT. 46.9)

                    JAN    FEB    MAR    APR    MAY    JUNE   JULY   AUG    SEP    OCT    NOV    DEC
HORIZ. RAD. (4)     4.83   8.24  12.85  15.98  17.77  19.86  19.86  16.88  12.41   7.74   4.18   3.61
KT                   .44    .50    .54    .50    .46    .48    .50    .49    .46    .42    .34    .38
AVE. TEMP.  (3)   -11.0  -10.0   -4.0    3.0   10.0   15.0   18.0   17.0   12.0    7.0     .0   -8.0
DEGREE-DAYS (3)    935.   811.   713.   472.   263.    94.    47.    68.   182.   365.   560.   842.


                    CASPER             WY          (LAT. 42.9)

                    JAN    FEB    MAR    APR    MAY    JUNE   JULY   AUG    SEP    OCT    NOV    DEC
HORIZ. RAD. (4)     7.75  11.50  16.36  20.96  25.01  28.39  28.77  25.26  19.86  13.83   8.68   6.74
KT                   .58    .62    .63    .63    .64    .69    .72    .71    .70    .67    .60    .57
AVE. TEMP.  (3)    -5.0   -3.0   -1.0    6.0   11.0   17.0   22.0   21.0   15.0    9.0    1.0   -3.0
DEGREE-DAYS (3)    720.   594.   586.   372.   216.    82.     7.     9.   127.   298.   518.   668.


                    CEDER CITY         UT          (LAT. 37.7)

                    JAN    FEB    MAR    APR    MAY    JUNE   JULY   AUG    SEP    OCT    NOV    DEC
HORIZ. RAD. (4)    10.01  13.39  18.56  23.75  28.00  30.71  28.41  25.44  22.34  16.57  11.26   8.91
KT                   .61    .62    .66    .68    .71    .74    .71    .70    .73    .70    .64    .59
AVE. TEMP.  (3)    -2.0    1.0    4.0    8.0   13.0   18.0   23.0   22.0   17.0   11.0    4.0   -1.0
DEGREE-DAYS (3)    625.   496.   458.   298.   156.    48.     0.     3.    63.   236.   437.   589.


                    CHARLESTON         SC          (LAT. 32.9)

                    JAN    FEB    MAR    APR    MAY    JUNE   JULY   AUG    SEP    OCT    NOV    DEC
HORIZ. RAD. (4)     8.45  11.30  15.19  19.66  21.11  20.93  20.42  17.99  15.82  13.54  10.60   8.18
KT                   .44    .47    .50    .55    .53    .51    .51    .48    .49    .52    .52    .46
AVE. TEMP.  (3)    10.0   10.0   14.0   18.0   22.0   25.0   27.0   26.0   24.0   19.0   14.0   10.0
DEGREE-DAYS (3)    267.   218.   157.    29.     0.     0.     0.     0.     0.    28.   117.   242.


                    CHARLESTON         WV          (LAT. 38.4)

                    JAN    FEB    MAR    APR    MAY    JUNE   JULY   AUG    SEP    OCT    NOV    DEC
HORIZ. RAD. (4)     5.66   8.02  11.46  15.39  18.61  20.15  19.10  17.19  14.44  11.03   6.96   5.00
KT                   .35    .38    .41    .44    .47    .49    .47    .47    .48    .48    .40    .34
AVE. TEMP.  (3)     1.0    3.0    7.0   13.0   18.0   22.0   24.0   23.0   20.0   14.0    7.0    2.0
DEGREE-DAYS (3)    526.   443.   357.   159.    63.     6.     0.     0.    26.   148.   327.   496.
```

CHARLOTTE   NC   (LAT. 35.2)

	JAN	FEB	MAR	APR	MAY	JUNE	JULY	AUG	SEP	OCT	NOV	DEC
HORIZ. RAD. (4)	8.16	11.02	14.95	19.24	21.06	21.80	20.78	19.24	16.07	13.32	9.82	9.63
KT.	.45	.48	.51	.54	.53	.53	.52	.52	.51	.54	.51	.58
AVE. TEMP. (5)	6.0	7.0	10.0	16.0	20.0	24.0	26.0	25.0	22.0	16.0	11.0	6.0
DEGREE-DAYS (5)	394.	327.	256.	81.	19.	0.	0.	0.	6.	84.	233.	388.

CHATTANOOGA   TN   (LAT. 35.0)

	JAN	FEB	MAR	APR	MAY	JUNE	JULY	AUG	SEP	OCT	NOV	DEC
HORIZ. RAD. (4)	7.16	9.74	13.35	17.59	19.65	20.78	19.69	18.50	15.16	12.58	8.77	6.59
KT	.40	.42	.46	.50	.50	.51	.49	.50	.48	.51	.46	.40
AVE. TEMP. (5)	5.0	6.0	10.0	16.0	20.0	24.0	26.0	26.0	22.0	16.0	9.0	5.0
DEGREE-DAYS (5)	427.	347.	268.	92.	28.	0.	0.	0.	5.	101.	268.	410.

CHERRY POINT   NC   (LAT. 34.9)

	JAN	FEB	MAR	APR	MAY	JUNE	JULY	AUG	SEP	OCT	NOV	DEC
HORIZ. RAD. (4)	8.59	11.63	15.74	20.36	21.84	22.00	20.76	18.55	16.27	13.27	10.29	8.15
KT	.47	.50	.54	.57	.55	.54	.52	.50	.51	.53	.53	.49
AVE. TEMP. (3)	7.0	8.0	10.0	15.0	19.0	23.0	26.0	25.0	23.0	18.0	13.0	9.0
DEGREE-DAYS (3)	339.	299.	254.	104.	26.	0.	0.	0.	0.	42.	154.	298.

CHEYENNE   WY   (LAT. 41.1)

	JAN	FEB	MAR	APR	MAY	JUNE	JULY	AUG	SEP	OCT	NOV	DEC
HORIZ. RAD. (4)	8.69	12.12	16.26	20.09	22.64	25.63	25.31	22.31	18.92	14.09	9.34	7.62
KT	.61	.62	.61	.59	.58	.62	.63	.62	.65	.65	.60	.9
AVE. TEMP. (3)	-3.0	-2.0	.0	6.0	11.0	16.0	21.0	20.0	15.0	9.0	2.0	-2.0
DEGREE-DAYS (3)	661.	560.	575.	372.	219.	87.	12.	17.	125.	294.	492.	617.

CHICAGO   IL   (LAT. 42.0)

	JAN	FEB	MAR	APR	MAY	JUNE	JULY	AUG	SEP	OCT	NOV	DEC
HORIZ. RAD. (4)	5.75	8.62	12.56	16.56	20.30	22.78	22.06	19.51	15.37	11.00	6.42	4.56
KT	.42	.45	.48	.49	.52	.55	.55	.55	.53	.52	.43	.37
AVE. TEMP. (3)	-3.0	-2.0	3.0	10.0	16.0	21.0	24.0	23.0	19.0	13.0	5.0	-1.0
DEGREE-DAYS (3)	701.	585.	486.	252.	116.	14.	0.	4.	32.	176.	410.	629.

CLEVELAND   OH   (LAT. 41.4)

	JAN	FEB	MAR	APR	MAY	JUNE	JULY	AUG	SEP	OCT	NOV	DEC
HORIZ. RAD. (4)	4.41	6.82	10.47	15.32	19.08	20.92	20.74	17.96	14.07	9.84	5.29	3.61
KT	.31	.35	.39	.45	.49	.51	.52	.50	.48	.46	.34	.28
AVE. TEMP. (3)	-2.0	-1.0	2.0	9.0	15.0	20.0	22.0	21.0	18.0	12.0	5.0	.0
DEGREE-DAYS (3)	656.	577.	498.	278.	136.	22.	5.	9.	53.	197.	390.	598.

COLORADO SPRINGS   CO   (LAT. 38.8)

	JAN	FEB	MAR	APR	MAY	JUNE	JULY	AUG	SEP	OCT	NOV	DEC
HORIZ. RAD. (4)	10.11	13.37	17.59	21.92	24.16	26.88	25.10	22.99	19.96	15.42	10.72	8.87
KT	.64	.64	.64	.63	.61	.65	.62	.63	.66	.67	.63	.62
AVE. TEMP. (3)	-2.0	.0	2.0	8.0	13.0	18.0	22.0	21.0	16.0	10.0	3.0	-1.0
DEGREE-DAYS (3)	627.	524.	512.	313.	167.	57.	5.	7.	86.	253.	458.	586.

COLUMBIA   MO   (LAT. 39.0)

	JAN	FEB	MAR	APR	MAY	JUNE	JULY	AUG	SEP	OCT	NOV	DEC
HORIZ. RAD. (4)	6.95	9.93	13.38	17.32	21.33	23.71	24.01	21.31	16.46	12.49	7.98	5.93
KT	.44	.47	.48	.50	.54	.57	.60	.59	.55	.55	.47	.42
AVE. TEMP. (3)	-1.0	.0	6.0	12.0	18.0	23.0	25.0	24.0	20.0	14.0	6.0	.0
DEGREE-DAYS (3)	615.	488.	406.	174.	61.	6.	0.	0.	23.	137.	352.	554.

COLUMBIA   SC   (LAT. 33.9)

	JAN	FEB	MAR	APR	MAY	JUNE	JULY	AUG	SEP	OCT	NOV	DEC
HORIZ. RAD. (4)	8.64	11.58	15.38	19.82	21.50	22.10	20.90	19.32	16.33	13.75	10.45	8.19
KT	.46	.49	.52	.56	.54	.54	.52	.52	.51	.54	.53	.47
AVE. TEMP. (3)	7.0	9.0	12.0	18.0	22.0	26.0	27.0	27.0	24.0	18.0	12.0	8.0
DEGREE-DAYS (3)	338.	274.	200.	46.	7.	0.	0.	0.	0.	62.	189.	327.

COLUMBUS   GA   (LAT. 40.0)

	JAN	FEB	MAR	APR	MAY	JUNE	JULY	AUG	SEP	OCT	NOV	DEC
HORIZ. RAD. (4)	5.21	7.68	11.12	15.35	18.69	20.57	19.92	18.62	14.54	10.73	6.10	4.39
KT	.35	.38	.41	.45	.48	.50	.49	.51	.49	.48	.38	.32
AVE. TEMP. (3)	-1.0	.0	4.0	11.0	16.0	21.0	23.0	22.0	18.0	12.0	5.0	.0
DEGREE-DAYS (3)	631.	540.	444.	232.	98.	7.	0.	4.	42.	190.	388.	591.

CONCORD   NH   (LAT. 43.2)

	JAN	FEB	MAR	APR	MAY	JUNE	JULY	AUG	SEP	OCT	NOV	DEC
HORIZ. RAD. (4)	5.21	7.79	11.05	14.95	17.96	19.35	19.00	16.52	12.94	9.27	5.25	4.11
KT	.40	.42	.43	.45	.46	.47	.47	.47	.46	.45	.37	.35
AVE. TEMP. (3)	-6.0	-5.0	.0	7.0	13.0	18.0	21.0	20.0	15.0	10.0	3.0	-4.0
DEGREE-DAYS (3)	764.	659.	563.	347.	175.	32.	9.	25.	101.	271.	450.	692.

CORPUS CHRISTI   TX   (LAT. 27.8)

	JAN	FEB	MAR	APR	MAY	JUNE	JULY	AUG	SEP	OCT	NOV	DEC
HORIZ. RAD. (4)	10.19	13.02	16.23	18.64	21.18	23.76	24.81	22.59	19.15	16.07	11.83	9.59
KT	.46	.49	.51	.51	.54	.59	.62	.60	.57	.57	.51	.46
AVE. TEMP. (3)	13.0	15.0	18.0	22.0	25.0	27.0	29.0	29.0	27.0	23.0	18.0	15.0
DEGREE-DAYS (3)	169.	111.	67.	0.	0.	0.	0.	0.	0.	4.	45.	122.

COVINGTON   KY   (LAT. 39.1)

	JAN	FEB	MAR	APR	MAY	JUNE	JULY	AUG	SEP	OCT	NOV	DEC
HORIZ. RAD. (4)	4.23	5.82	11.76	16.91	21.27	24.33	28.05	22.90	16.70	9.84	5.86	3.39
KT	.27	.28	.43	.49	.54	.59	.70	.63	.56	.43	.35	.24
AVE. TEMP. (3)	-1.0	1.0	5.0	12.0	17.0	22.0	24.0	24.0	20.0	14.0	7.0	1.0
DEGREE-DAYS (3)	584.	493.	401.	189.	77.	5.	0.	0.	24.	151.	353.	539.

DAGGETT   CA   (LAT. 34.9)

	JAN	FEB	MAR	APR	MAY	JUNE	JULY	AUG	SEP	OCT	NOV	DEC
HORIZ. RAD. (4)	10.87	14.53	20. 1	25.81	29.41	31.39	29.55	27.04	22.79	17.20	12.31	9.94
KT	.60	.63	.68	.73	.74	.76	.73	.73	.72	.69	.64	.59
AVE. TEMP. (3)	8.0	11.0	14.0	18.0	22.0	27.0	31.0	30.0	26.0	20.0	13.0	9.0
DEGREE-DAYS (3)	305.	206.	151.	66.	8.	0.	0.	0.	0.	32.	164.	293.

DALLAS   TX   (LAT. 32.8)

	JAN	FEB	MAR	APR	MAY	JUNE	JULY	AUG	SEP	OCT	NOV	DEC
HORIZ. RAD. (4)	9.32	12.16	16.14	18.46	21.43	24.23	24.08	22.13	18.01	14.48	10.63	8.85
KT	.48	.50	.53	.51	.54	.59	.60	.59	.56	.56	.52	.49
AVE. TEMP. (3)	7.0	10.0	13.0	19.0	23.0	28.0	30.0	30.0	26.0	20.0	13.0	9.0
DEGREE-DAYS (3)	338.	243.	174.	39.	0.	0.	0.	0.	0.	31.	158.	289.

DAYTON   OH   (LAT. 39.9)

	JAN	FEB	MAR	APR	MAY	JUNE	JULY	AUG	SEP	OCT	NOV	DEC
HORIZ. RAD. (4)	5.55	8.23	11.64	15.92	19.29	21.26	20.54	18.67	14.96	11.00	6.40	4.62
KT	.37	.40	.43	.46	.49	.51	.51	.52	.50	.49	.39	.34
AVE. TEMP. (5)	-2.0	-1.0	4.0	11.0	16.0	22.0	24.0	23.0	19.0	13.0	5.0	-1.0
DEGREE-DAYS (5)	636.	538.	448.	229.	92.	7.	0.	4.	35.	171.	387.	587.

DAYTONA BEACH   FL   (LAT. 29.2)

	JAN	FEB	MAR	APR	MAY	JUNE	JULY	AUG	SEP	OCT	NOV	DEC
HORIZ. RAD. (4)	10.88	13.77	17.57	21.38	22.33	20.72	20.25	19.09	16.77	14.20	11.75	9.88
KT	.51	.53	.56	.58	.56	.51	.51	.51	.50	.51	.52	.49
AVE. TEMP. (3)	15.0	15.0	18.0	21.0	24.0	26.0	27.0	27.0	26.0	23.0	18.0	15.0
DEGREE-DAYS (3)	134.	117.	67.	9.	0.	0.	0.	0.	0.	0.	54.	118.

DENVER   CO   (LAT. 39.7)

	JAN	FEB	MAR	APR	MAY	JUNE	JULY	AUG	SEP	OCT	NOV	DEC
HORIZ. RAD. (4)	10.16	13.71	18.71	22.97	26.09	28.63	27.71	24.87	21.05	15.81	10.68	8.82
KT (6)	.67	.67	.69	.67	.66	.69	.69	.69	.71	.70	.65	.64
AVE. TEMP. (5)	-1.0	.0	3.0	9.0	14.0	19.0	23.0	22.0	17.0	11.0	4.0	.0
DEGREE-DAYS (5)	604.	501.	482.	292.	141.	44.	0.	0.	67.	227.	427	558.

DES MOINES   IA   (LAT. 41.5)

	JAN	FEB	MAR	APR	MAY	JUNE	JULY	AUG	SEP	OCT	NOV	DEC
HORIZ. RAD. (4)	6.59	9.77	13.40	17.67	21.19	24.11	23.80	20.74	16.27	12.12	7.47	5.53
KT (6)	.47	.50	.51	.52	.54	.58	.59	.58	.56	.56	.49	.44
AVE. TEMP. (5)	-7.0	-4.0	1.0	10.0	16.0	21.0	24.0	23.0	18.0	12.0	3.0	-4.0
DEGREE-DAYS (5)	786.	634.	536.	258.	103.	14.	0.	7.	52.	194.	453.	689.

DETROIT   MI   (LAT. 42.2)

	JAN	FEB	MAR	APR	MAY	JUNE	JULY	AUG	SEP	OCT	NOV	DEC
HORIZ. RAD. (4)	4.74	7.72	11.35	15.88	19.47	21.18	20.83	17.88	14.22	9.94	5.42	3.90
KT (6)	.35	.41	.43	.47	.50	.51	.52	.50	.49	.47	.36	.32
AVE. TEMP. (5)	-4.0	-3.0	2.0	9.0	14.0	20.0	22.0	22.0	18.0	12.0	4.0	-2.0
DEGREE-DAYS (5)	696.	597.	512.	288.	136.	20.	3.	9.	53.	209.	415.	629.

```
                   DILLON          MT       (LAT. 45.2)

                JAN    FEB    MAR    APR    MAY   JUNE   JULY   AUG    SEP    OCT    NOV    DEC
HORIZ. RAD. (4) 5.97   9.60  14.52  18.60  22.58  24.33  27.15  22.96  17.26  11.61   6.83   5.11
KT          (6)  .50    .56    .59    .57    .58    .59    .68    .66    .63    .60    .52    .49
AVE. TEMP.  (3) -7.0   -4.0   -1.0    5.0   10.0   14.0   19.0   18.0   13.0    7.0     .0   -5.0
DEGREE-DAYS (3) 772.   614.   609.   398.   252.   132.    30.    47.   181.   344.   553.   708.


                   DODGE CITY      KA       (LAT. 37.8)

                JAN    FEB    MAR    APR    MAY   JUNE   JULY   AUG    SEP    OCT    NOV    DEC
HORIZ. RAD. (4) 9.38  12.73  16.76  21.40  23 72  26.76  26.05  23.33  19.14  14.76  10.14   8.31
KT          (6)  .57    .59    .60    .61    .60    .65    .65    .64    .65    .63    .58    .55
AVE. TEMP.  (3) -1.0    1.0    5.C   12.0   17.0   23.0   26.0   25.0   20.0   13.0    5.0     .0
DEGREE-DAYS (3) 589.   463.   410.   191.    64.    12.     0.     0.    23.   137.   370    544.


                   DULUTH          MN       (LAT. 46.8)

                JAN    FEB    MAR    APR    MAY   JUNE   JULY   AUG    SEP    OCT    NOV    DEC
HORIZ. RAD. (4) 4.41   7.64  11.74  15.58  18.64  20.06  21.04  17.56  12.43   8.23   4.32   3.31
KT          (6)  .40    .47    .49    .48    .48    .49    .53    .51    .46    .44    .36    .35
AVE. TEMP.  (5)-13.0  -11.0   -5.0    4.0   10.0   15.0   19.0   18.0   12.0    7.0   -2.0  -10.0
DEGREE-DAYS (5) 973.   823.   715.   440.   269.   108.    37.    58.   177.   339.   610.   872.


                   EAGLE           CO       (LAT. 39.6)

                JAN    FEB    MAR    APR    MAY   JUNE   JULY   AUG    SEP    OCT    NOV    DEC
HORIZ. RAD. (4) 8.56  12.23  17.04  21.93  25.60  28.47  27.06  23.65  20.05  14.84   9.86   7.84
KT          (6)  .56    .60    .62    .64    .65    .69    .67    .65    .67    .66    .60    .57
AVE. TEMP.  (5) -8.0   -5.0   -1.0    5.0   11.0   15.0   19.0   18.0   13.0    7.0   -1.0   -7.0
DEGREE-DAYS (5) 809.   649.   584.   385.   236.   106.    24.    44.   158.   348.   568.   770.


                   EAST LANSING    MI       (LAT. 42.7)

                 JAN    FEB    MAR    APR    MAY   JUNE   JULY   AUG    SEP    OCT    NOV    DEC
HORIZE. RAD. (2) 5.07   8.79  12.94  15.03  20.22  22.90  22.61  19.51  15.62  10.68   5.69   4.52
KT          (6)  .38    .47    .50    .45    .52    .55    .56    .55    .55    .51    .39    .38
AVE. TEMP.  (3) -5.0   -4.0     .0    8.0   13.0   19.0   21.0   20.0   16.0   10.0    3.0   -2.0
DEGREE-DAYS (3) 730.   638.   553.   308.   156.    27.     5.    15.    74.   234.   443.   653.


                   EAU CLAIRE      WI       (LAT. 44.9)

                JAN    FEB    MAR    APR    MAY   JUNE   JULY   AUG    SEP    OCT    NOV    DEC
HORIZ. RAD. (4) 5.13   8.47  12.37  16.18  19.08  21.24  21.41  18.39  13.58   9.38   5.11   3.87
KT          (6)  .42    .48    .50    .49    .49    .51    .53    .52    .49    .48    .38    .36
AVE. TEMP.  (3)-11.0   -9.0   -3.0    7.0   13.0   19.0   21.0   20.0   15.0    9.0     .0   -8.0
DEGREE-DAYS (3) 918.   772.   649.   342.   163.    36.     8.    21.   112.   281.   550.   809.


                   ELKO            NV       (LAT. 40.8)

                JAN    FEB    MAR    APR    MAY   JUNE   JULY   AUG    SEP    OCT    NOV    DEC
HORIZ. RAD. (4) 7.82  11.74  16.60   1.56  26.14  28.75  29.77  26.28  21.48  15.01   9.22   7.00
KT          (6)  .54    .59    .62    .63    .67    .70    .74    .73    .73    .69    .58    .53
AVE. TEMP.  (3) -5.0   -2.0    2.0    6.0   11.0   15.0   21.0   19.0   14.0    8.0    2.0   -3.0
DEGREE-DAYS (3) 720.   557.   517.   358.   226.   106.    15.    33.   138.   312.   503.   673.


                   EL PASO         TX       (LAT. 31.8)

                JAN    FEB    MAR    APR    MAY   JUNE   JULY   AUG    SEP    OCT    NOV    DEC
HORIZ. RAD. (4)12.77  16.80  21.67  26.82  29.51  30.44  27.81  25.93  22.55  18.60  14.12  11.70
KT          (6)  .64    .68    .71    .74    .74    .74    .69    .69    .69    .70    .67    .63
AVE. TEMP.  (3)  7.0    9.0   13.0   17.0   22.0   27.0   27.0   26.0   23.0   18.0   11.0    7.0
DEGREE-DAYS (3) 368.   258.   182.    49.     0.     0.     0.     0.     0.    51.   223.   355.


                   EL TORO         CA       (LAT. 33.7)

                JAN    FEB    MAR    APR    MAY   JUNE   JULY   AUG    SEP    OCT    NOV    DEC
HORIZ. RAD. (4)10.75  14.03  18.27  21.89  23.49  24.90  26.82  24.46  19.72  15.40  11.65   9.87
KT          (6)  .57    .59    .61    .61    .59    .61    .6     .66    .62    .60    .58    .56
AVE. TEMP.  (3) 12.0   13.0   13.0   15.0   17.0   19.0   22.0   22.0   21.0   18.0   15.0   12.0
DEGREE-DAYS (3) 207.   166.   155.    98.    52.    21.     0.     0.     5.    36.   108.   189.


                   ELY             NV       (LAT. 39.3)

                JAN    FEB    MAR    APR    MAY   JUNE   JULY   AUG    SEP    OCT    NOV    DEC
HORIZ. RAD. (4) 9.30  12.95  18.23  22.80  26.22  28.52  27.77  25.31  21.96  15.98  10.51   8.20
KT          (6)  .60    .62    .66    .66    .67    .69    .69     .0    .73    .70    .63    .58
AVE. TEMP.  (3) -4.0   -2.0     .0    5.0   10.0   14.0   19.0   18.0   13.0    7.0    1.0   -3.0
DEGREE-DAYS (3) 713.   577.   554.   395.   261.   134.    13.    34.    47.   327.   520.   557.
```

ERIE   PA   (LAT. 42.1)

		JAN	FEB	MAR	APR	MAY	JUNE	JULY	AUG	SEP	OCT	NOV	DEC
HORIZ. RAD.	(4)	3.92	6.55	10.44	15.42	18.68	20.96	20.80	16.51	13.63	9.39	4.72	3.15
KT	(6)	.28	.34	.40	.46	.48	.51	.52	.46	.47	.44	.31	.26
AVE. TEMP.	(3)	-4.0	-4.0	.0	7.0	13.0	18.0	20.0	20.0	16.0	11.0	4.0	-2.0
DEGREE-DAYS	(3)	687.	619.	553.	337.	187.	44.	13.	24.	78.	231.	415.	618.

EVANSVILLE   IN   (LAT. 38.0)

		JAN	FEB	MAR	APR	MAY	JUNE	JULY	AUG	SEP	OCT	NOV	DEC
HORIZ. RAD.	(4)	6.51	9.34	13.06	17.03	20.23	22.50	21.79	19.69	15.93	12.34	7.75	5.66
KT	(6)	.40	.44	.47	.49	.51	.55	.54	.54	.52	.53	.44	.38
AVE. TEMP.	(3)	.0	2.0	7.0	14.0	19.0	24.0	25.0	25.0	21.0	15.0	7.0	2.0
DEGREE-DAYS	(3)	558.	453.	363.	146.	53.	0.	0.	0.	19.	131	335.	512.

FARGO   ND   (LAT. 46.9)

		JAN	FEB	MAR	APR	MAY	JUNE	JULY	AUG	SEP	OCT	NOV	DEC
HORIZ. RAD.	(4)	4.71	8.01	12.46	16.75	20.82	22.63	24.06	20.71	14.80	9.92	5.19	3.83
KT	(6)	.43	.49	.52	.52	.54	.55	.60	.60	.55	.54	.43	.41
AVE. TEMP.	(5)	-14.0	-12.0	-4.0	6.0	13.0	18.0	21.0	21.0	14.0	8.0	-2.0	-11.0

FARMINGTON   NM   (LAT. 36.7)

		JAN	FEB	MAR	APR	MAY	JUNE	JULY	AUG	SEP	OCT	NOV	DEC
HORIZ. RAD.	(4)	10.72	14.54	19.22	24.21	27.82	30.25	28.13	25.56	21.95	16.78	11.88	9.50
KT	(6)	.63	.66	.67	.69	.70	.73	.70	.70	.71	.70	.65	.61
AVE. TEMP.	(3)	-2.0	2.0	5.0	10.0	15.0	20.0	24.0	23.0	18.0	12.0	4.0	-1.0
DEGREE-DAYS	(3)	627.	467.	420.	258.	102.	20.	0.	3.	37.	208.	430.	601.

FLINT   MI   (LAT. 43.0)

		JAN	FEB	MAR	APR	MAY	JUNE	JULY	AUG	SEP	OCT	NOV	DEC
HORIZ. RAD.	(4)	4.35	7.22	10.86	15.20	18.82	20.58	20.39	17.65	13.57	9.41	4.87	3.51
KT	(6)	.33	.39	.42	.45	.48	.50	.51	.50	.48	.45	.34	.30
AVE. TEMP.	(3)	-5.0	-5.0	.0	8.0	13.0	19.0	21.0	20.0	16.0	11.0	3.0	-3.0
DEGREE-DAYS	(3)	736.	641.	558.	318.	170.	36.	8.	20.	82.	241.	445.	658.

FORT SMITH   AR   (LAT. 35.3)

		JAN	FEB	MAR	APR	MAY	JUNE	JULY	AUG	SEP	OCT	NOV	DEC
HORIZ. RAD.	(4)	8.44	11.34	14.89	18.34	21.70	23.71	23.44	21.31	17.04	13.63	9.66	7.74
KT	(6)	.47	.49	.41	.52	.55	.58	.58	.58	.54	.55	.51	.47
AVE. TEMP.	(5)	4.0	6.0	10.0	17.0	21.0	26.0	28.0	27.0	23.0	17.0	10.0	5.0
DEGREE-DAYS	(5)	448.	238.	262.	73.	9.	0.	0.	0.	0.	75.	243.	405.

FORT WAYNE   IN   (LAT. 41.0)

		JAN	FEB	MAR	APR	MAY	JUNE	JULY	AUG	SEP	OCT	NOV	DEC
HORIZ. RAD.	(4)	5.17	7.92	11.14	15.44	18.97	20.90	20.28	18.09	14.45	10.49	5.86	4.19
KT	(6)	.36	.40	.42	.45	.48	.51	.50	.50	.49	.48	.37	.32
AVE. TEMP.	(5)	-4.0	-2.0	2.0	10.0	15.0	21.0	23.0	22.0	18.0	12.0	5.0	-2.0
DEGREE-DAYS	(5)	684.	582.	491.	262.	120.	13.	0.	7.	50.	202.	413.	627.

FORT WORTH   TX   (LAT. 32.8)

		JAN	FEB	MAR	APR	MAY	JUNE	JULY	AUG	SEP	OCT	NOV	DEC
HORIZ. RAD.	(4)	9.09	12.07	15.91	18.28	21.37	24.36	24.37	22.39	18.28	14.49	10.52	8.60
KT	(6)	.47	.50	.53	.51	.54	.59	.61	.60	.57	.56	.51	.48
AVE. TEMP.	(3)	7.0	9.0	13.0	18.0	22.0	27.0	29.0	29.0	25.0	19.0	13.0	8.0
DEGREE-DAYS	(3)	348.	253.	186.	49.	0.	0.	0.	0.	0.	33.	159.	294.

FRESNO   CA   (LAT. 36.8

		JAN	FEB	MAR	APR	MAY	JUNE	JULY	AUG	SEP	OCT	NOV	DEC
HORIZ. RAD.	(4)	7.32	11.35	17.31	22.32	26.21	28.72	28.66	25.82	21.25	15.39	9.55	6.24
KT	(6)	.43	.51	.61	.64	.66	.70	.71	.70	.69	.64	.53	.40
AVE. TEMP.	(3)	7.0	10.0	12.0	16.0	19.0	23.0	27.0	26.0	23.0	18.0	12.0	7.0
DEGREE-DAYS	(3)	339.	235.	191.	101.	28.	5.	0.	0.	0.	50.	192.	331.

GLASGOW   MT   (LAT. 48.2)

		JAN	FEB	MAR	APR	MAY	JUNE	JULY	AUG	SEP	OCT	NOV	DEC
HORIZ. RAD.	(4)	4.40	7.62	12.54	16.89	20.74	23.23	24.89	21.14	15.21	9.96	5.43	3.79
KT	(6)	.44	.49	.54	.53	.54	.56	.63	.62	.58	.56	.48	.44
AVE. TEMP.	(3)	-12.0	-8.0	-3.0	6.0	12.0	17.0	21.0	21.0	14.0	8.0	-1.0	-7.0
DEGREE-DAYS	(3)	961.	774.	686.	370.	191.	84.	8.	1.	146.	321.	600.	825.

GRAND ISLAND   NE   (LAT. 41.0)

		JAN	FEB	MAR	APR	MAY	JUNE	JULY	AUG	SEP	OCT	NOV	DEC
HORIZ. RAD.	(4)	7.50	10.41	14.36	19.21	22.38	25.45	25.14	22.01	17.13	12.91	8.38	6.46
KT	(6)	.52	.53	.54	.56	.47	.62	.63	.61	.58	.59	.53	.50
AVE. TEMP.	(3)	-5.0	-3.0	3.0	10.0	16.0	22.0	25.0	24.0	19.0	12.0	4.0	-3.0
DEGREE-DAYS	(3)	704.	582.	479.	232.	100.	20.	0.	0.	48.	198.	447.	625.

GRAND JUNCTION   CO   (LAT. 39.1)

		JAN	FEB	MAR	APR	MAY	JUNE	JULY	AUG	SEP	OCT	NOV	DEC
HORIZ. RAD.	(4)	8.98	12.70	17.63	22.54	27.01	29.49	27.98	24.76	20.82	15.26	10.42	8.30
KT	(6)	.58	.61	.64	.65	.69	.71	.69	.68	.69	.67	.62	.59
AVE. TEMP.	(3)	-3.0	.0	5.0	11.0	16.0	22.0	25.0	24.0	19.0	12.0	4.0	-1.0
DEGREE-DAYS	(3)	661.	488.	410.	224.	74.	12.	0.	0.	33.	180	420.	612.

GRAND LAKE   CO   (LAT. 40.3)

		JAN	FEB	MAR	APR	MAY	JUNE	JULY	AUG	SEP	OCT	NOV	DEC
HORIZ. RAD.	(2)	8.88	13.11	17.71	21.44	23.11	26.46	25.12	21.14	19.93	15.11	9.80	7.70
KT	(6)	.60	.65	.66	.63	.59	.64	.62	.59	.61	.68	.61	.57
AVE. TEMP.	(3)	-9.0	-7.0	-4.0	.0	6.0	10.0	13.0	12.0	8.0	3.0	-3.0	-8.0
DEGREE DAYS	(3)	864.	734.	720.	525.	381.	250.	153.	174.	280.	446.	653.	820.

GRAND RAPIDS   MI   (LAT. 42.9)

		JAN	FEB	MAR	APR	MAY	JUNE	JULY	AUG	SEP	OCT	NOV	DEC
HORIZ. RAD.	(4)	4.19	7.36	11.51	16.02	19.92	22.20	21.73	19.02	14.32	9.74	5.06	3.53
KT	(6)	.32	.3	.45	.48	.51	.54	.54	.53	.50	.47	.35	.30
AVE. TEMP.	(3)	-5.0	-4.0	1.0	8.0	14.0	20.0	22.0	21.0	17.0	11.0	4.0	-3.0
DEGREE-DAYS	(3)	720.	630.	549.	308.	150.	24.	4.	15.	63.	227.	438.	648.

GREAT FALLS   MT   (LAT. 47.5)

		JAN	FEB	MAR	APR	MAY	JUNE	JULY	AUG	SEP	OCT	NOV	DEC
HORIZ. RAD.	(4)	4.77	8.17	13.28	16.89	20.97	23.85	26.43	21.94	15.64	10.49	5.65	3.81
KT	(6)	.45	.51	.56	.53	.55	.58	.66	.64	.49	.58	.48	.42
AVE. TEMP.	(3)	-5.0	-2.0	.0	6.0	12.0	16.0	21.0	20.0	14.0	9.0	1.0	-2.0
DEGREE-DAYS	(3)	767.	597.	594.	360.	204.	90.	10.	23.	144.	291.	507.	663.

GREEN BAY   WI   (LAT. 44.5)

		JAN	FEB	MAR	APR	MAY	JUNE	JULY	AUG	SEP	OCT	NOV	DEC
HORIZ. RAD.	(4)	5.12	8.23	12.53	16.33	19.51	21.65	21.43	18.41	13.82	9.31	5.28	3.97
KT	(6)	.42	.46	.50	.50	.50	.52	.54	.52	.50	.47	.39	.36
AVE. TEMP.	(5)	-9.0	-8.0	-2.0	7.0	12.0	18.0	21.0	20.0	15.0	10.0	1.0	-6.0
DEGREE-DAYS	(5)	854.	731.	627.	353.	188.	51.	12.	30.	106.	272.	515.	759.

GREENSBORO   NC   (LAT. 36.1)

		JAN	FEB	MAR	APR	MAY	JUNE	JULY	AUG	SEP	OCT	NOV	DEC
HORIZ. RAD.	(4)	8.12	11.01	14.90	19.10	21.20	22.17	21.15	19.25	16.09	12.95	9.52	7.48
KT	(6)	.47	.49	.52	.54	.54	.54	.53	.53	.52	.53	.51	.47
AVE. TEMP.	(3)	3.0	4.0	8.0	14.0	19.0	23.0	25.0	24.0	21.0	14.0	8.0	4.0
DEGREE-DAYS	(3)	453.	379.	302.	113.	33.	0.	0.	0.	13.	116.	278.	437.

HARRISBURG   PA   (LAT. 40.2)

		JAN	FEB	MAR	APR	MAY	JUNE	JULY	AUG	SEP	OCT	NOV	DEC
HORIZ. RAD.	(4)	6.08	8.75	12.29	16.01	18.75	20.48	20.01	17.60	14.37	10.60	6.57	5.08
KT	(6)	.41	.43	.45	.47	.48	.50	.50	.49	.49	.48	.41	.38
AVE. TEMP.	(3)	-1.0	.0	5.0	12.0	17.0	22.0	24.0	23.0	19.0	13.0	7.0	.0
DEGREE-DAYS	(3)	601.	509.	413.	206.	71.	0.	0.	0.	28.	163.	353.	558.

HARTFORD   CT   (LAT. 41.9)

		JAN	FEB	MAR	APR	MAY	JUNE	JULY	AUG	SEP	OCT	NOV	DEC
HORIZ. RAD.	(4)	5.42	8.11	11.11	14.92	17.80	19.13	18.71	16.13	13.10	9.68	5.64	4.37
KT	(6)	.39	.42	.42	.44	.46	.46	.47	.45	.45	.46	.37	.35
AVE. TEMP.	(5)	-4.0	-3.0	2.0	9.0	15.0	20.0	23.0	21.0	17.0	11.0	5.0	-2.0
DEGREE-DAYS	(5)	692.	594.	506.	288.	126.	13.	0.	7.	59.	213.	395.	634.

HELENA   MT   (LAT. 46.6)

		JAN	FEB	MAR	APR	MAY	JUNE	JULY	AUG	SEP	OCT	NOV	DEC
HORIZ. RAD.	(4)	4.76	8.04	13.00	16.87	21.11	23.15	26.49	21.91	16.03	10.51	5.92	4.13
KT	(6)	.43	.49	.54	.52	.55	.56	.66	.63	.59	.56	.48	.43
AVE. TEMP.	(3)	-8.0	-4.0	-1.0	6.0	11.0	15.0	20.0	19.0	13.0	7.0	.0	-5.0
DEGREE-DAYS	(3)	808.	616.	592.	372.	223.	108.	18.	32.	169.	339.	555.	718.

HOUSTON TX (LAT. 30.0)

		JAN	FEB	MAR	APR	MAY	JUNE	JULY	AUG	SEP	OCT	NOV	DEC
HORIZ. RAD.	(4)	8.77	11.74	14.72	17.28	20.14	21.54	20.75	19.14	16.69	14.48	10.49	8.28
KT	(6)	.42	.46	.47	.47	.51	.53	.52	.51	.51	.53	.47	.42
AVE. TEMP.	(5)	11.0	13.0	16.0	21.0	24.0	27.0	28.0	29.0	26.0	22.0	16.0	13.0
DEGREE-DAYS	(5)	231.	163.	105.	13.	0.	0.	0.	0.	0.	13.	86.	185.

HURON SD (LAT. 44.4)

		JAN	FEB	MAR	APR	MAY	JUNE	JULY	AUG	SEP	OCT	NOV	DEC
HORIZ. RAD.	(4)	.54	8.45	12.64	17.36	21.24	23.84	24.77	21.48	16.09	11.22	6.55	4.60
KT	(6)	.45	.47	.50	.53	.55	.58	.62	.61	.58	.56	.48	.42
AVE. TEMP.	(3)	-11.0	-8.0	-2.0	8.0	14.0	19.0	23.0	22.0	16.0	10.0	.0	-7.0
DEGREE-DAYS	(3)	904.	733.	620.	320.	152.	40.	5.	7.	94.	268.	543.	789.

INDIANAPOLIS IN (LAT. 39.7)

		JAN	FEB	MAR	APR	MAY	JUNE	JULY	AUG	SEP	OCT	NOV	DEC
HORIZ. RAD.	(4)	5.62	8.48	11.77	15.87	19.16	21.20	20.50	18.65	15.03	11.09	6.57	4.73
KT	(6)	.37	.41	.43	.46	.49	.51	.51	.52	.50	.49	.40	.34
AVE. TEMP.	(3)	-1.0	.0	4.0	11.0	17.0	22.0	24.0	23.0	19.0	13.0	.0	.0
DEGREE-DAYS	(3)	639.	533.	436.	215.	88.	6.	0.	3.	35.	168.	388.	587.

INTRNTNAL FALLS MN (LAT. 48.6)

		JAN	FEB	MAR	APR	MAY	JUNE	JULY	AUG	SEP	OCT	NOV	DEC
HORIZ. RAD.	(4)	4.04	7.52	11.87	16.38	19.48	21.03	21.80	18.37	12.73	7.99	3.92	3.08
KT	(6)	.41	.49	.52	.52	.51	.51	.55	.54	.49	.45	.37	.37
AVE. TEMP.	(3)	-17.0	-14.0	-6.0	3.0	10.0	16.0	19.0	17.0	12.0	6.0	-4.0	-13.0
DEGREE-DAYS	(3)	1087.	902.	764.	447.	257.	93.	37.	62.	202.	371.	668.	969.

JACKSON MS (LAT. 32.3)

		JAN	FEB	MAR	APR	MAY	JUNE	JULY	AUG	SEP	OCT	NOV	DEC
HORIZ. RAD.	(4)	8.55	11.65	15.54	19.39	22.03	22.97	21.66	20.21	17.13	14.43	10.23	8.04
KT	(6)	.43	.48	.51	.54	.56	.56	.54	.54	.53	.55	.49	.44
AVE. TEMP.	(5)	8.0	10.0	13.0	19.0	23.0	26.0	28.0	27.0	24.0	19.0	13.0	9.0
DEGREE-DAYS	(5)	316.	246.	174.	41.	3.	0.	0.	0.	0.	51.	167.	280.

JACKSONVILLE FL (LAT. 30.4)

		JAN	FEB	MAR	APR	MAY	JUNE	JULY	AUG	SEP	OCT	NOV	DEC
HORIZ. RAD.	(4)	10.21	13.21	17.27	21.06	22.20	21.39	20.45	19.23	16.37	13.88	11.30	9.28
KT	(6)	.49	.52	.56	.58	.56	.52	.51	.51	.50	.51	.52	.48
AVE. TEMP.	(3)	12.0	13.0	16.0	20.0	23.0	26.0	27.0	27.0	25.0	21.0	16.0	12.0
DEGREE-DAYS	(3)	193.	157.	98.	13.	0.	0.	0.	0.	0.	11.	89.	176.

KANSAS CITY MO (LAT. 39.3)

		JAN	FEB	MAR	APR	MAY	JUNE	JULY	AUG	SEP	OCT	NOV	DEC
HORIZ. RAD.	(4)	7.35	10.15	13.65	17.87	21.25	23.60	23.86	21.14	16.48	12.40	8.37	6.37
KT	(6)	.47	.49	.50	.52	.54	.57	.59	.58	.55	.55	.0	.45
AVE. TEMP.	(5)	-2.0	1.0	5.0	13.0	18.0	23.0	26.0	25.0	20.0	15.0	6.0	.0
DEGREE-DAYS	(5)	641.	496.	414.	174.	62.	7.	0.	0.	23.	131.	357.	563.

KINGSVILLE TX (LAT. 27.5)

		JAN	FEB	MAR	APR	MAY	JUNE	JULY	AUG	SEP	OCT	NOV	DEC
HORIZ. RAD.	(4)	10.35	13.18	16.28	18.87	21.16	23.11	23.96	21.81	18.44	15.77	11.74	9.64
KT	(6)	.46	.49	.51	.51	.54	.57	.60	.58	.55	.56	.50	.46
AVE. TEMP.	(3)	13.0	15.0	18.0	23.0	26.0	29.0	30.0	30.0	27.0	23.0	18.0	15.0
DEGREE-DAYS	(3)	177.	111.	67.	4.	0.	0.	0.	0.	0.	6.	56.	140.

LA CROSSE WI (LAT. 43.9)

		JAN	FEB	MAR	APR	MAY	JUNE	JULY	AUG	SEP	OCT	NOV	DEC
HORIZ. RAD.	(4)	5.46	8.68	12.49	16.19	19.44	21.62	21.57	18.91	14.10	9.80	5.61	4.19
KT	(6)	.43	.48	.49	.49	.50	.52	.54	.53	.50	.48	.40	.37
AVE. TEMP.	(3)	-9.0	-7.0	-1.0	9.0	15.0	20.0	23.0	22.0	17.0	11.0	2.0	-6.0
DEGREE-DAYS	(3)	842.	700.	584.	290.	124.	22.	6.	9.	72.	234.	493.	744.

LAKE CHARLES LA (LAT. 30.2)

		JAN	FEB	MAR	APR	MAY	JUNE	JULY	AUG	SEP	OCT	NOV	DEC
HORIZ. RAD.	(4)	8.27	11.46	14.91	17.82	20.99	22.36	20.29	18.81	16.86	15.67	10.40	8.01
KT	(6)	.40	.45	.48	.49	.53	.55	.51	.50	.51	.58	.47	.41
AVE. TEMP.	(3)	11.0	12.0	15.0	20.0	23.0	27.0	28.0	27.0	25.0	21.0	15.0	12.0
DEGREE-DAYS	(3)	231.	170.	111.	14.	0.	0.	0.	0.	0.	20.	98.	188.

LAKEHURST　　　NJ　　　(LAT. 40.0)

		JAN	FEB	MAR	APR	MAY	JUNE	JULY	AUG	SEP	OCT	NOV	DEC
HORIZ. RAD.	(4)	6.35	9.04	12.59	16.52	18.98	20.14	19.33	17.39	14.31	10.85	7.05	5.39
KT	(6)	.42	.45	.46	.48	.48	.49	.48	.48	.48	.49	.43	.40
AVE. TEMP.	(3)	-1.0	.0	5.0	11.0	16.0	21.0	24.0	23.0	19.0	13.0	7.0	1.0
DEGREE-DAYS	(3)	584.	504.	424.	228.	87.	5.	0.	0.	29.	158.	327.	539.

LAS VEGAS　　　NV　　　(LAT. 36.1)

		JAN	FEB	MAR	APR	MAY	JUNE	JULY	AUG	SEP	OCT	NOV	DEC
HORIZ. RAD.	(4)	11.10	15.20	20.69	26.32	30.03	31.53	29.38	26.72	23.12	17.47	12.32	9.99
KT	(6)	.64	.68	.72	.75	.76	.77	.73	.73	.74	.72	.66	.62
AVE. TEMP.	(3)	6.0	9.0	12.0	17.0	23.0	28.0	31.0	30.0	26.0	19.0	11.0	7.0
DEGREE-DAYS	(3)	358.	257.	180.	70.	6.	0.	0.	0.	0.	41.	198.	341.

LEWISTOWN　　　MT　　　(LAT. 47.0)

		JAN	FEB	MAR	APR	MAY	JUNE	JULY	AUG	SEP	OCT	NOV	DEC
HORIZ. RAD.	(4)	4.77	7.86	12.81	16.39	20.51	23.37	25.96	21.58	15.57	10.27	5.70	4.12
KT	(6)	.44	.48	.54	.51	.53	.57	.65	.62	.58	.56	.47	.44
AVE. TEMP.	(3)	-7.0	-5.0	-3.0	4.0	10.0	14.0	19.0	18.0	12.0	8.0	.0	-4.0
DEGREE-DAYS	(3)	791.	641.	646.	415.	265.	147.	39.	52.	193.	336.	547.	698.

LEXINGTON　　　KY　　　(LAT. 38.0)

		JAN	FEB	MAR	APR	MAY	JUNE	JULY	AUG	SEP	OCT	NOV	DEC
HORIZ. RAD.	(4)	6.19	8.85	12.48	16.79	19.83	21.53	21.00	19.13	15.46	11.85	7.46	5.51
KT	(6)	.38	.41	.45	.48	.50	.52	.52	.52	.51	.51	.43	.37
AVE. TEMP.	(3)	.0	2.0	6.0	13.0	18.0	23.0	25.0	24.0	20.0	14.0	7.0	2.0
DEGREE-DAYS	(3)	553.	462.	374.	168.	59.	4.	0.	0.	22.	137.	340.	508.

LITTLE ROCK　　　AR　　　(LAT. 34.7)

		JAN	FEB	MAR	APR	MAY	JUNE	JULY	AUG	SEP	OCT	NOV	DEC
HORIZ. RAD.	(4)	8.30	11.38	14.90	18.28	21.89	23.91	23.06	21.12	17.23	13.94	9.62	7.65
KT	(6)	.45	.49	.51	.51	.55	.58	.57	.57	.55	.56	.50	.45
AVE. TEMP.	(3)	4.0	6.0	10.0	16.0	21.0	26.0	27.0	27.0	23.0	17.0	10.0	5.0
DEGREE-DAYS	(3)	439.	344.	264.	77.	12.	0.	0.	0.	3.	79.	245.	403.

LONGBEACH　　　CA　　　(LAT. 33.8)

		JAN	FEB	MAR	APR	MAY	JUNE	JULY	AUG	SEP	OCT	NOV	DEC
HORIZ. RAD.	(4)	10.53	13.79	18.27	21.99	23.43	24.29	26.10	23.83	19.30	15.05	11.39	9.61
KT	(6)	.56	.58	.61	.62	.59	.59	.65	.64	.61	.59	.57	.55
AVE. TEMP.	(3)	12.0	13.0	14.0	16.0	18.0	20.0	22.0	23.0	22.0	19.0	16.0	13.0
DEGREE-DAYS	(3)	188.	152.	137.	82.	39.	13.	0.	0.	4.	27.	86.	164.

LOS ANGELES　　　CA　　　(LAT. 33.9)

		JAN	FEB	MAR	APR	MAY	JUNE	JULY	AUG	SEP	OCT	NOV	DEC
HORIZ. RAD.	(4)	10.51	13.78	18.37	22.14	23.37	24.05	26.19	23.60	19.08	14.95	11.39	9.63
KT	(6)	.56	.58	.62	.62	.59	.59	.65	.64	.60	.59	.47	.56
AVE. TEMP.	(3)	12.0	13.0	13.0	15.0	16.0	18.0	20.0	20.0	20.0	18.0	15.0	13.0
DEGREE-DAYS	(3)	184.	150.	148.	108.	63.	39.	.11	8.	13.	43.	88.	155.

LOUISVILLE　　　KY　　　(LAT. 38.2)

		JAN	FEB	MAR	APR	MAY	JUNE	JULY	AUG	SEP	OCT	NOV	DEC
HORIZ. RAD.	(4)	6.19	8.96	12.51	16.65	19.52	12.60	20.85	19.07	15.45	11.83	7.41	5.54
KT	(6)	.38	.42	.45	.48	.50	.52	.52	.52	.51	.51	.43	.38
AVE. TEMP.	(3)	.0	2.0	6.0	13.0	18.0	22.0	24.0	24.0	20.0	14.0	7.0	1.0
DEGREE-DAYS	(3)	546.	454.	367.	159.	58.	0.	0.	0.	19.	134.	333.	506.

LOVELOCK　　　NV　　　(LAT. 40.1)

		JAN	FEB	MAR	APR	MAY	JUNE	JULY	AUG	SEP	OCT	NOV	DEC
HORIZ. RAD.	(4)	9.12	13.23	18.80	24.57	28.99	31.20	31.59	28.19	23.01	16.47	10.55	8.11
KT	(6)	.61	.65	.69	.72	.74	.75	.78	.78	.78	.74	.65	.60
AVE. TEMP.	(3)	-2.0	2.0	4.0	9.0	14.0	19.0	23.0	22.0	17.0	11.0	4.0	-1.0
DEGREE-DAYS	(3)	582.	426.	393.	250.	124.	38.	0.	6.	55.	202.	398.	433.

LUBBOCK　　　TX　　　(LAT. 33.6)

		JAN	FEB	MAR	APR	MAY	JUNE	JULY	AUG	SEP	OCT	NOV	DEC
HORIZ. RAD.	(4)	11.70	15.11	20.00	24.60	27.19	28.88	27.37	25.06	20.66	16.66	12.67	10.61
KT	(6)	.62	.63	.67	.69	.69	.70	.68	.67	.75	.65	.63	.61
AVE. TEMP.	(3)	4.0	6.0	9.0	16.0	20.0	25.0	27.0	26.0	22.0	16.0	9.0	5.0
DEGREE-DAYS	(3)	446.	347.	282.	106.	16.	0.	0.	0.	4.	90.	270.	408.

LUFKIN   TX   (LAT. 31.2)

		JAN	FEB	MAR	APR	MAY	JUNE	JULY	AUG	SEP	OCT	NOV	DEC
HORIZ. RAD.	(4)	9.01	12.13	15.62	18.43	21.18	22.33	22.77	21.15	17.37	15.31	10.93	8.71
KT	(6)	.44	.48	.51	.51	.53	.57	.57	.57	.53	.57	.51	.46
AVE. TEMP.	(3)	9.0	11.0	14.0	20.0	23.0	27.0	28.0	28.0	25.0	20.0	14.0	10.0
DEGREE-DAYS	(3)	283.	206.	142.	31.	0.	0.	0.	0.	0.	29.	142.	244.

MACON   GA   (LAT. 32.7)

		JAN	FEB	MAR	APR	MAY	JUNE	JULY	AUG	SEP	OCT	NOV	DEC
HORIZ. RAD.	(4)	8.73	11.57	15.47	19.70	21.39	21.78	20.26	19.49	16.33	14.15	10.66	8.27
KT	(6)	.45	.48	.51	.55	.54	.53	.50	.52	.51	.54	.52	.46
AVE. TEMP.	(5)	9.0	10.0	14.0	19.0	23.0	26.0	27.0	27.0	24.0	19.0	13.0	9.0
DEGREE-DAYS	(5)	302.	235.	166.	37.	3.	0.	0.	0.	0.	46.	169.	288.

MADISON   WI   (LAT. 43.1)

		JAN	FEB	MAR	APR	MAY	JUNE	JULY	AUG	SEP	OCT	NOV	DEC
HORIZ. RAD.	(4)	5.85	9.12	12.89	15.87	19.78	22.11	21.95	19.38	14.75	10.34	5.72	4.41
KT	(6)	.44	.49	.50	.48	.51	.53	.55	.55	.52	.50	.40	.38
AVE. TEMP.	(3)	-7.0	-6.0	.0	7.0	13.0	19.0	21.0	20.0	15.0	10.0	1.0	-5.0
DEGREE-DAYS	(3)	830.	696.	599.	328.	165.	40.	8.	22.	96.	263.	505.	742.

MANHATTAN   KA   (LAT. 39.2)

		JAN	FEB	MAR	APR	MAY	JUNE	JULY	AUG	SEP	OCT	NOV	DEC
HORIZ. RAD.	(2)	8.04	11.05	14.44	18.13	22.06	23.07	22.23	22.02	17.17	12.23	9.50	6.53
KT	(6)	.52	.53	.52	.53	.56	.56	.55	.61	.57	.54	.57	.46
AVE. TEMP.	(3)	-1.0	1.0	5.0	13.0	18.0	23.0	26.0	25.0	20.0	14.0	6.0	.0
DEGREE-DAYS	(3)	625.	482.	401.	169.	61.	7.	0.	0.	26.	134.	358.	562.

MASON CITY   IA   (LAT. 43.1)

		JAN	FEB	MAR	APR	MAY	JUNE	JULY	AUG	SEP	OCT	NOV	DEC
HORIZ. RAD.	(4)	6.28	9.49	13.25	17.23	21.51	23.99	23.65	20.80	15.95	11.47	6.81	5.03
KT	(6)	.48	.51	.52	.52	.55	.58	.59	.59	.56	.56	.47	.43
AVE. TEMP.	(3)	-10.0	-8.0	-2.0	8.0	14.0	20.0	22.0	21.0	16.0	10.0	1.0	-7.0
DEGREE-DAYS	(3)	875.	723.	620.	322.	147.	36.	7.	17.	92.	254.	523.	773.

MASSENA   NY   (LAT. 44.9)

		JAN	FEB	MAR	APR	MAY	JUNE	JULY	AUG	SEP	OCT	NOV	DEC
HORIZ. RAD.	(4)	4.90	7.86	12.43	17.04	20.54	22.78	22.57	19.16	14.39	9.43	5.09	3.75
KT	(6)	.41	.45	.50	.52	.53	.55	.56	.55	.52	.48	.38	.35
AVE. TEMP.	(3)	-10.0	-9.0	-2.0	6.0	12.0	18.0	21.0	19.0	15.0	9.0	2.0	-7.0
DEGREE-DAYS	(3)	870.	751.	644.	380.	194.	43.	12.	32.	107.	284.	485.	773.

MATANUSKA   AK   (LAT. 61.6)

		JAN	FEB	MAR	APR	MAY	JUNE	JULY	AUG	SEP	OCT	NOV	DEC
HORIZ. RAD.	(2)	1.34	3.85	10.13	14.90	18.25	19.34	17.12	13.15	8.29	4.19	1.59	.63
KT	(6)	.41	.41	.64	.56	.51	.48	.45	.44	.42	.43	.43	.40
AVE. TEMP.	(3)	-11.0	-7.0	-3.0	2.0	8.0	12.0	14.0	12.0	8.0	1.0	-6.0	-10.0
DEGREE-DAYS	(3)	914.	714.	689.	477.	310.	168.	129.	169.	288.	526.	738.	904.

MEDFORD   OR   (LAT. 42.4)

		JAN	FEB	MAR	APR	MAY	JUNE	JULY	AUG	SEP	OCT	NOV	DEC
HORIZ. RAD.	(4)	4.62	8.37	12.85	18.60	23.08	25.85	28.09	24.07	18.03	11.14	5.72	3.82
KT	(6)	.34	.44	.49	.55	.59	.63	.70	.67	.63	.53	.39	.31
AVE. TEMP.	(3)	3.0	5.0	7.0	10.0	14.0	18.0	22.0	21.0	18.0	12.0	6.0	3.0
DEGREE-DAYS	(3)	489.	369.	348.	274.	139.	52.	6.	12.	49.	200.	358.	470.

MEMPHIS   TN   (LAT. 35.0)

		JAN	FEB	MAR	APR	MAY	JUNE	JULY	AUG	SEP	OCT	NOV	DEC
HORIZ. RAD.	(4)	7.75	10.72	14.50	18.60	21.39	23.20	22.38	20.70	16.69	13.67	9.27	7.13
KT	(6)	.43	.46	.50	.52	.54	.56	.56	.56	.53	.55	.48	.43
AVE. TEMP.	(5)	5.0	7.0	11.0	17.0	22.0	26.0	28.0	27.0	23.0	17.0	10.0	6.0
DEGREE-DAYS	(5)	422.	330.	254.	73.	12.	0.	0.	0.	4.	79.	235.	384.

MERIDIAN   MS   (LAT. 32.3)

		JAN	FEB	MAR	APR	MAY	JUNE	JULY	AUG	SEP	OCT	NOV	DEC
HORIZ. RAD.	(4)	8.44	11.49	15.07	18.86	21.11	22.27	20.69	19.74	16.50	14.27	10.18	7.94
KT	(6)	.43	.47	.50	.52	.53	.54	.52	.53	.51	.55	.49	.43
AVE. TEMP.	(3)	8.0	10.0	13.0	19.0	22.0	26.0	27.0	27.0	24.0	18.0	12.0	9.0
DEGREE-DAYS	(3)	319.	246.	173.	44.	4.	0.	0.	0.	0.	62.	184.	294.

```
                    MIAMI           FL        (LAT. 25.8)

                 JAN    FEB    MAR    APR    MAY   JUNE   JULY   AUG    SEP    OCT    NOV    DEC
HORIZ. RAD. (4) 12.00 14.91 18.20 21.10 20.92 19.38 20.01 18.50 16.53 14.78 12.69 11.57
KT          (6)   .51    .54    .56    .57    .53    .48    .50    .49    .48    .51    .52    .52
AVE. TEMP.  (3) 19.0   19.0   12.0   23.0   25.0   27.0   27.0   28.0   27.0   25.0   22.0   20.0
DEGREE-DAYS (3) 29.    37.    9.     0.     0.     0.     0.     0.     0.     0.     7.    31.


                    MIDLAND         TX        (LAT. 31.9)

                 JAN    FEB    MAR    APR    MAY   JUNE   JULY   AUG    SEP    OCT    NOV    DEC
HORIZ. RAD. (4) 12.27 15.69 20.87 24.88 27.58 29.08 27.12 25.08 20.93 17.27 13.35 11.35
KT          (6)   .62    .63    .58    .69    .70    .71    .68    .67    .64    .65    .64    .61
AVE. TEMP.  (3)  6.0    8.0   12.0   17.0   22.0   26.0   27.0   27.0   23.0   18.0   11.0    7.0
DEGREE-DAYS (3) 368.   268.   194.   54.    0.     0.     0.     0.     0.    45.   198.   329.


                    MILES CITY      MT        (LAT. 46.4)

                 JAN    FEB    MAR    APR    MAY   JUNE   JULY   AUG    SEP    OCT    NOV    DEC
HORIZ. RAD. (4)  5.19   8.46 13.45 17.50 21.51 24.35 26.07 22.44 16.39 10.90  6.25  4.53
KT          (6)   .47    .51    .56    .54    .56    .59    .65    .65    .61    .58    .50    .47
AVE. TEMP.  (3) -9.0   -6.0   -1.0   7.0   13.0   18.0   24.0   23.0   15.0   9.0    .0   -6.0
DEGREE-DAYS (3) 854.   675.   599.   328.   160.   65.    5.     9.    121.   282.   543.   741.


                    MILWAUKEE       WI        (LAT. 42.9)

                 JAN    FEB    MAR    APR    MAY   JUNE   JULY   AUG    SEP    OCT    NOV    DEC
HORIZ. RAD. (4)  5.44   8.36 12.36 16.37 20.07 22.44 22.26 19.51 14.87 10.30  5.95  4.29
KT          (6)   .41    .45    .48    .49    .51    .54    .55    .55    .52    .50    .41    .36
AVE. TEMP.  (5) -7.0   -5.0    .0    7.0   12.0   18.0   21.0   21.0   16.0   11.0   2.0   -4.0
DEGREE-DAYS (5) 786.   661.   579.   338.   193.   50.    8.    20.    78.   244.   475.   703.


                    MINN-ST. PAUL   MN        (LAT. 44.9)

                 JAN    FEB    MAR    APR    MAY   JUNE   JULY   AUG    SEP    OCT    NOV    DEC
HORIZ. RAD. (4)  5.27   8.67 12.52 16.36 19.72 21.88 22.36 19.15 14.24  9.76  5.45  4.01
KT          (6)   .44    .49    .50    .50    .51    .53    .56    .54    .51    .50    .41    .38
AVE. TEMP.  (5) -11.0  -9.0   -2.0   7.0   14.0   19.0   22.0   21.0   16.0   10.0   .0   -7.0
DEGREE-DAYS (5) 916.   759.   637.   340.   159.   42.    8.    14.   108.   276.   552.   806.


                    MINOT           ND        (LAT. 48.2)

                 JAN    FEB    MAR    APR    MAY   JUNE   JULY   AUG    SEP    OCT    NOV    DEC
HORIZ. RAD. (4)  4.35   7.44 11.85 16.58 20.95 22.41 23.81 20.43 14.49  9.64  4.98  3.52
KT          (6)   .43    .48    .51    .52    .55    .54    .60    .60    .55    .54    .44    .41
AVE. TEMP.  (3) -13.0 -11.0   -5.0   5.0   12.0   17.0   20.0   20.0   13.0   8.0   -2.0 -10.0
DEGREE-DAYS (3) 983.   812.   713.   398.   213.   83.   15.    39.   159.   326.   618.   866.


                    MISSOULA        MT        (LAT. 46.9)

                 JAN    FEB    MAR    APR    MAY   JUNE   JULY   AUG    SEP    OCT    NOV    DEC
HORIZ. RAD. (4)  3.54   6.52 11.14 15.69 20.23 21.94 26.41 21.35 15.41  9.22  4.65  3.03
KT          (6)   .33    .40    .47    .49    .52    .53    .66    .62    .57    .50    .38    .32
AVE. TEMP.  (3) -6.0   -3.0    1.0   7.0   11.0   15.0   19.0   18.0   13.0   7.0    .0   -4.0
DEGREE-DAYS (3) 761.   588.   546.   352.   221.   112.   22.    39.   167.   360.   545.   694.


                    MOBILE          AL        (LAT. 30.7)

                 JAN    FEB    MAR    APR    MAY   JUNE   JULY   AUG    SEP    OCT    NOV    DEC
HORIZ. RAD. (4)  9.40 12.48 15.97 19.54 21.25 21.20 19.47 18.63 16.45 14.74 10.84  8.62
KT          (6)   .46    .49    .52    .54    .54    .52    .49    .50    .50    .55    .50    .45
AVE. TEMP.  (3) 11.0   12.0   15.0   20.0   24.0   27.0   28.0   28.0   25.0   20.0   15.0   12.0
DEGREE-DAYS (3) 251.   187.   123.   22.    0.     0.     0.     0.     0.    22.   117.   214.


                    MOLINE          IL        (LAT. 41.4)

                 JAN    FEB    MAR    APR    MAY   JUNE   JULY   AUG    SEP    OCT    NOV    DEC
HORIZ. RAD. (4)  6.07   9.22 12.69 16.56 19.90 22.35 22.00 19.46 15.40 11.30  6.75  4.91
KT          (6)   .43    .47    .48    .49    .51    .54    .55    .54    .53    .53    .44    .39
AVE. TEMP.  (3) -6.0   -4.0    2.0  10.0   16.0   22.0   24.0   23.0   18.0   12.0   4.0   -3.0
DEGREE-DAYS (3) 749.   611.   504.   242.   102.   11.    0.     6.    44.   191.   430.   661.


                    MONTGOMERY      AL        (LAT. 32.3)

                 JAN    FEB    MAR    APR    MAY   JUNE   JULY   AUG    SEP    OCT    NOV    DEC
HORIZ. RAD. (4)  8.53 11.50 15.21 19.62 21.53 22.38 20.89 19.81 16.66 14.32 10.39  8.16
KT          (6)   .43    .47    .50    .54    .54    .55    .52    .53    .51    .55    .50    .45
AVE. TEMP.  (3)  9.0   10.0   14.0   18.0   22.0   26.0   27.0   27.0   24.0   19.0   13.0    9.0
DEGREE-DAYS (3) 309.   233.   166.   42.    4.     0.     0.     0.     0.    52.   170.   284.
```

MT. SHASTA   CA   (LAT. 41.3)

		JAN	FEB	MAR	APR	MAY	JUNE	JULY	AUG	SEP	OCT	NOV	DEC
HORIZ. RAD.	(4)	6.36	9.73	14.19	19.93	24.81	27.65	29.25	25.12	19.69	13.11	7.48	5.73
KT	(6)	.45	.50	.53	.59	.63	.67	.73	.70	.67	.61	.48	.45
AVE. TEMP.	(3)	1.0	3.0	5.0	8.0	12.0	16.0	20.0	19.0	16.0	11.0	5.0	2.0
DEGREE-DAYS	(3)	541.	423.	424.	321.	206.	99.	21.	36.	81.	234.	388.	508.

MT. WEATHER   VA   (LAT. 39.1)

		JAN	FEB	MAR	APR	MAY	JUNE	JULY	AUG	SEP	OCT	NOV	DEC
HORIZ. RAD.	(2)	7.20	11.47	14.15	17.33	21.27	21.98	21.35	18.00	15.70	11.76	8.46	7.03
KT	(6)	.46	.55	.51	.50	.54	.53	.53	.50	.52	.52	.50	.50
AVE. TEMP.	(3)	-1.0	.0	3.0	10.0	15.0	20.0	22.0	21.0	18.0	12.0	5.0	.0
DEGREE-DAYS	(3)	615.	535.	453.	251.	102.	13.	0.	3.	44.	189.	370.	573.

NASHVILLE   TN   (LAT. 36.1)

		JAN	FEB	MAR	APR	MAY	JUNE	JULY	AUG	SEP	OCT	NOV	DEC
HORIZ. RAD.	(4)	6.58	9.35	12.82	17.52	20.71	22.28	21.46	19.71	15.87	12.64	8.07	5.91
KT	(6)	.38	.42	.44	.50	.52	.-4	.53	.54	.51	.52	.43	.37
AVE. TEMP.	(3)	3.0	5.0	9.0	15.0	20.0	24.0	26.0	25.0	22.0	16.0	9.0	5.0
DEGREE-DAYS	(3)	460.	373.	291.	98.	25.	0.	0.	0.	6.	100.	277.	424.

NEEDLES   CA   (LAT. 34.8)

		JAN	FEB	MAR	APR	MAY	JUNE	JULY	AUG	SEP	OCT	NOV	DEC
HORIZ. RAD.	(4)	11.18	15.36	20.71	26.30	30.10	31.68	28.84	25.85	22.86	17.45	12.75	10.37
KT	(6)	.62	.66	.71	.74	.76	.77	.72	.70	.72	.30	.66	.62
AVE. TEMP.	(3)	11.0	14.0	16.0	21.0	26.0	31.0	35.0	34.0	31.0	23.0	16.0	11.0
DEGREE-DAYS	(3)	234.	145.	83.	23.	0.	0.	0.	0.	0.	6.	91.	212.

NEWARK   NJ   (LAT. 40.7)

		JAN	FEB	MAR	APR	MAY	JUNE	JULY	AUG	SEP	OCT	NOV	DEC
HORIZ. RAD.	(4)	6.26	9.00	12.58	16.44	19.15	20.37	19.97	17.76	14.45	10.79	6.77	5.16
KT	(6)	.43	.45	.47	.48	.49	.49	.50	.49	.49	.49	.43	.39
AVE. TEMP.	(3)	.0	.0	5.0	11.0	17.0	22.0	25.0	24.0	20.0	14.0	8.0	1.0
DEGREE-DAYS	(3)	479.	504.	420.	222.	79.	0.	0.	0.	19.	135.	313.	526.

NEW ORLEANS   LA   (LAT. 30.0)

		JAN	FEB	MAR	APR	MAY	JUNE	JULY	AUG	SEP	OCT	NOV	DEC
HORIZ. RAD.	(4)	9.47	12.62	16.06	20.20	22.33	22.74	20.58	19.48	17.18	15.15	11.04	8.85
KT	(6)	.45	.49	.51	.55	.56	.56	.51	.52	.52	.56	.50	.45
AVE. TEMP.	(3)	11.0	13.0	15.0	20.0	23.0	26.0	27.0	27.0	25.0	20.0	15.0	12.0
DEGREE-DAYS	(3)	224.	166.	104.	16.	0.	0.	0.	0.	0.	22.	179.	327.

NEW YORK   NY   (LAT. 40.8)

		JAN	FEB	MAR	APR	MAY	JUNE	JULY	AUG	SEP	OCT	NOV	DEC
HORIZ. RAD.	(4)	5.68	8.18	11.77	15.48	18.57	19.41	19.15	16.83	13.77	10.16	6.05	4.58
KT	(6)	.39	.41	.44	.45	.47	.47	.48	.47	.47	.46	.38	.35
AVE. TEMP.	(3)	.0	1.0	5.0	11.0	17.0	22.0	25.0	24.0	20.0	15.0	9.0	2.0
DEGREE-DAYS	(3)	579.	510.	443.	252.	104.	5.	0.	0.	23.	137.	308.	518.

NORFOLK   VA   (LAT. 36.9)

		JAN	FEB	MAR	APR	MAY	JUNE	JULY	AUG	SEP	OCT	NOV	DEC
HORIZ. RAD.	(4)	7.70	10.58	14.54	19.03	21.42	22.70	21.03	19.07	15.84	12.29	9.21	7.08
KT	(6)	.45	.48	.51	.54	.54	.55	.52	.52	.51	.51	.51	46
AVE. TEMP.	(3)	4.0	5.0	8.0	14.0	19.0	23.0	25.0	24.0	22.0	16.0	10.0	5.0
DEGREE-DAYS	(3)	422.	367.	296.	126.	29.	0.	0.	0.	5.	78.	223.	391.

NORTH BEND   OR   (LAT. 43.4)

		JAN	FEB	MAR	APR	MAY	JUNE	JULY	AUG	SEP	OCT	NOV	DEC
HORIZ. RAD.	(4)	4.98	8.00	12.01	17.13	21.08	22.63	23.92	20.27	15.63	10.13	5.95	4.32
KT	(6)	.38	.44	.47	.51	.54	.55	.60	.57	.55	.50	.42	.37
AVE. TEMP.	(3)	7.0	8.0	-.0	10.0	12.0	14.0	15.0	15.0	15.0	13.0	10.0	8.0
DEGREE-DAYS	(3)	351.	286.	312.	265.	205.	135.	104.	93.	112.	174.	248.	319.

NORTH OMAHA   NE   (LAT. 41.4)

		JAN	FEB	MAR	APR	MAY	JUNE	JULY	AUG	SEP	OCT	NOV	DEC
HORIZ. RAD.	(4)	7.19	10.12	13.87	17.69	21.25	24.09	23.91	21.09	15.58	11.91	7.31	5.80
KT	(6)	.41	.52	.52	.52	.54	.58	.59	.59	.53	.55	.57	.45
AVE. TEMP.	(2)	-5.0	-3.0	3.0	10.0	17.0	22.0	25.0	23.0	19.0	12.0	4.0	-2.0
DEGREE-DAYS	(3)	730.	576.	481.	217.	82.	11.	0.	3.	39.	167.	417.	637.

NORTH PLATTE   NE   (LAT. 41.1)

		JAN	FEB	MAR	APR	MAY	JUNE	JULY	AUG	SEP	OCT	NOV	DEC
HORIZ. RAD.	(4)	7.86	10.88	15.13	19.57	22.56	25.72	25.84	22.58	17.76	13.36	8.62	6.87
KT	(6)	.55	.55	.57	.58	.58	.62	.64	.63	.61	.62	.55	.53
AVE. TEMP.	(3)	-5.0	-2.0	2.0	9.0	15.0	20.0	24.0	23.0	18.0	11.0	3.0	-3.0
DEGREE-DAYS	(3)	717.	574.	529.	290.	132.	36.	4.	4.	78.	244.	258.	658.

OAKLAND   CA   (LAT. 37.7)

		JAN	FEB	MAR	APR	MAY	JUNE	JULY	AUG	SEP	OCT	NOV	DEC
HORIZ. RAD.	(4)	8.03	11.55	16.53	21.81	25.10	26.67	26.36	23.30	19.31	13.75	9.33	7.34
KT	(6)	.49	.53	.59	.63	.64	.65	.65	.64	.63	.59	.53	.49
AVE. TEMP.	(3)	9.0	11.0	12.0	13.0	15.0	17.0	17.0	18.0	18.0	16.0	13.0	10.0
DEGREE-DAYS	(3)	282.	204.	194.	150.	107.	63.	44.	41.	33.	75.	162.	260.

OKLAHOMA CITY   OK   (LAT. 35.4)

		JAN	FEB	MAR	APR	MAY	JUNE	JULY	AUG	SEP	OCT	NOV	DEC
HORIZ. RAD.	(4)	9.09	11.97	15.89	19.58	21.77	24.33	24.15	22.13	17.64	13.90	10.23	8.23
KT	(6)	.51	.52	.55	.55	.55	.59	.60	.60	.56	.57	.54	.50
AVE. TEMP.	(3)	2.0	4.0	9.0	15.0	20.0	25.0	27.0	27.0	23.0	16.0	9.0	4.0
DEGREE-DAYS	(3)	486.	369.	296.	100.	20.	0.	0.	0.	7.	82.	263.	431.

OLYMPIA   WA   (LAT. 47.0)

		JAN	FEB	MAR	APR	MAY	JUNE	JULY	AUG	SEP	OCT	NOV	DEC
HORIZ. RAD.	(4)	3.05	5.71	9.59	14.25	18.52	19.22	21.71	17.58	13.13	7.22	3.85	2.51
KT	(6)	.28	.35	.40	.44	.48	.47	.54	.51	.49	.39	.32	.27
AVE. TEMP.	(3)	3.0	5.0	6.0	9.0	12.0	15.0	18.0	17.0	15.0	10.0	6.0	4.0
DEGREE-DAYS	(3)	479.	373.	376.	280.	189.	109.	49.	47.	110.	248.	362.	439.

ORLANDO   FL   (LAT. 28.5)

		JAN	FEB	MAR	APR	MAY	JUNE	JULY	AUG	SEP	OCT	NOV	DEC
HORIZ. RAD.	(4)	11.34	14.11	17.96	21.54	22.57	20.78	20.44	18.99	16.98	14.80	12.44	10.51
KT	(6)	.52	.53	.57	.59	.57	.51	.51	.50	.51	.53	.54	.51
AVE. TEMP.	(3)	16.0	17.0	19.0	22.0	25.0	27.0	28.0	28.0	27.0	24.0	19.0	17.0
DEGREE-DAYS	(3)	108.	99.	49.	5.	0.	0.	0.	0.	0.	0.	38.	92.

PAGE   AZ   (LAT. 36.6)

		JAN	FEB	MAR	APR	MAY	JUNE	JULY	AUG	SEP	OCT	NOV	DEC
HORIZ. RAD.	(2)	12.56	15.99	22.02	25.87	29.10	29.60	28.47	24.95	21.60	16.83	12.98	10.17
KT	(6)	.73	.72	.77	.74	.74	.72	.71	.68	.70	.70	.71	.65
AVE. TEMP.	(3)	.0	2.0	5.0	10.0	15.0	20.0	24.0	22.0	18.0	12.0	5.0	.0
DEGREE-DAYS	(3)	591.	447.	396.	240.	107.	21.	0.	6.	41.	189.	390.	562.

PASEDENA   CA   (LAT. 34.1)

		JAN	FEB	MAR	APR	MAY	JUNE	JULY	AUG	SEP	OCT	NOV	DEC
HORIZ. RAD.	(1)	10.51	13.94	18.38	21.31	23.82	24.28	26.54	25.08	20.18	15.32	11.35	9.88
KT	(6)	.57	.59	.62	.60	.60	.59	.66	.68	.64	.61	.58	.57
AVE. TEMP.	(3)	191.	151.	141.	94.	47.	26.	0.	0.	6.	29.	91.	166.

PATUXENT RIVER   MD   (LAT. 39.2)

		JAN	FEB	MAR	APR	MAY	JUNE	JULY	AUG	SEP	OCT	NOV	DEC
HORIZ. RAD.	(4)	6.90	9.78	13.40	17.46	20.01	21.48	20.62	18.47	15.40	11.59	8.02	6.09
KT	(6)	.44	.47	.49	.41	.51	.52	.51	.51	.51	.51	.48	.43
AVE. TEMP.	(3)	1.0	2.0	6.0	12.0	18.0	23.0	25.0	24.0	20.0	15.0	9.0	3.0
DEGREE-DAYS	(3)	526.	454.	376.	183.	58.	0.	0.	0.	16.	137.	293.	484.

PENDLETON   OR   (LAT. 45.7)

		JAN	FEB	MAR	APR	MAY	JUNE	JULY	AUG	SEP	OCT	NOV	DEC
HORIZ. RAD.	(4)	3.95	6.96	11.84	17.05	21.85	24.34	27.19	22.63	17.05	10.31	4.97	3.31
KT	(6)	.34	.41	.48	.52	.56	.59	.68	.65	.62	.54	.39	.33
AVE. TEMP.	(3)	.0	4.0	7.0	10.0	15.0	19.0	23.0	22.0	18.0	11.0	5.0	2.0
DEGREE-DAYS	(3)	568.	406.	365.	235.	122.	39.	3.	7.	54.	213.	393.	504.

PENSACOLA   FL   (LAT. 30.5)

		JAN	FEB	MAR	APR	MAY	JUNE	JULY	AUG	SEP	OCT	NOV	DEC
HORIZ. RAD.	(1)	10.47	13.44	16.96	21.31	23.53	23.78	22.48	21.31	18.00	16.49	11.64	9.38
KT	(6)	.40	.53	.55	.59	.59	.58	.56	.57	.55	.61	.53	.48
AVE. TEMP.	(3)	11.0	12.0	15.0	20.0	23.0	26.0	27.0	27.0	25.0	21.0	15.0	12.0
DEGREE-DAYS	(3)	237.	179.	117.	21.	0.	0.	0.	0.	0.	18.	105.	199.

```
                 PHOENIX          AZ          (LAT. 33.4)

                 JAN    FEB    MAR    APR    MAY    JUNE   JULY   AUG    SEP    OCT    NOV    DEC
HORIZ. RAD. (4)  11.59  15.60  20.59  26.72  30.37  21.09  28.22  26.02  22.87  17.89  13.06  10.58
KT          (6)   .61    .65    .69    .75    .77    .76    .70    .70    .71    .70    .65    .60
AVE. TEMP.  (3)  10.0   13.0   15.0   19.0   24.0   29.0   32.0   31.0   28.0   22.0   15.0   11.0
DEGREE-DAYS (3)  238.   162.   103.    33.     0.     0.     0.     0.     0.     9.   101.   216.


                 PHILADELPHIA     PA          (LAT. 39.9)

                 JAN    FEB    MAR    APR    MAY    JUNE   JULY   AUG    SEP    OCT    NOV    DEC
HORIZ. RAD. (1)   6.30   9.02  12.58  16.27  18.84  20.56  19.95  17.87  14.54  10.88   7.03   5.34
KT          (6)   .42    .44    .46    .47    .48    .50    .50    .49    .49    .49    .43    .39
AVE. TEMP.  (3)    .0    1.0    5.0   11.0   17.0   22.0   24.0   23.0   20.0   14.0    8.0    2.0
DEGREE-DAYS (3)  563.   484.   398.   204.    68.     0.     0.     0.    21.   138.   313.   513.


                 PIERRE           SD          (LAT. 44.4)

                 JAN    FEB    MAR    APR    MAY    JUNE   JULY   AUG    SEP    OCT    NOV    DEC
HORIZ. RAD. (4)   6.01   9.02  13.69  18.32  22.32  24.91  25.85  22.61  16.98  11.94   7.07   5.02
KT          (6)   .49    .51    .55    .55    .57    .60    .65    .64    .61    .60    .52    .45
AVE. TEMP.  (3)  -9.0   -6.0   -1.0    8.0   14.0   20.0   24.0   23.0   17.0   10.0    1.0   -6.0
DEGREE-DAYS (3)  851.   694.   606.   312.   148.    41.     3.     6.    84.   251.   520.   749.


                 POCATELLO        ID          (LAT. 42.9)

                 JAN    FEB    MAR    APR    MAY    JUNE   JULY   AUG    SEP    OCT    NOV    DEC
HORIZ. RAD. (4)   6.12  10.01  15.56  20.66  25.88  28.14  29.50  25.41  20.08  13.65   7.82   5.41
KT          (6)   .46    .54    .60    .62    .66    .68    .74    .71    .70    .66    .54    .46
AVE. TEMP.  (3)  -4.0     .0    2.0    7.0   12.0   17.0   22.0   21.0   15.0    9.0    2.0   -2.0
DEGREE-DAYS (3)  720.   554.   510.   328.   187.    77.     0.    11.   107.   286.   488.   656.


                 POINT MUGU       CA          (LAT. 34.1)

                 JAN    FEB    MAR    APR    MAY    JUNE   JULY   AUG    SEP    OCT    NOV    DEC
HORIZ. RAD. (4)  10.52  13.85  18.56  22.14  22.90  23.32  24.04  21.96  18.25  14.71  11.42   9.72
KT          (6)   .57    .59    .63    .62    .58    .57    .60    .59    .57    .58    .58    .56
AVE. TEMP.  (3)  11.0   11.0   11.0   11.0   12.0   13.0   14.0   14.0   14.0   14.0   13.0   12.0
DEGREE-DAYS (3)  222.   199.   224.   215.   198.   167.   148.   143.   124.   138.   153.   202.


                 PORT ARTHUR      TX          (LAT. 29.9)

                 JAN    FEB    MAR    APR    MAY    JUNE   JULY   AUG    SEP    OCT    NOV    DEC
HORIZ. RAD. (4)   9.08  12.15  15.36  18.27  21.23  22.82  20.95  19.70  17.33  15.00  10.81   8.56
KT          (6)   .43    .47    .49    .50    .54    .56    .52    .52    .52    .55    .49    .43
AVE. TEMP.  (5)  11.0   13.0   16.0   20.0   24.0   27.0   28.0   28.0   26.0   21.0   16.0   12.0
DEGREE-DAYS (5)  233.   168.   112.    18.     0.     0.     0.     0.     0.    19.   102.   190.


                 PORTLAND         ME          (LAT. 43.6)

                 JAN    FEB    MAR    APR    MAY    JUNE   JULY   AUG    SEP    OCT    NOV    DEC
HORIZ. RAD. (4)   5.11   7.74  11.00  14.80  17.79  19.42  18.83  16.58  13.14   9.33   5.21   4.12
KT          (6)   .40    .42    .43    .45    .46    .47    .47    .47    .47    .46    .37    .36
AVE. TEMP.  (3)  -5.0   -4.0     .0    6.0   11.0   17.0   20.0   19.0   15.0    9.0    4.0   -3.0
DEGREE DAYS (3)  777.   655.   572.   594.   212.    59.    15.    31.   111.   274.   440.   677.


                 PORTLAND         OR          (LAT. 45.6)

                 JAN    FEB    MAR    APR    MAY    JUNE   JULY   AUG    SEP    OCT    NOV    DEC
HORIZ. RAD. (4)   3.52   6.29  10.16  14.84  18.88  20.12  23.12  18.99  13.81   8.21   4.40   2.95
KT          (6)   .30    .37    .41    .45    .49    .49    .58    .54    .50    .43    .34    .29
AVE. TEMP.  (5)   3.0    6.0    8.0   10.0   14.0   17.0   19.0   19.0   17.0   12.0    7.0    5.0
DEGREE-DAYS (5)  463.   346.   332.   240.   147.    71.    27.    31.    66.   193.   328.   418.


                 PRESCOTT         AZ          (LAT. 34.6)

                 JAN    FEB    MAR    APR    MAY    JUNE   JULY   AUG    SEP    OCT    NOV    DEC
HORIZ. RAD. (4)  11.53  15.15  20.17  25.82  29.84  31.35  26.21  23.74  22.18  17.51  12.94  10.52
KT          (6)   .63    .65    .68    .73    .75    .76    .65    .64    .70    .70    .67    .62
AVE. TEMP.  (3)   3.0    5.0    7.0   11.0   16.0   21.0   24.0   23.0   20.0   14.0    8.0    4.0
DEGREE-DAYS (3)  481.   381.   357.   219.    92.    18.     0.     0.    13.   141.   320.   454.


                 PROSSER          WA          (LAT. 46.2)

                 JAN    FEB    MAR    APR    MAY    JUNE   JULY   AUG    SEP    OCT    NOV    DEC
HORIZ. RAD. (2)   4.90   9.29  14.70  21.81  25.79  28.47  29.60  25.29  19.18  11.47   5.69   4.19
KT          (6)   .44    .56    .61    .67    .67    .69    .74    .73    .71    .61    .46    .43
AVE. TEMP.  (3)  -1.0    4.0    6.0   10.0   14.0   18.0   21.0   20.0   16.0   10.0    4.0     .0
DEGREE-DAYS (3)  603.   428.   376.   240.   127.    47.     7.    16.    66.   231.   413.   543.
```

```
                PROVIDENCE        RI          (LAT. 41.7)

                JAN    FEB    MAR    APR    MAY    JUNE   JULY   AUG    SEP    OCT    NOV    DEC
HORIZ. RAD.  (4)  5.75   8.38  11.71  15.59  18.78  20.15  19.24  17.01  13.72  10.29   6.10   4.75
KT           (6)   .41    .43    .44    .46    .48    .49    .48    .47    .47    .48    .40    .38
AVE. TEMP.   (3)  -2.0   -1.0    3.0    8.0   14.0   19.0   22.0   21.0   17.0   12.0    6.0     .0
DEGREE-DAYS  (3)  631.   554.   484.   295.   144.    20.     0.     6.    52.   194.   362.   577.


                PUEBLO            CO          (LAT. 38.3)

                JAN    FEB    MAR    APR    MAY    JUNE   JULY   AUG    SEP    OCT    NOV    DEC
HORIZ. RAD.  (4) 10.15  13.30  17.75  22.20  24.54  27.63  26.23  23.86  20.19  15.45  10.82   8.88
KT           (6)   .63    .62    .64    .64    .62    .67    .65    .65    .67    .67    .63    .61
AVE. TEMP.   (5)  -1.0    1.0    4.0   11.0   16.0   21.0   25.0   24.0   19.0   12.0    5.0    1.0
DEGREE-DAYS  (5)  601.   471.   431.   225.    82.    16.     0.     0.    31.   186.   403.   551.


                RALEIGH           NC          (LAT. 35.8)

                JAN    FEB    MAR    APR    MAY    JUNE   JULY   AUG    SEP    OCT    NOV    DEC
HORIZ. RAD.  (4)  7.87  10.70  14.48  18.66  20.52  21.16  20.15  18.29  15.63  12.54   9.22   7.21
KT           (6)   .45    .47    .50    .53    .52    .51    .50    .50    .50    .51    .49    .45
AVE. TEMP.   (3)   5.0    6.0   10.0   15.0   20.0   24.0   25.0   25.0   22.0   16.0   10.0    5.0
DEGREE-DAYS  (3)  422.   354.   279.   100.    27.     0.     0.     0.     7.   103.   250.   410.


                RAPID CITY        SD          (LAT. 44.1)

                JAN    FEB    MAR    APR    MAY    JUNE   JULY   AUG    SEP    OCT    NOV    DEC
HORIZ. RAD.  (4)  6.16   9.38  13.95  18.03  21.41  24.19  25.23  22.27  17.23  12.07   7.34   5.41
KT           (6)   .49    .52    .55    .55    .55    .59    .63    .63    .61    .60    .53    .49
AVE. TEMP.   (3)  -5.0   -2.0     .0    7.0   13.0   18.0   23.0   22.0   16.0   10.0    2.0   -2.0
DEGREE-DAYS  (3)  742.   610.   582.   340.   177.    74.     7.     6.   106.   263.   493.   663.


                REDMOND           OR          (LAT. 44.3)

                JAN    FEB    MAR    APR    MAY    JUNE   JULY   AUG    SEP    OCT    NOV    DEC
HORIZ. RAD.  (4)  5.57   8.79  13.51  19.10  23.60  25.96  27.76  23.48  17.98  11.34   6.49   4.82
KT           (6)   .45    .49    .54    .58    .51    .63    .69    .67    .64    .57    .47    .44
AVE. TEMP.   (3)  -1.0    2.0    4.0    7.0   11.0   15.0   19.0   18.0   14.0    9.0    4.0    1.0
DEGREE-DAYS  (3)  599.   454.   454.   343.   236.   122.    31.    57.   129.   286.   433.   544.


                RENO              NV          (LAT. 39.5)

                JAN    FEB    MAR    APR    MAY    JUNE   JULY   AUG    SEP    OCT    NOV    DEC
HORIZ. RAD.  (4)  9.08  13.05  18.72  24.51  28.63  30.66  30.55  27.30  22.67  16.24  10.35   8.01
KT           (6)   .59    .63    .68    .7o    .73    .74    .76    .75    .76    .72    .62    .58
AVE. TEMP.   (3)   .0     2.0    4.0    8.0   12.0   16.0   20.0   19.0   15.0   10.0    4.0     .0
DEGREE-DAYS  (3)  570.   434.   426.   303.   182.    81.     9.    28.    93.   253.   415.   551.


                RICHMOND          VA          (LAT. 37.8)

                JAN    FEB    MAR    APR    MAY    JUNE   JULY   AUG    SEP    OCT    NOV    DEC
HORIZ. RAD.  (4)  7.17   9.95  13.74  17.77  20.00  21.25  20.14  18.16  15.30  11.72   8.32   6.43
KT           (6)   .44    .46    .49    .51    .51    .51    .50    .50    .50    .50    .47    .43
AVE. TEMP.   (5)   3.0    4.0    8.0   14.0   19.0   23.0   25.0   25.0   21.0   15.0    9.0    4.0
DEGREE-DAYS  (5)  474.   398.   216.   126.    36.     0.     0.     0.    12.   113.   267.   448.


                RIVERSIDE         CA          (LAT. 33.9)

                JAN    FEB    MAR    APR    MAY    JUNE   JULY   AUG    SEP    OCT    NOV    DEC
HORIZ. RAD.  (2) 11.51  15.37  20.01  22.65  26.08  28.47  28.18  25.87  22.40  17.04  13.36  11.30
KT           (6)   .62    .65    .67    .63    .66    .69    .70    .70    .70    .67    .67    .65
AVE. TEMP.   (3)  11.0   12.0   13.0   15.0   18.0   20.0   24.0   24.0   22.0   18.0   14.0   11.0
DEGREE-DAYS  (3)  226.   173.   157.    93.    41.    12.     0.     0.     3.    34.   118.   208.


                ROANOKE           VA          (LAT. 37.3)

                JAN    FEB    MAR    APR    MAY    JUNE   JULY   AUG    SEP    OCT    NOV    DEC
HORIZ. RAD.  (4)  7.50  10.21  14.03  17.95  20.02  21.36  20.38  18.39  15.41  12.26   8.68   6.70
KT           (6)   .45    .47    .49    .51    .51    .52    .51    .50    .50    .52    .49    .44
AVE. TEMP.   (3)   2.0    3.0    7.0   13.0   18.0   22.0   24.0   23.0   20.0   14.0    8.0    3.0
DEGREE-DAYS  (3)  493.   418.   339.   157.    56.     0.     0.     0.    18.   131.   305.   470.


                ROCHESTER         MN          (LAT. 43.9)

                JAN    FEB    MAR    APR    MAY    JUNE   JULY   AUG    SEP    OCT    NOV    DEC
HORIZ. RAD.  (4)  5.41   8.54  12.28  16.00  19.24  21.58  21.66  18.86  14.19   9.87   5.61   4.20
KT           (6)   .43    .47    .48    .48    .49    .52    .54    .53    .50    .50    .40    .37
AVE. TEMP.   (3) -11.0   -8.0   -2.0    7.0   13.0    9.0   21.0   20.0   15.0   10.0     .0   -7.0
DEGREE-DAYS  (3)  897.   748.   641.   342.   162.    43.    12.    19.   103.   269.   540.   794.
```

ROCHESTER   NY   (LAT. 43.1)

		JAN	FEB	MAR	APR	MAY	JUNE	JULY	AUG	SEP	OCT	NOV	DEC
HORIZ. RAD.	(4)	4.13	6.35	10.25	15.20	18.23	20.62	20.21	17.24	13.16	8.87	4.58	3.19
KT	(6)	.31	.34	.40	.46	.47	.50	.50	.49	.46	.43	.32	.27
AVE. TEMP.	(5)	-4.0	-4.0	1.0	8.0	14.0	19.0	22.0	21.0	17.0	11.0	5.0	-2.0
DEGREE-DAYS	(5)	706.	626.	551.	315.	158.	26.	5.	14.	70.	221.	408.	632.

ROCK SPRINGS   WY   (LAT. 41.6)

		JAN	FEB	MAR	APR	MAY	JUNE	JULY	AUG	SEP	OCT	NOV	DEC
HORIZ. RAD.	(4)	8.34	12.36	17.37	22.06	26.61	29.22	28.91	25.42	20.80	14.82	9.38	7.38
KT	(6)	.59	.64	.66	.65	.68	.71	.72	.71	.72	.69	.61	.58
AVE. TEMP.	(3)	-7.0	-5.0	-2.0	4.0	10.0	15.0	20.0	19.0	14.0	7.0	-1.0	-5.0
DEGREE-DAYS	(3)	789.	647.	622.	415.	252.	110.	10.	27.	149.	349.	472.	730.

ROSWELL   NM   (LAT. 33.4)

		JAN	FEB	MAR	APR	MAY	JUNE	JULY	AUG	SEP	OCT	NOV	DEC
HORIZ. RAD.	(4)	11.88	15.58	20.51	25.17	27.91	29.62	27.70	25.44	21.71	17.33	12.84	10.80
KT	(6)	.62	.65	.68	.70	.70	.72	.69	.68	.68	.68	.64	.61
AVE. TEMP.	(3)	3.0	6.0	10.0	15.0	20.0	25.0	26.0	25.0	21.0	15.0	8.0	4.0
DEGREE-DAYS	(3)	643.	344.	271.	103.	11.	0.	0.	0.	9.	108.	302.	443.

SACRAMENTO   CA   (LAT. 38.5)

		JAN	FEB	MAR	APR	MAY	JUNE	JULY	AUG	SEP	OCT	NOV	DEC
HORIZ. RAD.	(4)	6.77	10.66	16.55	22.74	27.63	30.46	30.51	26.88	21.64	14.92	8.87	6.11
KT	(6)	.42	.50	.59	.66	.70	.74	.76	. 4	.72	.65	.52	.42
AVE. TEMP.	(5)	7.0	10.0	12.0	15.0	18.0	21.0	24.0	23.0	22.0	17.0	12.0	8.0
DEGREE-DAYS	(5)	343.	237.	207.	126.	67.	11.	0.	0.	3.	56.	200.	331.

ST. LOUIS   MO   (LAT. 38.7)

		JAN	FEB	MAR	APR	MAY	JUNE	JULY	AUG	SEP	OCT	NOV	DEC
HORIZ. RAD.	(4)	7.12	10.05	13.67	17.75	21.24	23.75	23.26	20.62	16.56	12.48	8.15	6.02
KT	(6)	.45	.48	.49	.51	.54	.57	.58	.57	.55	.54	.48	.42
AVE. TEMP.	(5)	.0	2.0	6.0	14.0	19.0	24.0	26.0	25.0	21.0	15.0	7.0	1.0
DEGREE-DAYS	(5)	581.	465.	379.	151.	57.	6.	0.	0.	19.	124.	333.	523.

SALEM   OR   (LAT. 44.9)

		JAN	FEB	MAR	APR	MAY	JUNE	JULY	AUG	SEP	OCT	NOV	DEC
HORIZ. RAD.	(4)	3.77	6.67	10.75	15.55	19.72	20.08	24.31	20.14	15.08	8.73	4.66	3.15
KT	(6)	.31	.38	.43	.47	.51	.51	.61	.57	.54	.44	.35	.30
AVE. TEMP.	(3)	4.0	6.0	7.0	10.0	13.0	16.0	19.0	19.0	17.0	12.0	7.0	5.0
DEGREE-DAYS	(3)	451.	344.	341.	253.	164.	74.	24.	29.	67.	203.	330.	415.

SALT LAKE CITY   UT   (LAT. 40.8)

		JAN	FEB	MAR	APR	MAY	JUNE	JULY	AUG	SEP	OCT	NOV	DEC
HORIZ. RAD.	(4)	7.25	11.22	16.50	21.50	26.81	29.06	29.39	25.58	20.92	14.68	8.94	6.47
KT	(6)	.50	.56	.62	.63	.68	.70	.73	.71	.71	.67	.57	.49
AVE. TEMP.	(3)	-1.0	1.0	4.0	10.0	15.0	19.0	25.0	24.0	18.0	11.0	4.0	.0
DEGREE-DAYS	(3)	637.	492.	437.	263.	132.	49.	0.	3.	58.	223.	432.	498.

SAN ANGELO   TX   (LAT. 31.4)

		JAN	FEB	MAR	APR	MAY	JUNE	JULY	AUG	SEP	OCT	NOV	DEC
HORIZ. RAD.	(4)	10.91	13.71	18.23	21.00	23.05	24.81	24.09	22.31	18.24	15.17	11.85	10.15
KT	(6)	.54	.55	.59	.58	.48	.61	.60	.60	.56	.57	.56	.54
AVE. TEMP.	(3)	8.0	10.0	14.0	20.0	24.0	28.0	29.0	29.0	25.0	20.0	13.0	9.0
DEGREE-DAYS	(3)	321.	229.	159.	41.	0.	0.	0.	0.	0.	41.	166.	288.

SAN ANTONIO   TX   (LAT. 29.5)

		JAN	FEB	MAR	APR	MAY	JUNE	JULY	AUG	SEP	OCT	NOV	DEC
HORIZ. RAD.	(4)	10.16	13.10	16.46	18.30	21.50	23.48	24.07	22.10	18.59	15.32	11.45	9.61
KT	(6)	.48	.50	.52	.50	.54	.58	.60	.59	.56	.56	.51	.48
AVE. TEMP.	(3)	11.0	12.0	16.0	20.0	24.0	27.0	28.0	28.0	26.0	21.0	15.0	12.0
DEGREE-DAYS	(3)	257.	172.	108.	17.	0.	0.	0.	0.	0.	18.	99.	207.

SAN DIEGO   CA   (LAT. 32.7)

		JAN	FEB	MAR	APR	MAY	JUNE	JULY	AUG	SEP	OCT	NOV	DEC
HORIZ. RAD.	(4)	11.07	14.37	18.52	21.98	2 .73	23.40	24.81	23.35	19.49	15.59	12.06	10.26
KT	(6)	.57	.59	.61	.61	.57	.57	.62	.63	.60	.60	.59	.57
AVE. TEMP.	(3)	12.0	13.0	14.0	15.0	17.0	18.0	20.0	21.0	21.0	18.0	15.0	13.0
DEGREE-DAYS	(3)	174.	132.	122.	80.	44.	29,	3.	0.	9.	24.	78.	143.

SAN FRANCISCO   CA   (LAT. 37.8)

	JAN	FEB	MAR	APR	MAY	JUNE	JULY	AUG	SEP	OCT	NOV	DEC
HORIZ. RAD. (4)	8.03	11.45	16.51	21.79	25.26	26.97	27.14	24.02	19.77	13.91	9.32	7.29
KT (6)	.49	.53	.59	.63	.64	.65	.67	.66	.65	.59	.53	.49
AVE. TEMP. (5)	10.0	12.0	12.0	13.0	14.0	15.0	15.0	15.0	17.0	16.0	14.0	11.0
DEGREE-DAYS (5)	288.	214.	207.	162.	117.	67.	52.	47.	37.	76.	162.	263.

SANTA MARIA   CA   (LAT. 34.9)

	JAN	FEB	MAR	APR	MAY	JUNE	JULY	AUG	SEP	OCT	NOV	DEC
HORIZ. RAD. (4)	9.69	12.95	17.95	21.80	24.29	26.65	26.57	23.90	19.64	15.36	11.05	9.12
KT (6)	.53	.56	.61	.61	.61	.65	.66	.65	.62	.62	.57	.55
AVE. TEMP. (3)	10.0	11.0	11.0	12.0	13.0	15.0	16.0	16.0	17.0	15.0	13.0	10.0
DEGREE-DAYS (3)	250.	202.	210.	168.	136.	92.	62.	57.	53.	88.	150.	227.

SAULT ST. MARIE   MI   (LAT. 46.5)

	JAN	FEB	MAR	APR	MAY	JUNE	JULY	AUG	SEP	OCT	NOV	DEC
HORIZ. RAD. (4)	3.69	6.85	11.67	15.70	19.16	20.55	20.83	17.28	11.91	7.64	3.76	2.87
KT (6)	.33	.41	.49	.49	.50	.50	.52	.50	.44	.41	.30	.30
AVE. TEMP. (3)	-9.0	-10.0	-4.0	3.0	9.0	14.0	17.0	17.0	13.0	7.0	.0	-6.0
DEGREE-DAYS (3)	875.	774.	706.	447.	276.	112.	53.	69.	162.	324.	537.	773.

SCOTTS BLUFF   NE   (LAT. 41.9)

	JAN	FEB	MAR	APR	MAY	JUNE	JULY	AUG	SEP	OCT	NOV	DEC
HORIZ. RAD. (4)	7.67	10.79	14.84	18.93	21.94	25.38	25.92	22.69	18.15	13.00	8.21	6.53
KT (6)	.55	.56	.56	.56	.56	.61	.64	.63	.63	.61	.54	.52
AVE. TEMP. (3)	-4.0	-2.0	2.0	8.0	14.0	19.0	23.0	22.0	16.0	10.0	3.0	-2.0
DEGREE-DAYS (3)	691.	552.	529.	313.	156.	51.	0.	4.	89.	255.	480.	644.

SHERIDAN   WY   (LAT. 44.8)

	JAN	FEB	MAR	APR	MAY	JUNE	JULY	AUG	SEP	OCT	NOV	DEC
HORIZ. RAD. (4)	5.87	8.94	13.67	17.45	21.37	24.47	26.43	22.77	17.04	11.41	6.70	5.01
KT (6)	.49	.51	.55	.53	.55	.59	.66	.65	.61	.58	.50	.47
AVE. TEMP. (3)	-6.0	-3.0	-1.0	6.0	12.0	16.0	21.0	21.0	14.0	9.0	1.0	-4.0
DEGREE-DAYS (3)	758.	608.	586.	357.	208.	93.	16.	17.	136.	296.	527.	681.

SIOUX CITY   IA   (LAT. 42.4)

	JAN	FEB	MAR	APR	MAY	JUNE	JULY	AUG	SEP	OCT	NOV	DEC
HORIZ. RAD. (4)	6.45	9.55	13.28	17.91	21.58	24.10	24108	20.94	16.13	11.78	7.29	5.33
KT (6)	.47	.50	.51	.53	.55	.58	.60	.59	.56	.56	.49	.44
AVE. TEMP. (3)	-8.0	-5.0	1.0	10.0	16.0	21.0	24.0	23.0	17.0	12.0	2.0	-5.0
DEGREE-DAYS (3)	809.	647.	548.	263.	105.	18.	0.	6.	63.	210.	478.	715.

SIOUX FALLS   SD   (LAT. 43.6)

	JAN	FEB	MAR	APR	MAY	JUNE	JULY	AUG	SEP	OCT	NOV	DEC
HORIZ. RAD. (4)	6.04	9.10	13.08	17.51	21.49	23.83	24.40	20.93	16.00	11.41	6.80	5.01
KT (6)	.47	.50	.51	.53	.55	.58	.61	.59	.57	.56	.49	.44
AVE. TEMP. (3)	-10.0	-7.0	-1.0	8.0	14.0	20.0	23.0	22.0	16.0	10.0	1.0	-7.0
DEGREE-DAYS (3)	875.	709.	603.	315.	144.	36.	6.	10.	92.	258.	532.	775.

SOUTH BEND   IN   (LAT. 41.7)

	JAN	FEB	MAR	APR	MAY	JUNE	JULY	AUG	SEP	OCT	NOV	DEC
HORIZ. RAD. (4)	4.72	7.49	11.26	15.75	19.55	21.81	21.02	18.91	14.65	10.32	5.64	3.86
KT (6)	.34	.39	.43	.47	.50	.53	.52	.53	.50	.48	.37	.31
AVE. TEMP. (3)	-4.0	-3.0	2.0	9.0	15.0	20.0	22.0	22.0	18.0	12.0	4.0	-2.0
DEGREE-DAYS (3)	706.	602.	512.	282.	136.	19.	3.	13.	54.	204.	423.	634.

SPOKANE   WA   (LAT. 47.7)

	JAN	FEB	MAR	APR	MAY	JUNE	JULY	AUG	SEP	OCT	NOV	DEC
HORIZ. RAD. (4)	3.57	6.88	11.81	16.97	21.77	23.64	26.75	22.04	16.29	9.54	4.51	2.90
KT (6)	.34	.43	.50	.53	.47	.57	.67	.64	.61	.53	.39	.32
AVE. TEMP. (3)	-3.0	.0	3.0	8.0	13.0	16.0	21.0	20.0	15.0	9.0	2.0	-1.0
DEGREE-DAYS (3)	682.	510.	474.	315.	182.	80.	12.	26.	109.	296.	492.	620.

SPRINGFIELD   IL   (LAT. 39.8)

	JAN	FEB	MAR	APR	MAY	JUNE	JULY	AUG	SEP	OCT	NOV	DEC
HORIZ. RAD. (4)	6.64	9.77	12.97	17.19	21.17	23.80	23.36	20.49	16.50	12.12	7.68	5.56
KT (6)	.44	.48	.48	.50	.54	.58	.58	.57	.55	.54	.47	.41
AVE. TEMP. (3)	-3.0	-1.0	4.0	12.0	17.0	23.0	24.0	24.0	20.0	14.0	5.0	-1.0
DEGREE-DAYS (3)	659.	538.	441.	202.	73.	7.	0.	4.	27.	157.	385.	594.

```
              SPRINGFIELD       MO           (LAT. 37.2)

              JAN    FEB    MAR    APR    MAY    JUNE   JULY   AUG    SEP    OCT    NOV    DEC
HORIZ. RAD. (4)   7.76  10.51  14.02  18.21  21.36  23.55  23.41  21.26  16.80  12.98   8.80   6.84
KT          (6)    .46    .48    .49    .51    .52    .57    .58    .58    .55    .55    .59    .45
AVE. TEMP.  (5)    1.0    3.0    7.0   14.0   18.0   23.0   25.0   25.0   21.0   15.0    7.0    2.0
DEGREE-DAYS (5)   553.   436.   367.   153.    52.     6.     0.     3.    19.   126.   325.   499.


              SUMMIT            MT           (LAT. 48.3)

              JAN    FEB    MAR    APR    MAY    JUNE   JULY   AUG    SEP    OCT    NOV    DEC
HORIZ. RAD. (2)   5.11   6.78  11.22  17.33  19.34  20.64  23.45  21.35  14.82   9.04   4.27   3.18
KT          (6)    .51    .44    .49    .49    .50    .50    .59    .62    .57    .57    .51    .37
AVE. TEMP.  (3)   -8.0   -5.0   -4.0    1.0    6.0   10.0   14.0   13.0    8.0    4.0   -2.0   -6.0
DEGREE-DAYS (3)   854.   691.   715.   518.   377.   253.   143.   171.   302.   457.   647.   777.


              SUNNYVALE         CA           (LAT. 37.4)

              JAN    FEB    MAR    APR    MAY    JUNE   JULY   AUG    SEP    OCT    NOV    DEC
HORIZ. RAD. (4)   8.37  11.77  16.86  22.06  25.84  27.84  27.71  24.59  19.97  14.17   9.57   7.49
KT          (6)    .50    .54    .60    .63    .65    .67    .69    .67    .65    .60    .54    .49
AVE. TEMP.  (3)   10.0   11.0   13.0   14.0   17.0   19.0   20.0   20.0   20.0   17.0   13.0   10.0
DEGREE-DAYS (3)   267.   194.   179.   127.    68.    28.     7.     8.     7.    50.   153.   253.


              SYRACUSE          NY           (LAT. 43.1)

              JAN    FEB    MAR    APR    MAY    JUNE   JULY   AUG    SEP    OCT    NOV    DEC
HORIZ. RAD. (4)   4.37   6.48  10.11  15.02  17.91  20.18  19.95  17.06  13.23   8.82   4.52   3.24
KT          (6)    .33    .35    .39    .45    .46    .49    .50    .58    .46    .43    .31    .28
AVE. TEMP.  (5)   -5.0   -4.0    1.0    8.0   14.0   19.0   22.0   21.0   17.0   11.0    5.0   -2.0
DEGREE-DAYS (5)   713.   628.   548.   308.   151.    26.     6.     0.    67.   218.   400.   636.


              TACOMA            WA           (LAT. 47.2)

              JAN    FEB    MAR    APR    MAY    JUNE   JULY   AUG    SEP    OCT    NOV    DEC
HORIZ. RAD. (4)   2.97   5.62   9.64  14.68  19.45  20.45  25.51  18.34  13.02   7.45   3.83   2.40
KT          (6)    .28    .35    .41    .46     .1    .50    .64    .53    .49    .41    .32    .26
AVE. TEMP.  (3)    4.0    6.0    7.0   10.0   13.0   16.0   18.0   18.0   16.0   12.0    8.0    5.0
DEGREE-DAYS (3)   432.   333.   341.   250.   159.    83.    36.    41.    88.   207.   318.   398.


              TALLAHASSEE       FL           (LAT. 30.4)

              JAN    FEB    MAR    APR    MAY    JUNE   JULY   AUG    SEP    OCT    NOV    DEC
HORIZ. RAD. (4)   9.95  12.91  16.79  20.69  21.97  21.37  19.84  19.01  16.94  14.96  11.44   9.23
KT          (6)    .48    .51    .54    .57    .55    .52    .50    .51    .51    .55    .52    .47
AVE. TEMP.  (3)   11.0   12.0   16.0   19.0   23.0   26.0   27.0   27.0   25.0   20.0   15.0   12.0
DEGREE-DAYS (3)   227.   179.   104.    19.     0.     0.     0.     0.     0.    17.   113.   209.


              TAMPA             FL           (LAT. 28.0)

              JAN    FEB    MAR    APR    MAY    JUNE   JULY   AUG    SEP    OCT    NOV    DEC
HORIZ. RAD. (4)  11.47  14.29  18.09  21.66  22.68  20.97  19.89  18.76  16.93  15.28  12.57  10.62
KT          (6)    .52    .53    .57    .59    .57    .52    .50    .50    .50    .54    .54    .51
AVE. TEMP.  (3)   16.0   16.0   19.0   21.0   24.0   26.0   27.0   27.0   26.0   23.0   19.0   16.0
DEGREE-DAYS (3)   113.    98.    50.     5.     0.     0.     0.     0.     0.     0.    39.    94.


              TOLEDO            OH           (LAT. 41.6)

              JAN    FEB    MAR    APR    MAY    JUNE   JULY   AUG    SEP    OCT    NOV    DEC
HORIZ. RAD. (4)   4.94   7.72  11.31  15.71  19.48  21.32  20.99  18.34  14.48  10.34   5.65   4.03
KT          (6)    .35    .50    .43    .46    .50    .52    .52    .51    .50    .48    .37    .32
AVE. TEMP.  (3)   -4.0   -3.0    2.0    9.0   15.0   20.0   22.0   22.0   18.0   12.0    4.0   -2.0
DEGREE-DAYS (3)   692.   589.   503.   277.   127.    18.     3.    10.    55.   211.   423.   637.


              TONOPAH           NV           (LAT. 38.1)

              JAN    FEB    MAR    APR    MAY    JUNE   JULY   AUG    SEP    OCT    NOV    DEC
HORIZ. RAD. (4)  10.42  14.46  20.17  25.55  29.25  31.64  30.67  27.67  23.18  17.26  11.70   9.38
KT          (6)    .64    .68    .72    .77    .77    .76    .76    .76    .76    .74    .67    .63
AVE. TEMP.  (3)   -1.0    1.0    4.0    .90   14.0   18.0   23.0   22.0   18.0   11.0    4.0     .0
DEGREE-DAYS (3)   599.   473.   437.   284.   149.    51.     0.     7.    60.   226.   420.   470.


              TOPEKA            KA           (LAT. 39.1)

              JAN    FEB    MAR    APR    MAY    JUNE   JULY   AUG    SEP    OCT    NOV    DEC
HORIZ. RAD. (4)   7.73  10.68  14.26  18.63  21.74  24.13  24.15  21.68  17.21  13.01   8.76   6.62
KT          (6)    .49    .51    .52    .54    .55    .58    .60    .60    .57    .57    .52    .47
AVE. TEMP.  (3)   -2.0    1.0    5.0   13.0   18.0   23.0   26.0   25.0   20.0   14.0    6.0     .0
DEGREE-DAYS (3)   637.   492.   414.   183.    66.     7.     0.     0.    31.   144.   368.   572.
```

TRAVERSE CITY   MI   (LAT. 44.7)

		JAN	FEB	MAR	APR	MAY	JUNE	JULY	AUG	SEP	OCT	NOV	DEC
HORIZ. RAD.	(4)	3.55	6.44	11.36	15.95	19.62	21.70	21.67	18.26	13.23	8.56	4.28	2.91
KT	(6)	.29	.37	.46	.48	.41	.53	.54	.52	.48	.43	.32	.27
AVE. TEMP.	(3)	-6.0	-6.0	-2.0	6.0	12.0	18.0	20.0	20.0	15.0	10.0	3.0	-3.0
DEGREE-DAYS	(3)	761.	689.	625.	372.	215.	58.	18.	37.	99.	262.	468.	673.

TRENTON   NJ   (LAT. 40.2)

		JAN	FEB	MAR	APR	MAY	JUNE	JULY	AUG	SEP	OCT	NOV	DEC
HORIZ. RAD.	(1)	7.24	10.22	14.36	17.75	20.56	22.86	22.61	19.64	16.29	12.31	8.16	6.49
KT	(6)	.49	.51	.53	.52	.52	.55	.56	.54	.55	.55	.51	.48
AVE. TEMP.	(3)	.0	.0	5.0	11.0	16.0	21.0	24.0	23.0	19.0	13.0	7.0	1.0
DEGREE-DAYS	(3)	567.	492.	410.	213.	75.	0.	0.	0.	22.	140	312.	518.

TUCSON   AZ   (LAT. 32.1)

		JAN	FEB	MAR	APR	MAY	JUNE	JULY	AUG	SEP	OCT	NOV	DEC
HORIZ. RAD.	(4)	12.47	16.25	21.16	26.82	30.32	30.98	26.57	24.77	22.46	18.18	13.71	11.30
KT	(6)	.63	.66	.69	.74	.77	.76	.66	.66	.69	.69	.66	.61
AVE. TEMP.	(3)	10.0	11.0	14.0	18.0	22.0	27.0	30.0	28.0	26.0	20.0	14.0	10.0
DEGREE-DAYS	(3)	242.	182.	131.	42.	0.	0.	0.	0.	0.	14.	116.	221.

TULSA   OK   (LAT. 36.2)

		JAN	FEB	MAR	APR	MAY	JUNE	JULY	AUG	SEP	OCT	NOV	DEC
HORIZ. RAD.	(4)	8.30	11.10	14.82	18.19	20.68	22.93	23.04	21.17	16.71	13.21	9.39	7.48
KT	(6)	.48	.49	.51	.52	.52	.56	.57	.58	.54	.54	.51	.47
AVE. TEMP.	(5)	3.0	5.0	9.0	16.0	20.0	25.0	28.0	27.0	23.0	17.0	10.0	4.0
DEGREE-DAYS	(5)	489.	370.	293.	98.	16.	0.	0.	0.	6.	79.	260.	434.

TWIN FALLS   ID   (LAT. 40.6)

		JAN	FEB	MAR	APR	MAY	JUNE	JULY	AUG	SEP	OCT	NOV	DEC
HORIZ. RAD.	(2)	6.82	10.05	14.86	19.34	23.11	24.79	25.20	22.61	18.09	11.97	7.37	5.49
KT	(6)	.46	.50	.55	.57	.59	.60	.63	.63	.61	.54	.46	.41
AVE. TEMP.	(3)	-1.0	1.0	4.0	9.0	13.0	17.0	22.0	21.0	16.0	10.0	4.0	.0
DEGREE-DAYS	(3)	644.	490.	454.	290.	161.	73.	0.	12.	99.	260.	442.	589.

WACO   TX   (LAT. 31.6)

		JAN	FEB	MAR	APR	MAY	JUNE	JULY	AUG	SEP	OCT	NOV	DEC
HORIZ. RAD.	(4)	9.45	12.44	16.20	18.30	20.13	23.97	24.18	22.22	18.17	14.77	10.86	9.11
KT	(6)	.47	.50	.53	.51	.51	.59	.60	.59	.56	.56	.51	.49
AVE. TEMP.	(3)	8.0	10.0	14.0	20.0	24.0	28.0	30.0	30.0	26.0	21.0	14.0	10.0
DEGREE-DAYS	(3)	310.	223.	156.	31.	0.	0.	0.	0.	0.	28.	134.	262.

WASHINGTON   DC   (LAT. 38.8)

		JAN	FEB	MAR	APR	MAY	JUNE	JULY	AUG	SEP	OCT	NOV	DEC
HORIZ. RAD.	(4)	6.49	9.25	12.77	16.56	19.50	21.57	20.63	18.36	15.21	11.39	7.39	5.46
KT	(6)	.41	.44	.46	.48	.50	.52	.51	.50	.50	.50	.44	.38
AVE. TEMP.	(3)	2.0	3.0	7.0	13.0	18.0	23.0	25.0	24.0	21.0	15.0	9.0	3.0
DEGREE-DAYS	(3)	484.	423.	348.	160.	41.	0.	0.	0.	18.	121.	288.	463.

WHIDBEY ISLAND   WA   (LAT. 48.3)

		JAN	FEB	MAR	APR	MAY	JUNE	JULY	AUG	SEP	OCT	NOV	DEC
HORIZ. RAD.	(4)	8.90	12.01	15.95	20.23	23.10	25.70	25.40	23.06	18.34	14.18	9.88	7.83
KT	(6)	.89	.78	.69	.64	.60	.62	.64	.67	.70	.80	.88	.91
AVE. TEMP.	(3)	3.0	5.0	6.0	9.0	12.0	14.0	16.0	16.0	14.0	10.0	7.0	5.0
DEGREE-DAYS	(3)	462.	367.	371.	280.	200.	124.	77.	74.	132.	252.	353.	424.

WICHITA   KA   (LAT. 37.6)

		JAN	FEB	MAR	APR	MAY	JUNE	JULY	AUG	SEP	OCT	NOV	DEC
HORIZ. RAD.	(4)	3.21	6.04	10.42	15.26	19.98	20.66	22.48	18.07	13.32	7.43	4.05	2.64
KT	(6)	.19	.28	.37	.44	.51	.50	.56	.49	.44	.32	.23	.18
AVE. TEMP.	(5)	.0	2.0	6.0	14.0	19.0	24.0	27.0	26.0	21.0	15.0	7.0	1.0
DEGREE-DAYS	(5)	581.	447.	373.	153.	50.	4.	0.	0.	18.	117.	337.	526.

WICHITA FALLS   TX   (LAT. 34.0)

		JAN	FEB	MAR	APR	MAY	JUNE	JULY	AUG	SEP	OCT	NOV	DEC
HORIZ. RAD.	(4)	9.78	12.74	16.70	20.01	22.89	25.21	24.59	22.35	18.18	14.66	10.86	9.07
KT	(6)	.52	.54	.56	.56	.58	.61	.61	.60	.57	.58	.55	.52
AVE. TEMP.	(3)	5.0	8.0	11.0	18.0	22.0	27.0	30.0	30.0	25.0	19.0	12.0	7.0
DEGREE-DAYS	(3)	405.	297.	227.	62.	7.	0.	0.	0.	0.	51.	205.	358.

WILMINGTON   DE   (LAT. 39.7)

		JAN	FEB	MAR	APR	MAY	JUNE	JULY	AUG	SEP	OCT	NOV	DEC
HORIZ. RAD.	(4)	6.49	9.39	13.04	16.80	19.41	21.37	20.69	18.32	14.96	11.17	7.31	5.55
KT	(6)	.43	.46	.48	.49	.49	.52	.51	.51	.50	.50	.44	.40
AVE. TEMP.	(3)	.0	1.0	5.0	11.0	17.0	22.0	24.0	23.0	20.0	14.0	8.0	1.0
DEGREE-DAYS	(3)	568.	488.	403.	212.	71.	0.	0.	0.	18.	141.	322.	522.

WINNEMUCCA   NV   (LAT. 40.9)

		JAN	FEB	MAR	APR	MAY	JUNE	JULY	AUG	SEP	OCT	NOV	DEC
HORIZ. RAD.	(4)	7.84	11.66	16.71	22.33	26.80	29.16	30.39	26.64	21.64	15.00	9.19	7.02
KT	(6)	.54	.59	.62	.66	.68	.71	.76	.74	.74	.69	.58	.54
AVE. TEMP.	(3)	-2.0	1.0	3.0	7.0	12.0	17.0	22.0	20.0	15.0	9.0	3.0	-1.0
DEGREE-DAYS	(3)	634.	481.	472.	332.	199.	83.	3.	23.	111.	288.	462.	596.

WINSLOW   AZ   (LAT. 35.0)

		JAN	FEB	MAR	APR	MAY	JUNE	JULY	AUG	SEP	OCT	NOV	DEC
HORIZ. RAD.	(4)	11.17	15.06	20.20	25.91	29.45	30.77	26.63	24.29	21.88	17.17	12.70	10.15
KT	(6)	.62	.65	.69	.73	.74	.75	.66	.66	.70	.69	.66	.61
AVE. TEMP.	(3)	.0	4.0	7.0	12.0	17.0	22.0	26.0	24.0	21.0	14.0	6.0	1.0
DEGREE-DAYS	(3)	558.	403.	348.	193.	69.	8.	0.	0.	11.	140.	363.	537.

YOUNGSTOWN   OH   (LAT. 41.3)

		JAN	FEB	MAR	APR	MAY	JUNE	JULY	AUG	SEP	OCT	NOV	DEC
HORIZ. RAD.	(4)	4.37	6.66	10.10	14.51	18.00	19.96	19.68	17.10	13.55	9.66	5.18	3.58
KT	(6)	.31	.34	.38	.43	.46	.48	.49	.48	.46	.45	.33	.28
AVE. TEMP.	(3)	-4.0	-3.0	2.0	9.0	14.0	19.0	22.0	21.0	17.0	11.0	5.0	-2.0
DEGREE-DAYS	(3)	677.	596.	512.	288.	143.	23.	5.	12.	66.	213.	412.	623.

YUMA   AZ   (LAT. 32.7)

		JAN	FEB	MAR	APR	MAY	JUNE	JULY	AUG	SEP	OCT	NOV	DEC
HORIZ. RAD.	(4)	12.44	16.38	21.78	27.38	30.96	31.93	27.84	26.43	23.28	18.42	13.79	11.35
KT	(6)	.64	.67	.72	.76	.78	.78	.69	.71	.72	.71	.67	.63
AVE. TEMP.	(3)	12.0	15.0	17.0	21.0	25.0	29.0	34.0	33.0	30.0	24.0	17.0	13.0
DEGREE-DAYS	(3)	171.	107.	54.	13.	0.	0.	0.	0.	0.	0.	60.	153.

ZUNI   NM   (LAT. 35.1)

		JAN	FEB	MAR	APR	MAY	JUNE	JULY	AUG	SEP	OCT	NOV	DEC
HORIZ. RAD.	(4)	11.19	14.72	19.15	24.59	28.07	29.53	25.70	23.59	21.50	16.98	21.35	10.13
KT	(6)	.62	.64	.65	.69	.71	.72	.64	.64	.68	.68	.64	.61
AVE. TEMP.	(3)	-1.0	1.0	4.0	9.0	14.0	19.0	22.0	21.0	17.0	11.0	4.0	.0
DEGREE-DAYS	(3)	598.	473.	437.	282.	147.	38.	0.	7.	51.	216.	415.	568.

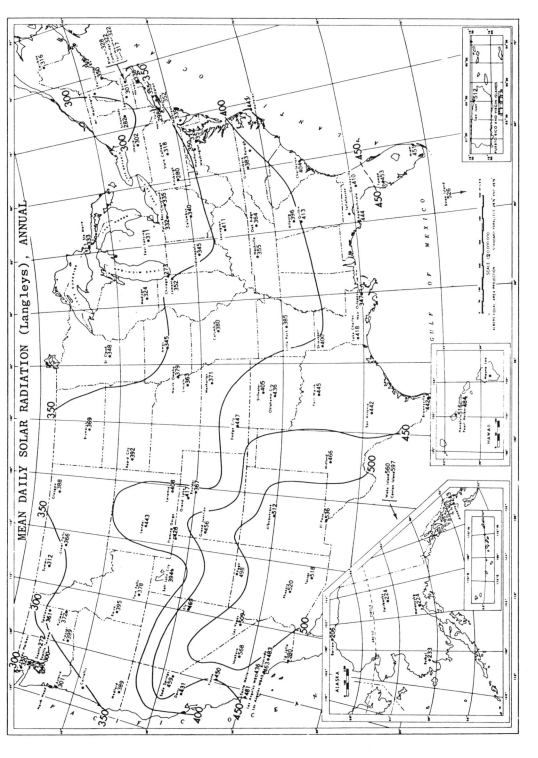

*U. S. Department of Commerce, *Climatic Atlas of the United States,* Asheville, North Carolina: National Oceanic and Atmospheric Administration (NOAA), 1977.

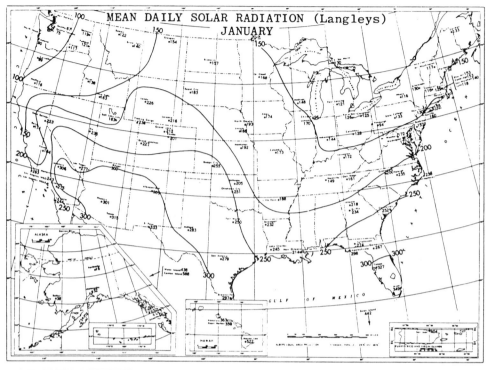

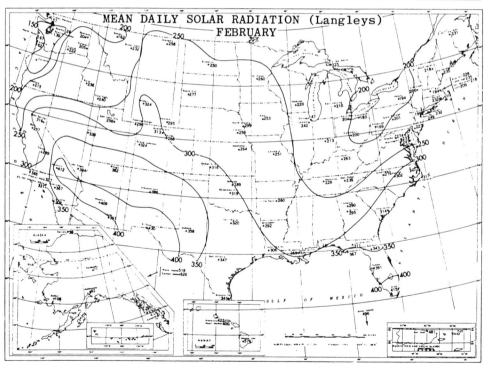

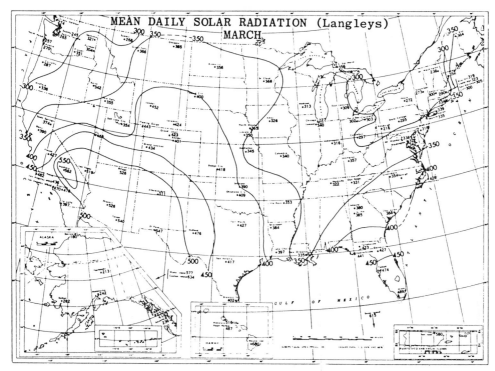

MEAN DAILY SOLAR RADIATION (Langleys)
MARCH

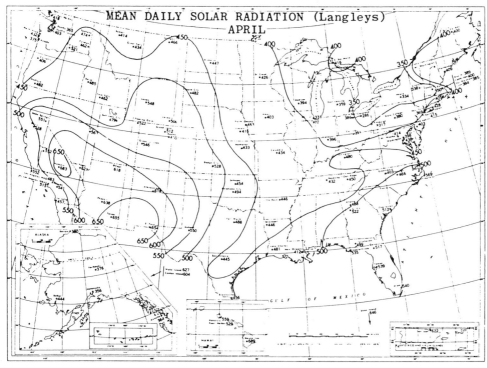

MEAN DAILY SOLAR RADIATION (Langleys)
APRIL

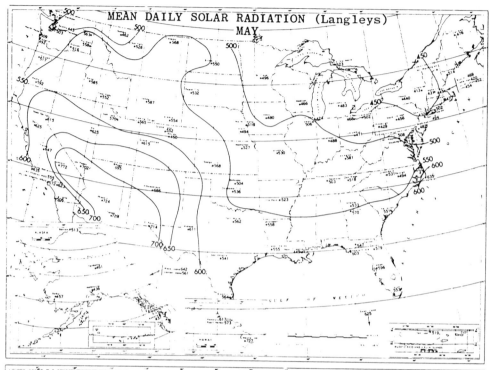

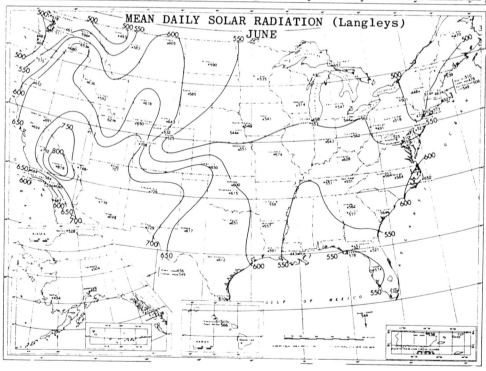

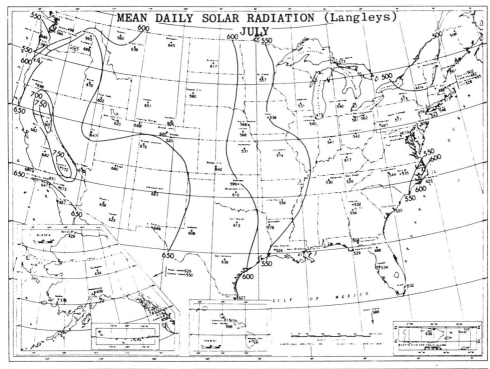

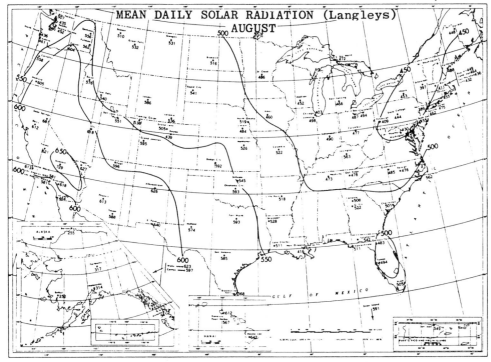

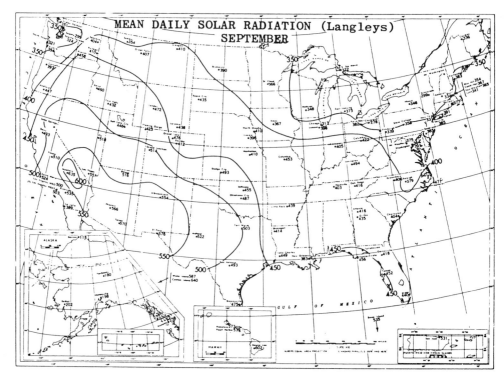

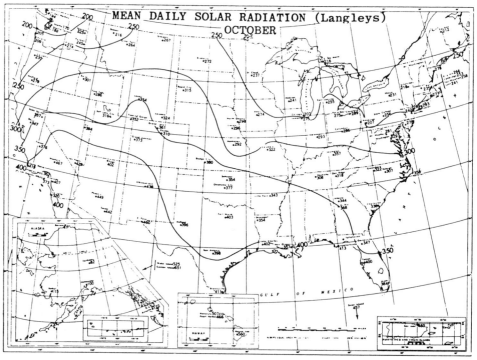

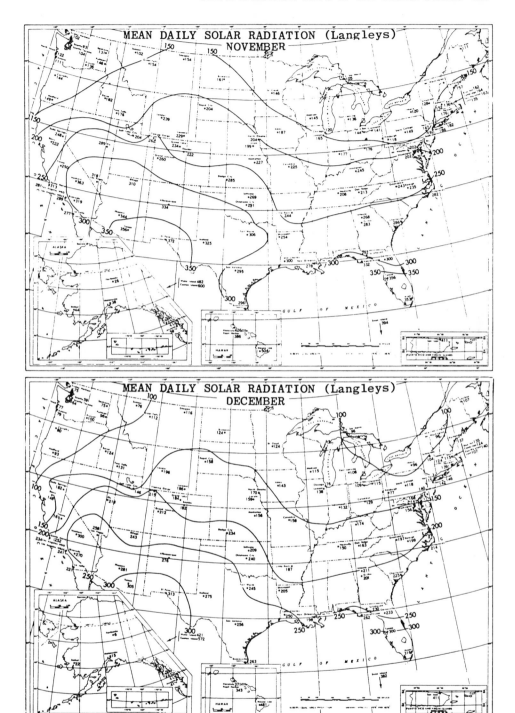

APPENDIX IV
HEAT GAIN COMPUTER PROGRAM

The following computer program calculates the heat gain for a building. The program is written for the Texas Instruments TI 59 programmable calculator used in conjunction with the Texas Instruments PC 100A printer. The program is stored on both sides of a magnetic card. A second card is used to store the 160 variables used in the program. The 160 variables relate to both the type construction of the building and the geographic location. Tables 1, 2 and 3 list steps in preparing the data and procedures for running the program. In running the program, The ASHRAE 1977 *Handbook of Fundamentals* is used to obtain actual values for the variables used in the heat gain program. For further information on this heat gain program, the reader should consult Sutch, Harry C., "Heat Gain Calculations for Programmable Calculators," *Heating/Piping/Air Conditioning,* March, **51**, 3:85–91, (1980).

Operation	Step	Code	Key	Operation	Step	Code	Key
Enter roof area	000	76	LBL		059	00	00
	001	11	A		060	70	70
	002	42	STO		061	43	RCL
	003	49	49		062	56	56
	004	91	R/S		063	42	STO
Enter glass area	005	76	LBL		064	00	00
	006	12	B		065	73	RC*
	007	42	STO		066	56	56
	008	50	50		067	61	GTO
	009	91	R/S		068	00	00
Enter Gross wall	010	76	LBL		069	76	76
area	011	13	C	Facing direction	070	43	RCL
	012	42	STO	is west	071	57	57
	013	51	51		072	42	STO
Set facing	014	43	RCL		073	00	00
direction	015	00	00		074	73	RC*
Set indirect	016	01	1		075	57	57
registers	017	42	STO	Calculate roof	076	65	×
	018	54	54	gain	077	01	1
	019	01	1		078	00	0
	020	01	1		079	00	0
	021	42	STO		080	95	=
	022	55	55		081	42	STO
	023	02	2		082	69	69
	024	01	1		083	59	INT
	025	42	STO		084	65	×
	026	56	56		085	43	RCL
	027	03	3		086	60	60
	028	01	1		087	65	×
	029	42	STO		088	43	RCL
	030	57	57		089	49	49
Is facing direction	031	97	DSZ		090	95	=
north?	032	00	00		091	42	STO
	033	00	00		092	52	52
	034	44	44	Calculate wall	093	71	SBR
	035	43	RCL	gain	094	01	01
	036	54	54		095	67	67
	037	42	STO		096	43	RCL
	038	00	00		097	61	61
	039	73	RC*		098	65	×
	040	54	54		099	53	(
	041	61	GTO		100	43	RCL
	042	00	00		101	51	51
	043	76	76		102	75	–
Is facing direction	044	97	DSZ		103	43	RCL
east?	045	00	00		104	50	50
	046	00	00		105	54	)
	047	57	57		106	95	=
	048	43	RCL		107	44	SUM
	049	55	55		108	52	52
	050	42	STO	Calculate glass	109	71	SBR
	051	00	00	conduction gain	110	01	01
	052	73	RC*		111	67	67
	053	55	55		112	43	RCL
	054	61	GTO		113	62	62
	055	00	00		114	65	×
	056	76	76		115	43	RCL
Is facing direction	057	97	DSZ		116	50	50
south?	058	00	00		117	95	=

Operation	Step	Code	Key
	118	44	SUM
	119	52	52
Calculate glass	120	43	RCL
solar gain	121	69	69
	122	22	INV
	123	59	INT
	124	65	x
	125	43	RCL
	126	50	50
	127	65	x
	128	43	RCL
	129	44	44
	130	65	x
	131	43	RCL
	132	45	45
	133	95	=
	134	44	SUM
	135	52	52
Is this hour gain	136	43	RCL
greater than previous	137	52	52
hour gain?	138	77	GE
	139	01	01
	140	52	52
"No"	141	01	1
Go to step 076	142	94	+/-
calculate room gain	143	44	SUM
for previous hour	144	00	00
	145	73	RC*
	146	00	00
	147	86	STF
	148	01	01
	149	61	GTO
	150	00	00
	151	76	76
Answer to step 138	152	87	IFF
"Yes"	153	01	01
Go to step 076	154	01	01
calculate room gain	155	81	81
for next hour	156	01	1
	157	44	SUM
	158	00	00
	159	43	RCL
	160	52	52
	161	32	XIT
	162	73	RC*
	163	00	00
	164	61	GTO
	165	00	00
	166	76	76
SBR 167 wall	167	43	RCL
& glass gain	168	69	69
	169	22	INV
	170	59	INT
	171	65	x
	172	01	1
	173	00	0
	174	00	0
	175	95	=
	176	42	STO
	177	69	69
	178	59	INT
	179	65	x
	180	92	RTN
Calculate lights	181	22	INV
gain	182	86	STF
	183	01	01
	184	43	RCL
	185	58	58
	186	65	x
	187	43	RCL
	188	46	46
	189	95	=
	190	44	SUM
	191	52	52
Calculate people	192	43	RCL
sensible gain	193	47	47
	194	65	x
	195	43	RCL
	196	65	65
	197	65	x
	198	43	RCL
	199	48	48
	200	95	=
	201	44	SUM
	202	52	52
Calculate people	203	43	RCL
latent gain	204	47	47
	205	65	x
	206	43	RCL

Operation	Step	Code	Key
	207	66	66
	208	95	=
	209	42	STO
	210	53	53
Calculate ventilation	211	53	(
& infiltration	212	43	RCL
sensible gain	213	42	42
	214	85	+
	215	43	RCL
	216	43	43
	217	54	)
	218	65	x
	219	01	1
	220	93	.
	221	01	1
	222	65	x
	223	43	RCL
	224	63	63
	225	95	=
	226	44	SUM
	227	52	52
Calculate ventilation	228	53	(
& infiltration	229	43	RCL
latent gain	230	42	42
	231	85	+
	232	43	RCL
	233	43	43
	234	54	)
	235	65	x
	236	04	4
	237	08	8
	238	04	4
	239	00	0
	240	65	x
	241	43	RCL
	242	64	64
	243	95	=
	244	44	SUM
	245	53	53
Sum zone sensible	246	43	RCL
& latent gains	247	52	52
	248	44	SUM
	249	67	67
	250	43	RCL
	251	53	53
	252	44	SUM
	253	68	68
Print room	254	69	OP
number	255	00	00
	256	03	3
	257	05	5
	258	03	3
	259	02	2
	260	03	3
	261	02	2
	262	03	3
	263	00	0
	264	69	OP
	265	01	01
	266	69	OP
	267	05	05
	268	43	RCL
	269	59	59
	270	99	PRT
Print room	271	71	SBR
sensible gain	272	03	03
	273	63	63
	274	43	RCL
	275	52	52
	276	58	FIX
	277	00	00
	278	99	PRT
	279	22	INV
	280	59	FIX
Print room	281	71	SBR
latent gain	282	03	03
	283	80	80
	284	43	RCL
	285	53	53
	286	58	FIX
	287	00	00
	288	99	PRT
	289	22	INV
	290	58	FIX
Print solar time	291	69	OP
of maximum gain	292	00	00
(24 hr clock)	293	03	3
	294	07	7
	295	02	2

Operation	Step	Code	Key
	296	04	4
	297	03	3
	298	00	0
	299	01	1
	300	07	7
	301	69	OP
	302	02	02
	303	69	OP
	304	05	05
	305	43	RCL
	306	00	00
	307	55	÷
	308	01	1
	309	00	0
	310	95	=
	311	22	INV
	312	59	INT
	313	65	x
	314	01	1
	315	00	0
	316	85	+
	317	08	8
	318	95	=
	319	99	PRT
	320	98	ADV
	321	98	ADV
	322	98	ADV
	323	29	CP
	324	91	R/S
SBR 325 print	325	69	OP
"Zone"	326	00	00
	327	04	4
	328	06	6
	329	03	3
	330	02	2
	331	03	3
	332	01	1
	333	01	1
	334	07	7
	335	69	OP
	336	01	01
	337	69	OP
	338	05	05
SBR 339 print	339	71	SBR
zone sensible	340	03	03
gain	341	63	63
	342	43	RCL
	343	67	67
	344	58	FIX
	345	00	00
	346	99	PRT
	347	22	INV

Operation	Step	Code	Key
	348	58	FIX
SBR 349	349	71	SBR
Print zone	350	03	03
latent gain	351	80	80
	352	43	RCL
	353	68	68
	354	58	FIX
	355	00	00
	356	99	PRT
	357	22	INV
	358	58	FIX
	359	98	ADV
	360	98	ADV
	361	98	ADV
	362	91	R/S
SBR 363 print	363	69	OP
"Sens."	364	00	00
	365	03	3
	366	06	6
	367	01	1
	368	07	7
	369	03	3
	370	01	1
	371	03	3
	372	06	6
	373	00	0
	374	00	0
	375	69	OP
	376	02	02
	377	69	OP
	378	05	05
	379	92	RTN
SBR 380 print	380	69	OP
"Latent"	381	00	00
	382	02	2
	383	07	7
	384	01	1
	385	03	3
	386	03	3
	387	07	7
	388	02	2
	389	09	9
	390	03	3
	391	07	7
	392	69	OP
	393	02	02
	394	69	OP
	395	05	05
	396	92	RTN
	397	00	0
	398	00	0
	399	00	0

Table A1 Preparing and Entering Heat Gain Data

1. It will be a great time-saver and will reduce the possibilities of error if a form similar to the sample heat gain data sheet is used to record the data.

2. The following data are taken from the ASHRAE *Handbook & Product Directory, 1977 Fundamentals*, Chapter 25. Solar times are from 9 to 18 inclusive.

   Roof CLTDs are in Table 5 of the handbook. The example uses Roof No. 9 with suspended ceiling and 4 in. h.w. concrete with 1 in. (or 2 in.) insulation. Note: the roof CLTDs are the same for all four facing directions.

   Wall CLTDs are in Table 7. The example uses group D walls.

   Glass CLTDs are interpolated from table 9 and are the same for the four facing directions.

   Glass cooling load factors are found in the appropriate line of Table 11 or Table 12. Note that the factors are decimals. Do NOT enter the decimal point in the data card. The program does it. Also, if the factor ends in zero, change the zero to 1.

3. Note that the data sheet form has a decimal point to the left of the roof CLTD. This is mandatory for proper operation of the program. Omitting the decimal point will produce impossible answers.

4. The numbers must be entered in the TI 59 programmable calculator in the following manner:
   12345678 STO nn

Table A2 Heat Gain Program

Operation	Press
Step 1 Turn on calculator	
Step 2 Set calculator partition.	7, OP, 17 = 399.69
Step 3 Select data card for building construction or record new data card as shown in Table 1.	
Step 4 Enter sides 1 and 2.	
Step 5 Enter sides 3 and 4.	
Step 6 Fill out heat gain calculations sheet with the building variables and room variables. If you have more than seven rooms in one zone, add more sheets. If any item is zero (such as roof area), show the zero in the heat gain calculations sheet. From the heat gain calculations sheet enter in the calculator:	
A) Building variables starting with the roof U factor.	STO 60
Then proceed on through the entire seven building variables.	
B) Room variables starting with the room number (which can be up to 8 digits).	STO 59
Then go on down the column for that particular room.	
Step 7 Enter gross wall area.	Lbl C
Calculations start.	
Step 8 If a printer is used, after about 30 seconds it will print out the following: 1) room number; 2) room sensible gain in Btuh; 3) room latent gain in Btuh, and 4) the 24 hour clock solar time of the maximum gain.	
Step 9 After the printout is complete for a room, repeat the process in Step 6 B above for the next room.	
Step 10 After all room gains have been calculated:	R/S
After about 60 seconds, the printer will print the total zone sensible gain and the total zone latent gain in Btuh, which will be the sum of the maximum room gains. Diversity must then be determined by the engineer using the calculations.	
Step 11 If the printer is not used, after each room calculation the display will show the 24 hour clock solar time of the maximum room gain	
to display the room sensible gain:	RCL 52
To display the room latent gain:	RCL 53
To display the zone sensible gain:	RCL 67
To display the zone latent gain:	RCL 68

Table A3 Obtaining Other Data
(All References are to the ASHRAE
Handbook and Product Directory: Fundamentals, 1977)

1. U factors should be found in Chapter 22.
2. ΔT is the outside-inside design temperature difference found in Chapter 23.
3. ΔW is the inside-outside air humidity ratio difference lb H_2O per lb air, and can be found from a psychrometric chart.
4. Rates of heat gain for people, both sensible and latent, can be found in Table 16, page 25.17.
5. Glass shading coefficient should be selected from Table 28, page 26.27.
6. Glass solar heat gain factors are in Table 10, page 25.12.
7. Lights cooling load factor should be selected from Table 15, page 15.16. This is a judgment selection. The sample problem uses 0.85.
8. People cooling load factor is also a judgment selection from Table 17, page 25.17. The sample problem uses 0.84.
9. Items not covered here are obtained in a normal manner from plans or appropriate codes or standards.

APPENDIX V
RSVP COMPUTER PROGRAM PRINTOUT

A NOTE ON ASSUMPTIONS

The tables on the following pages were prepared from *averaged* data furnished by RSVP, the U. S. Department of Housing and Urban Development's computer program for estimating solar economics. Annual energy savings (TABLE A) and suggested collector sizes (TABLE C) are based on:

Weather data for a *typical* year at each of the cities listed. Varying conditions from year to year will affect the fuel actually saved.

Assumed daily use of 20 gallons of water heated to 140°F by the first two occupants of a house, with each additional person using 15 gallons. Fuel savings will be greater if conservation measures are taken, such as setting the hot water thermostat at 120° to 130°F.

Average performance of a *typical* system—deemed for this purpose to be an open loop, two-tank, draindown system with storage capacity of 1.8 gallons for each square foot of collector area. Assumed collector performance (performance curve slope of 83 BTU/hr./sq ft degree F; y-intercept of .73) resembles that provided by flat steel absorber plates with a selective black coating under one layer of glazing. However, some collectors on the market may vary as much as 20% from this average performance. Annual Energy Savings figures can be adjusted to reflect data on the performance of a specific collector when this data is available (see TABLE K, page 122).

The *average* operating efficiency of conventional water heaters according to fuel used—specifically, electricity, 100%; natural gas, 60%; and fuel oil, 50%.

Maintenance and any additional property tax or insurance charges due to a solar water heating system have not been included in costs. These tend to be offset by Federal tax deductions for interest payments.

Table A. Annual Energy Savings from
a Typical Solar Water Heater

1. Killowatt-Hours of Electricity Saved

Number of Occupants	2		4			6 (or more)			
Collector Area (sq. ft.)	40	60	40	60	80	40	60	80	100
Location									
ALABAMA									
BIRMINGHAM	1900	2300	2300	3100	3600	2500	3400	4200	4800
ALASKA									
FAIRBANKS	2000	2600	2300	3200	3900	2500	3500	4300	5100
ARIZONA									
TUCSON	2500	2600	3300	4100	4400	3700	4900	5700	6100
ARKANSAS									
LITTLE ROCK	2000	2400	2500	3300	3800	2700	3700	4500	5100
CALIFORNIA									
LOS ANGELES	2300	2700	2900	3800	4300	3200	4300	5200	5900
SACRAMENTO	2300	2500	3000	3800	4100	3300	4400	5200	5700
SAN FRANCISCO	2300	2700	2900	3800	4400	3200	4300	5200	5900
COLORADO									
DENVER	2900	3100	3600	4600	5200	3900	5300	6400	7200
GRAND JUNCTION	2700	2900	3400	4400	4900	3700	5000	6000	6700
CONNECTICUT									
HARTFORD	1600	2100	1900	2600	3200	2000	2800	3500	4100
DELAWARE									
WILMINGTON	1900	2300	2200	3000	3500	2300	3200	4000	4700
DISTRICT OF COLUMBIA									
WASHINGTON	1800	2200	2100	2900	3400	2300	3100	3900	4500
FLORIDA									
JACKSONVILLE	1900	2300	2400	3100	3600	2600	3600	4300	4900
MIAMI	1800	2100	2300	3000	3400	2600	3400	4100	4600
TALLAHASSEE	2000	2300	2400	3200	3600	2600	3600	4300	4900
TAMPA	1900	2200	2500	3200	3600	2700	3600	4300	4900
GEORGIA									
ATLANTA	1900	2400	2300	3100	3700	2500	3500	4200	4900
SAVANNAH	2100	2400	2600	3300	3800	2800	3800	4600	5200
HAWAII									
HILO	1800	2100	2300	3000	3400	2500	3400	4100	4600
HONOLULU	2200	2300	3000	3700	3900	3400	4400	5100	5500

Table A-1 (Continued)

Number of Occupants	2		4			6 (or more)			
Collector Area (sq. ft.)	40	60	40	60	80	40	60	80	100
Location									
IDAHO									
BOISE	2400	2700	3000	3900	4400	3200	4400	5300	6000
POCATELLO	2500	2900	3100	4100	4700	3400	4600	5600	6400
ILLINOIS									
CHICAGO	1900	2400	2200	3000	3600	2400	3300	4100	4800
PEORIA	2100	2600	2500	3400	4100	2700	3800	4700	5400
INDIANA									
INDIANAPOLIS	1800	2200	2000	2800	3300	2200	3000	3800	4400
IOWA									
DES MOINES	2200	2600	2600	3400	4100	2700	3800	4700	5400
KANSAS									
WICHITA	2300	2700	2900	3800	4400	3100	4300	5200	5900
KENTUCKY									
LEXINGTON	1800	2300	2100	2900	3400	2300	3200	3900	4500
LOUISVILLE	1800	2300	2100	2900	3500	2300	3200	3900	4600
LOUISIANA									
NEW ORLEANS	1900	2300	2400	3100	3600	2600	3500	4300	4800
SHREVEPORT	2000	2300	2500	3200	3700	2700	3700	4400	5000
MAINE									
CARIBOU	1900	2400	2100	2900	3600	2200	3100	3900	4600
PORTLAND	1700	2200	1900	2700	3300	2000	2900	3600	4200
MARYLAND									
BALTIMORE	1900	2300	2200	2900	3500	2300	3200	4000	4700
MASSACHUSETTS									
AMHERST	1800	2300	2100	2900	3500	2200	3100	3900	4600
BOSTON	1700	2200	2000	2700	3300	2100	2900	3700	4300
MICHIGAN									
LANSING	2000	2500	2300	3100	3800	2500	3400	4300	5000
SAULT ST MARIE	1700	2200	1900	2600	3200	2000	2800	3600	4200
MINNESOTA									
MINN-ST PAUL	2000	2500	2300	3100	3800	2400	3400	4300	5000
MISSISSIPPI									
JACKSON	1900	2300	2400	3100	3600	2600	3500	4300	4900
MISSOURI									
KANSAS CITY	2100	2500	2500	3300	4000	2700	3700	4600	5300
ST. LOUIS	2000	2400	2400	3200	3800	2600	3600	4400	5100
MONTANA									
BILLINGS	2300	2700	2700	3700	4300	2900	4100	5000	5800
GREAT FALLS	2200	2700	2600	3500	4200	2800	3900	4800	5500
NEBRASKA									
LINCOLN	2300	2800	2800	3700	4400	3000	4100	5000	5800
NEVADA									
LAS VEGAS	2500	2600	3500	4200	4400	3900	5100	5900	6200
RENO	2900	3100	3700	4700	5200	4000	5400	6500	7200
NEW HAMPSHIRE									
CONCORD	1700	2200	1900	2600	3200	2000	2800	3600	4200
NEW JERSEY									
ATLANTIC CITY	2200	2600	2600	3500	4100	2800	3900	4700	5400
NEW MEXICO									
ALBUQUERQUE	2800	3000	3600	4600	5000	4000	5300	6400	7000
NEW YORK									
ALBANY	1900	2400	2300	3100	3700	2400	3400	4200	4900
NEW YORK	1600	2100	1900	2600	3100	2000	2800	3500	4100
ROCHESTER	1500	2000	1800	2400	2900	1800	2600	3300	3800
SYRACUSE	1500	2000	1700	2400	2900	1800	2600	3300	3800
NORTH CAROLINA									
CAPE HATTERAS	2000	2400	2400	3200	3700	2600	3600	4400	5000
RALEIGH	1900	2300	2300	3100	3600	2500	3400	4200	4800
NORTH DAKOTA									
BISMARCK	2300	2800	2700	3600	4300	2800	3900	4900	5700
OHIO									
CLEVELAND	1600	2000	1800	2500	3000	1900	2700	3400	4000
COLUMBUS	1700	2100	1900	2600	3200	2000	2900	3600	4200
OKLAHOMA									
OKLAHOMA CITY	2200	2600	2700	3500	4100	2900	4000	4800	5500
TULSA	2000	2400	2500	3300	3800	2700	3700	4500	5100
OREGON									
MEDFORD	2000	2300	2400	3200	3700	2600	3600	4400	5000
PORTLAND	1500	1900	1800	2400	2900	1900	2700	3300	3900
PENNSYLVANIA									
PHILADELPHIA	1800	2300	2100	2800	3400	2200	3100	3900	4500
PITTSBURGH	2000	2500	2400	3200	3900	2600	3600	4400	5100
STATE COLLEGE	1900	2400	2200	3000	3600	2300	3300	4100	4700
RHODE ISLAND									
PROVIDENCE	1800	2200	2000	2800	3400	2100	3000	3700	4400
SOUTH CAROLINA									
CHARLESTON	1900	2300	2300	3000	3500	2500	3400	4100	4700
SOUTH DAKOTA									
RAPID CITY	2300	2800	2800	3700	4400	3000	4100	5100	5800
TENNESSEE									
NASHVILLE	1800	2200	2200	2900	3400	2300	3200	3900	4500
TEXAS									
AMARILLO	2600	2800	3200	4100	4700	3500	4700	5700	6400
DALLAS	2000	2300	2500	3300	3800	2700	3700	4500	5100
EL PASO	2600	2700	3500	4300	4600	3900	5100	6000	6400
HOUSTON	1800	2100	2200	2800	3300	2300	3200	3900	4400
UTAH									
SALT LAKE CITY	2600	2800	3200	4200	4600	3500	4700	5700	6400
VERMONT									
BURLINGTON	1900	2400	2200	2900	3600	2300	3200	4000	4700
VIRGINIA									
MT. WEATHER	2100	2600	2500	3300	4000	2600	3700	4600	5300
NORFOLK	2000	2400	2400	3200	3800	2600	3500	4300	5000
RICHMOND	1900	2300	2200	3000	3600	2400	3300	4100	4700
WASHINGTON									
SEATTLE	1600	2000	1900	2600	3100	2100	2900	3500	4100
SPOKANE	2000	2400	2300	3100	3700	2500	3500	4300	4900
WEST VIRGINIA									
CHARLESTON	1600	2100	1900	2600	3100	2000	2800	3500	4100
WISCONSIN									
MADISON	2000	2500	2300	3100	3800	2400	3400	4300	5000
WYOMING									
CASPER	2800	3100	3400	4500	5100	3700	5000	6100	6900
CHEYENNE	2700	3100	3200	4300	5000	3500	4800	5800	6700

Table A. Annual Energy Savings from a Typical Solar Water Heater

2. Therms of Natural Gas Saved

Number of Occupants	2		4			6 (or more)			
Collector Area (sq. ft.)	40	60	40	60	80	40	60	80	100
Location									
ALABAMA									
BIRMINGHAM	110	130	130	170	200	140	190	240	270
ALASKA									
FAIRBANKS	120	150	130	180	220	140	200	250	290
ARIZONA									
TUCSON	140	150	190	230	250	210	280	330	350
ARKANSAS									
LITTLE ROCK	120	140	140	190	220	150	210	250	290
CALIFORNIA									
LOS ANGELES	130	150	170	220	250	180	250	300	330
SACRAMENTO	130	140	170	210	230	190	250	290	320
SAN FRANCISCO	130	150	170	220	250	180	250	300	330
COLORADO									
DENVER	160	180	200	260	300	220	300	360	410
GRAND JUNCTION	150	170	190	250	280	210	280	340	380
CONNECTICUT									
HARTFORD	90	120	110	150	180	110	160	200	230
DELAWARE									
WILMINGTON	110	130	120	170	200	130	180	230	270
DISTRICT OF COLUMBIA									
WASHINGTON	100	130	120	160	190	130	180	220	260
FLORIDA									
JACKSONVILLE	110	130	140	180	210	150	200	240	280
MIAMI	100	120	130	170	190	150	200	230	260
TALLAHASSEE	110	130	140	180	210	150	200	250	280
TAMPA	110	130	140	180	200	150	210	250	280
GEORGIA									
ATLANTA	110	130	130	180	210	140	200	240	280
SAVANNAH	120	140	150	190	220	160	220	260	290
HAWAII									
HILO	110	120	130	170	190	140	200	230	260
HONOLULU	120	130	170	210	220	190	250	290	310
IDAHO									
BOISE	140	150	170	220	250	180	250	300	340
POCATELLO	140	160	180	230	270	190	260	320	360
ILLINOIS									
CHICAGO	110	140	130	170	210	140	190	230	270
PEORIA	120	150	140	190	230	150	210	260	310
INDIANA									
INDIANAPOLIS	100	130	120	160	190	120	170	210	250
IOWA									
DES MOINES	120	150	150	200	230	160	220	270	310
KANSAS									
WICHITA	130	150	160	210	250	180	240	290	330
KENTUCKY									
LEXINGTON	100	130	120	160	200	130	180	220	260
LOUISVILLE	100	130	120	160	200	130	180	220	260
LOUISIANA									
NEW ORLEANS	110	130	140	180	200	150	200	240	270
SHREVEPORT	110	130	140	180	210	150	210	250	290
MAINE									
CARIBOU	110	140	120	170	200	130	180	220	260
PORTLAND	100	130	110	150	190	110	160	200	240
MARYLAND									
BALTIMORE	110	130	120	170	200	130	180	230	260
MASSACHUSETTS									
AMHERST	100	130	120	160	200	130	180	220	260
BOSTON	100	120	110	150	190	120	170	210	240
MICHIGAN									
LANSING	110	140	130	180	220	140	200	240	280
SAULT ST. MARIE	100	120	110	150	180	110	160	200	240
MINNESOTA									
MINN-ST. PAUL	110	140	130	180	220	140	190	240	280
MISSISSIPPI									
JACKSON	110	130	140	180	210	150	200	240	280
MISSOURI									
KANSAS CITY	120	140	140	190	230	150	210	260	300
ST. LOUIS	120	140	140	180	220	150	200	250	290
MONTANA									
BILLINGS	130	160	160	210	250	170	230	290	330
GREAT FALLS	130	150	150	200	240	160	220	270	320
NEBRASKA									
LINCOLN	130	160	160	210	250	170	230	290	330
NEVADA									
LAS VEGAS	140	150	200	240	250	220	290	340	360
RENO	160	180	210	270	300	230	310	370	410
NEW HAMPSHIRE									
CONCORD	100	120	110	150	180	110	160	200	240
NEW JERSEY									
ATLANTIC CITY	120	150	150	200	230	160	220	270	310
NEW MEXICO									
ALBUQUERQUE	160	170	210	260	290	230	300	360	400
NEW YORK									
ALBANY	110	140	130	170	210	140	190	240	280
NEW YORK	90	120	110	150	180	110	160	200	230
ROCHESTER	90	110	100	140	170	110	150	190	220
SYRACUSE	90	110	100	140	170	100	150	190	220
NORTH CAROLINA									
CAPE HATTERAS	110	140	140	180	210	150	200	250	280
RALEIGH	110	130	130	170	210	140	190	240	270
NORTH DAKOTA									
BISMARCK	130	160	150	200	250	160	220	280	320
OHIO									
CLEVELAND	90	120	100	140	170	110	150	190	230
COLUMBUS	100	120	110	150	180	120	160	200	240
OKLAHOMA									
OKLAHOMA CITY	120	150	150	200	230	160	230	270	310
TULSA	120	140	140	190	220	150	210	250	290
OREGON									
MEDFORD	110	130	140	180	210	150	200	250	280
PORTLAND	90	110	100	140	170	110	150	190	220
PENNSYLVANIA									
PHILADELPHIA	100	130	120	160	190	130	180	220	260
PITTSBURGH	120	140	140	180	220	150	200	250	290
STATE COLLEGE	110	140	130	170	210	130	190	230	270
RHODE ISLAND									
PROVIDENCE	100	130	110	160	190	120	170	210	250
SOUTH CAROLINA									
CHARLESTON	110	130	130	170	200	140	190	230	270

Table A-2 (Continued)

Number of Occupants	2		4			6 (or more)			
Collector Area (sq. ft.)	40	60	40	60	80	40	60	80	100
Location)									
SOUTH DAKOTA									
RAPID CITY	130	160	160	210	250	170	230	290	330
TENNESSEE									
NASHVILLE	100	130	120	160	200	130	180	220	260
TEXAS									
AMARILLO	150	160	180	240	270	200	270	320	360
DALLAS	120	130	140	190	210	160	210	250	290
EL PASO	150	150	200	250	260	220	290	340	370
HOUSTON	100	120	120	160	190	130	180	220	250
UTAH									
SALT LAKE CITY	150	160	180	240	260	200	270	330	360
VERMONT									
BURLINGTON	110	130	120	170	200	130	180	230	270
VIRGINIA									
MT. WEATHER	120	150	140	190	230	150	210	260	300
NORFOLK	110	140	140	180	210	150	200	250	280
RICHMOND	110	130	130	170	200	140	190	230	270
WASHINGTON									
SEATTLE	90	120	110	150	180	120	160	200	230
SPOKANE	110	130	130	180	210	140	200	240	280
WEST VIRGINIA									
CHARLESTON	90	120	110	150	180	110	160	200	230
WISCONSIN									
MADISON	110	140	130	180	220	140	190	240	280
WYOMING									
CASPER	160	180	190	250	290	210	290	350	390
CHEYENNE	150	180	180	240	280	200	270	330	380

Table A. Annual Energy Savings from a Typical Solar Water Heater

3. Gallons of Fuel Oil Saved

Number of Occupants	2		4			6 (or more)			
Collector Area (sq. ft.)	40	60	40	60	80	40	60	80	100
Location									
ALABAMA									
BIRMINGHAM	90	110	110	150	180	120	170	200	230
ALASKA									
FAIRBANKS	100	130	110	160	190	120	170	210	250
ARIZONA									
TUCSON	120	120	160	200	210	180	240	280	300
ARKANSAS									
LITTLE ROCK	100	120	120	160	190	130	180	220	250
CALIFORNIA									
LOS ANGELES	110	130	140	180	210	160	210	250	290
SACRAMENTO	110	120	140	180	200	160	210	250	280
SAN FRANCISCO	110	130	140	180	210	150	210	250	290
COLORADO									
DENVER	140	150	180	230	250	190	260	310	350
GRAND JUNCTION	130	140	160	210	240	180	240	290	330
CONNECTICUT									
HARTFORD	80	100	90	130	150	100	140	170	200
DELAWARE									
WILMINGTON	90	110	110	140	170	110	160	200	230
DISTRICT OF COLUMBIA									
WASHINGTON	90	110	100	140	170	110	150	190	220
FLORIDA									
JACKSONVILLE	90	110	120	150	180	130	170	210	240
MIAMI	90	100	110	150	170	120	170	200	220
TALLAHASSEE	100	110	120	150	180	130	170	210	240
TAMPA	90	110	120	150	180	130	180	210	240
GEORGIA									
ATLANTA	90	110	110	150	180	120	170	210	240
SAVANNAH	100	120	130	160	190	140	190	220	250
HAWAII									
HILO	90	100	110	150	170	120	170	200	230
HONOLULU	110	110	150	180	190	170	220	250	270
IDAHO									
BOISE	120	130	140	190	210	160	210	260	290
POCATELLO	120	140	150	200	230	170	230	270	310
ILLINOIS									
CHICAGO	90	120	110	150	180	120	160	200	230
PEORIA	100	130	120	170	200	130	180	230	260
INDIANA									
INDIANAPOLIS	90	110	100	140	160	110	150	180	210
IOWA									
DES MOINES	110	130	120	170	200	130	190	230	260
KANSAS									
WICHITA	110	130	140	180	210	150	210	250	290
KENTUCKY									
LEXINGTON	90	110	100	140	170	110	150	190	220
LOUISVILLE	90	110	100	140	170	110	150	190	220
LOUISIANA									
NEW ORLEANS	90	110	120	150	180	130	170	210	240
SHREVEPORT	100	110	120	160	180	130	180	220	250
MAINE									
CARIBOU	90	120	100	140	170	110	150	190	230
PORTLAND	80	110	90	130	160	100	140	180	210
MARYLAND									
BALTIMORE	90	110	110	140	170	110	160	200	230
MASSACHUSETTS									
AMHERST	90	110	100	140	170	110	150	190	220
BOSTON	80	110	100	130	160	100	140	180	210
MICHIGAN									
LANSING	100	120	110	150	180	120	170	210	240
SAULT ST MARIE	80	110	90	130	160	100	140	170	210

Table A-3 (Continued)

Number of Occupants	2		4			6 (or more)			
Collector Area (sq. ft.)	40	60	40	60	80	40	60	80	100
Location									
MINNESOTA									
MINN-ST. PAUL	100	120	110	150	190	120	170	210	240
MISSISSIPPI									
JACKSON	90	110	120	150	180	130	170	210	240
MISSOURI									
KANSAS CITY	100	120	120	160	190	130	180	220	260
ST LOUIS	100	120	120	160	190	130	170	210	250
MONTANA									
BILLINGS	110	130	130	180	210	140	200	240	280
GREAT FALLS	110	130	130	170	200	140	190	230	270
NEBRASKA									
LINCOLN	110	140	130	180	210	140	200	250	280
NEVADA									
LAS VEGAS	120	130	170	210	220	190	250	290	300
RENO	140	150	180	230	250	200	260	320	350
NEW HAMPSHIRE									
CONCORD	80	110	90	130	160	100	140	170	210
NEW JERSEY									
ATLANTIC CITY	110	130	130	170	200	140	190	230	270
NEW MEXICO									
ALBUQUERQUE	140	140	180	220	240	190	260	310	340
NEW YORK									
ALBANY	100	120	110	150	180	120	160	200	240
NEW YORK	80	100	90	130	150	100	140	170	200
ROCHESTER	70	100	90	120	140	90	130	160	190
SYRACUSE	70	100	80	120	140	90	130	160	190
NORTH CAROLINA									
CAPE HATTERAS	100	120	120	160	180	130	170	210	240
RALEIGH	90	110	110	150	180	120	170	200	230
NORTH DAKOTA									
BISMARCK	110	140	130	180	210	140	190	240	280
OHIO									
CLEVELAND	80	100	90	120	150	90	130	170	190
COLUMBUS	80	100	90	130	160	100	140	170	200
OKLAHOMA									
OKLAHOMA CITY	110	120	130	170	200	140	190	230	270
TULSA	100	120	120	160	190	130	180	220	250
OREGON									
MEDFORD	100	110	120	160	180	130	180	210	240
PORTLAND	80	90	90	120	140	90	130	160	190
PENNSYLVANIA									
PHILADELPHIA	90	110	100	140	170	110	150	190	220
PITTSBURGH	100	120	120	160	190	130	180	220	250
STATE COLLEGE	90	120	110	150	180	110	160	200	230
RHODE ISLAND									
PROVIDENCE	90	110	100	130	160	100	150	180	210
SOUTH CAROLINA									
CHARLESTON	90	110	110	150	170	120	160	200	230
SOUTH DAKOTA									
RAPID CITY	110	140	140	180	210	140	200	250	280
TENNESSEE									
NASHVILLE	90	110	110	140	170	110	160	190	220
TEXAS									
AMARILLO	120	140	160	200	230	170	230	280	310
DALLAS	100	1,10	120	160	180	130	180	220	250
EL PASO	130	130	170	210	220	190	250	290	310
HOUSTON	90	100	110	140	160	110	160	190	220
UTAH									
SALT LAKE CITY	120	140	160	200	230	170	230	280	310
VERMONT									
BURLINGTON	90	120	110	140	170	110	160	200	230
VIRGINIA									
MT. WEATHER	100	130	120	160	200	130	180	220	260
NORFOLK	100	120	120	150	180	120	170	210	240
RICHMOND	90	110	110	150	170	120	160	200	230
WASHINGTON									
SEATTLE	80	100	90	130	150	100	140	170	200
SPOKANE	100	110	110	150	180	120	170	210	240
WEST VIRGINIA									
CHARLESTON	80	100	90	130	150	100	140	170	200
WISCONSIN									
MADISON	100	120	110	150	180	120	170	210	240
WYOMING									
CASPER	130	150	170	220	250	180,	250	300	340
CHEYENNE	130	150	160	210	240	170	230	280	330

Table C. Suggested Collector Sizes for Solar Water Heating Systems

Note: The collector area indicated in each case can be expected to provide, at the lowest possible cost per unit of heat produced, at least 50 percent of the hot water used annually by the average household of that size. A collector panel typically has 20 square feet of collecting surface.

Location	Number of Users			Location	Number of Users		
	2	4	6		2	4	6
	Square Feet				Square Feet		
BIRMINGHAM, AL	40	60	80	BILLINGS, MT	40	60	80
FAIRBANKS, AK	60	80	100	GREAT FALLS, MT	40	60	80
TUCSON, AZ	40	60	60	LINCOLN, NE	40	60	80
LITTLE ROCK, AR	40	60	80	LAS VEGAS, NV	40	60	60
LOS ANGELES, CA	40	60	80	RENO, NV	40	60	80
SACRAMENTO, CA	40	60	60	CONCORD, NH	60	80	100
SAN FRANCISCO, CA	40	60	80	ATLANTIC CITY, NJ	40	60	80
DENVER, CO	40	60	80	ALBUQUERQUE, NM	40	60	60
GRAND JUNCTION, CO	40	60	80	ALBANY, NY	60	80	100
HARTFORD, CT	60	80	100	NEW YORK, NY	60	80	100
WILMINGTON, DE	60	80	100	ROCHESTER, NY	60	80	100
WASHINGTON, DC	60	80	100	SYRACUSE, NY	60	80	100
JACKSONVILLE, FL	40	60	80	CAPE HATTERAS, NC	40	60	80
MIAMI, FL	40	60	80	RALEIGH, NC	40	60	80
TALLAHASSEE, FL	40	60	80	BISMARCK, ND	40	80	100
TAMPA, FL	40	60	80	CLEVELAND, OH	60	80	100
ATLANTA, GA	40	60	80	COLUMBUS, OH	60	80	100
SAVANNAH, GA	40	60	80	OKLAHOMA CITY, OK	40	60	80
HILO, HI	40	60	80	TULSA, OK	40	60	80
HONOLULU, HI	40	60	60	MEDFORD, OR	40	60	80
BOISE, ID	40	60	80	PORTLAND, OR	60	80	100
POCATELLO, ID	40	60	80	PHILADELPHIA, PA	60	80	100
CHICAGO, IL	60	80	100	PITTSBURGH, PA	40	80	80
PEORIA, IL	40	60	80	STATE COLLEGE, PA	60	80	100
INDIANAPOLIS, IN	60	80	100	PROVIDENCE, RI	60	80	100
DES MOINES, IA	40	80	80	CHARLESTON, SC	40	60	80
WICHITA, KS	40	60	80	RAPID CITY, SD	40	60	80
LEXINGTON, KY	60	80	100	NASHVILLE, TN	40	60	80
LOUISVILLE, KY	60	80	100	AMARILLO, TX	40	60	80
NEW ORLEANS, LA	40	60	80	DALLAS, TX	40	60	80
SHREVEPORT, LA	40	60	80	EL PASO, TX	40	60	60
CARIBOU, ME	60	80	100	HOUSTON, TX	40	60	80
PORTLAND, ME	60	80	100	SALT LAKE CITY, UT	40	60	80
BALTIMORE, MD	60	80	100	BURLINGTON, VT	60	80	100
AMHERST, MA	60	80	100	MT. WEATHER, VA	60	80	80
BOSTON, MA	60	80	100	NORFOLK, VA	40	60	80
LANSING, MI	60	80	100	RICHMOND, VA	60	80	80
SAULT STE. MARIE, MI	60	80	100	SEATTLE, WA	60	80	100
MINN.-ST. PAUL, MN	60	80	100	SPOKANE, WA	40	60	80
JACKSON, MS	40	60	80	CHARLESTON, WV	60	80	100
KANSAS CITY, MO	40	60	80	MADISON, WI	60	80	100
ST. LOUIS, MO	40	60	80	CASPER, WY	40	60	80
				CHEYENNE, WY	40	60	80

Table K. Collector Performance Adjustment Factors

Note: Annual Energy Savings figures in Table A are based on *averaged* collector performance data. To adjust for *specific* performance data: 1) In Table A, find the number of units of electricity, natural gas, or oil that a typical solar water heater would save you annually. 2) Consult the charted curve representing a specific collector's thermal performance when tested according to ASHRAE Standard 93-77. Note the collector efficiency value at the point where the performance curve intercepts the vertical axis (see example below). 3) In this table, find the adjustment factor corresponding to that efficiency value and to the size of your household and proposed collector area. 4) Multiply the figure selected from Table A by this adjustment factor to find the estimated number of kilowatt-hours, therms, or gallons that a specific collector would save annually.

Persons in Household	2 Persons		4 Persons			6 Persons		
Collector Size	40 Sq Ft	60 Sq Ft	40 Sq Ft	60 Sq Ft	80 Sq Ft	60 Sq Ft	80 Sq Ft	100 Sq Ft
Efficiency Value								
0.65	0.90	0.92	0.88	0.89	0.90	0.88	0.89	0.90
0.67	0.92	0.94	0.91	0.92	0.92	0.91	0.92	0.92
0.69	0.95	0.96	0.94	0.95	0.95	0.94	0.95	0.95
0.71	0.98	0.98	0.97	0.97	0.97	0.97	0.97	0.98
0.73	1.00	1.00	1.00	1.00	1.00	1.00	1.00	1.00
0.75	1.02	1.01	1.03	1.02	1.02	1.03	1.03	1.02
0.77	1.04	1.03	1.06	1.05	1.04	1.06	1.05	1.04
0.79	1.06	1.04	1.09	1.07	1.06	1.08	1.08	1.06
0.81	1.09	1.06	1.11	1.10	1.09	1.11	1.10	1.09
0.83	1.11	1.07	1.14	1.12	1.11	1.14	1.13	1.11
0.85	1.13	1.08	1.17	1.15	1.13	1.17	1.15	1.13

Example

The Wainwrights of Los Angeles, a family of six, are comparing the economics of two packaged solar water heating systems which differ in collector efficiency. Their dealer has supplied test data for both types of collectors, combining the two performance curves on a single chart (below). Following the procedure described in TABLE K, the Wainwrights find that:

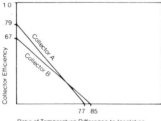

Ratio of Temperature Difference to Insolation
$$\left[\frac{T - Ta}{I} \right]$$

1) A typical solar water heater would save them 180 therms of natural gas, per TABLE A.

2) At the point where its performance curve intercepts the vertical axis, Collector A has an efficiency value of .79; Collector B has a value of .67.

3) For a six-person family planning 80 square feet of collector, an adjustment factor of 1.08 corresponds to Collector A's efficiency value; in the case of Collector B, the factor is .92.

4) Collector A could save about 195 therms annually (180 x 1.08 = 194.4), and Collector B about 165 therms (180 x .92 = 165.6).

REFERENCE

McPherson, Beth (of the Franklin Research Center), *Hot Water From the Sun*, U. S. Government Printing Office; Washington, D.C.: May, 1980, pp. 109–122.

APPENDIX VI
SOLAR DESIGN COMPUTER PROGRAM

```
10    DIM A[60],B[60],C[60],D[60],E[60],F[60],G[60]
20    DIM H[60],I[60],J[60],K[60],L[60],M[60],N[60]
30    DIM O$[30],P$[10]
40    REM
50    REM
60    REM         --------------------------------------------------
70    REM         ! PROGRAM WRITTEN BY JOE J. HARRELL, JR. !
80    REM         !           FOR THESIS PROJECT           !
90    REM         !           SUMMER QUARTER 1978          !
100   REM         --------------------------------------------------
110   REM
120   REM
130   REM
140   REM         ---------------------------------------------------------
150   REM         !       THIS PROGRAM IS BASED ON THE F-CHART METHOD     !
160   REM         ! ILLUSTRATED BY BECKMAN, KLEIN, AND DUFFIE IN THEIR    !
170   REM         ! BOOK; SOLAR HEATING DESIGN.                           !
180   REM         ---------------------------------------------------------
190   REM
200   REM
210   PRINT "          THIS PROGRAM WILL DETERMINE THE ANNUAL FRACTION OF"
220   PRINT "THERMAL ENERGY SUPPLIED TO A BUILDING BY SOLAR COLLECTORS"
230   PRINT "AND WILL THEN ECONOMICALLY DETERMINE, USING LIFE CYCLE"
240   PRINT "COSTING, THE MOST ECONOMICAL SIZE SOLAR COLLECTOR TO USE."
250   PRINT "(THIS PROGRAM WILL WORK FOR BOTH LIQUID AND AIR TYPE SOLAR"
260   PRINT "COLLECTORS AND ALSO FOR BOTH RESIDENTIAL AND COMMERCIAL"
270   PRINT "BUILDINGS.)"
280   PRINT
290   PRINT "          THE PROGRAM IS SET UP FOR THE PENSACOLA, FLORIDA AREA"
300   PRINT "(LATITUDE 30 DEGREES). FOR USE IN ANOTHER LOCATION BESIDES"
310   PRINT "PENSACOLA, NEW DATA WILL NEED TO BE ENTERED- STARTING IN LINE"
320   PRINT "2000. DATA FOR OTHER AREAS MAY BE FOUND IN THE BOOK WRITTEN BY"
330   PRINT "BECKMAN, KLEIN, AND DUFFIE ENTITLED; SOLAR HEATING DESIGN. (THIS"
340   PRINT "DATA MAY ALSO BE FOUND IN THIS PROGRAMMER'S THESIS.)"
350   PRINT LIN(3)
360   PRINT "What is the latitude (in degrees)";
370   INPUT A2
380   LET A1=A2+15
390   PRINT
400   PRINT "What is the ground reflectance (between 0.2 and 0.7)";
410   INPUT B1
420   PRINT
430   PRINT "What is the design space heating load (in Watts)";
440   INPUT C1
450   PRINT
460   PRINT "What is the design temperature difference (in degrees C.)";
470   INPUT D1
480   PRINT
490   PRINT "What is the water heating load (in liters/day)";
500   INPUT E1
510   PRINT
520   PRINT "What is the design hot water temperature (in degrees C.)";
530   INPUT F1
540   PRINT
550   PRINT "What is the average temperature of the water mains (in degrees C.)"
560   INPUT G1
570   PRINT
580   PRINT "What is the value for FRUL(F'R/FR)";
590   INPUT H1
600   PRINT
610   PRINT "What is the value for FR(Talpha)n(F'R/FR)";
620   INPUT I1
630   PRINT
640   PRINT "What is the storage size correction factor (X/Xo)";
650   INPUT J1
660   PRINT
670   PRINT "What is the load heat exchanger correction factor"
680   PRINT "(Y/Yo) [=1.0 for air systems]";
690   INPUT K1
700   PRINT
710   PRINT "What is the collector air flow rate correction factor"
720   PRINT "(X/Xo) [=1.0 for liquid systems]";
730   INPUT L1
```

```
740    PRINT
750    PRINT 'What are the three collector areas to compute (in square meters)';
760    INPUT M1,N1,O1
770    PRINT
780    PRINT 'Is this an air type or a liquid type solar collector (enter 1 or 2)
790    INPUT S1
800    PRINT LIN(2)
810    LET I3=E1*4190*(F1-G1)
820    LET P1=(1+COS(.01745*A1))/2
830    LET Q1=B1*(1-COS(.01745*A1))/2
840    LET R1=C1/D1
850    FOR I=1 TO 12
860    READ A[I],B[I],E[I]
870    NEXT I
880    FOR I=13 TO 24
890    READ A[I],B[I],C[I],F[I]
900    NEXT I
910    FOR I=25 TO 36
920    READ A[I],B[I]
930    NEXT I
940    FOR I=37 TO 48
950    READ A[I],G[I]
960    NEXT I
970    FOR P=1 TO 12
980    LET C[P]=1.39-4.03*B[P]+5.53*B[P]^2-3.11*B[P]^3
990    LET D[P]=1-C[P]
1000   LET F[P]=D[P]*E[P]
1010   LET G[P]=P1*C[P]
1020   LET H[P]=F[P]+G[P]+Q1
1030   LET I[P]=H[P]*A[P]
1040   NEXT P
1050   FOR R=1 TO 12
1060   LET D[R+12]=E[R]/H[R]
1070   LET E[R+12]=D[R]*B[R+12]*C[R+12]*D[R+12]
1080   LET G[R+12]=P1*C[R]/H[R]*.92*F[R+12]
1090   LET H[R+12]=Q1/H[R]*.92*F[R+12]
1100   LET I[R+12]=E[R+12]+G[R+12]+H[R+12]
1110   NEXT R
1120   FOR P=25 TO 36
1130   LET C[P]=86400.*R1*B[P]
1140   LET D[P]=I3*A[P]
1150   LET E[P]=C[P]+D[P]
1160   NEXT P
1170   FOR P=1 TO 12
1180   LET B[P+36]=100-G[P+36]
1190   LET C[P+36]=H1*B[P+36]*A[P+36]/E[P+24]
1200   LET D[P+36]=I[P+12]
1210   LET E[P+36]=A[P]*H[P]
1220   LET F[P+36]=I1*D[P+36]*E[P+36]*A[P+24]/E[P+24]
1230   NEXT P
1240   FOR U=49 TO 60
1250   LET A[U]=C[U-12]*J1*L1
1260   LET B[U]=F[U-12]*K1
1270   LET C[U]=A[U]*M1
1280   LET D[U]=B[U]*M1
1290   IF S1#1 THEN 1350
1300   E[U]=1.04*D[U]-.065*C[U]-.159*D[U]^2+.00187*C[U]^2-.0095*D[U]^3
1310   IF E[U]>1 THEN 1330
1320   GOTO 1390
1330   LET E[U]=1
1340   GOTO 1390
1350   E[U]=1.029*D[U]-.065*C[U]-.245*D[U]^2+.0018*C[U]^2+.0215*D[U]^3
1360   IF E[U]>1 THEN 1380
1370   GOTO 1390
1380   LET E[U]=1
1390   LET F[U]=E[U]*E[U-24]
1400   LET G[U]=A[U]*N1
1410   LET H[U]=B[U]*N1
1420   IF S1#1 THEN 1480
1430   I[U]=1.04*H[U]-.065*G[U]-.159*H[U]^2+.00187*G[U]^2-.0095*H[U]^3
1440   IF I[U]>1 THEN 1460
1450   GOTO 1520
1460   LET I[U]=1
1470   GOTO 1520
1480   I[U]=1.029*H[U]-.065*G[U]-.245*H[U]^2+.0018*G[U]^2+.0215*H[U]^3
1490   IF I[U]>1 THEN 1510
1500   GOTO 1520
1510   LET I[U]=1
```

```
1520    LET J[U]=I[U]*E[U-24]
1530    LET K[U]=A[U]*O1
1540    LET L[U]=B[U]*O1
1550    IF S1#1 THEN 1610
1560    M[U]=1.04*L[U]-.065*K[U]-.159*L[U]^2+.00187*K[U]^2-.0095*L[U]^3
1570    IF M[U]>1 THEN 1590
1580    GOTO 1650
1590    LET M[U]=1
1600    GOTO 1650
1610    M[U]=1.029*L[U]-.065*K[U]-.245*L[U]^2+.0018*K[U]^2+.0215*L[U]^3
1620    IF M[U]>1 THEN 1640
1630    GOTO 1650
1640    LET M[U]=1
1650    LET N[U]=M[U]*E[U-24]
1660    NEXT U
1670    T1=F[49]+F[50]+F[51]+F[52]+F[53]+F[54]+F[55]+F[56]+F[57]+F[58]+F[59]+F[60]
1680    U1=J[49]+J[50]+J[51]+J[52]+J[53]+J[54]+J[55]+J[56]+J[57]+J[58]+J[59]+J[60]
1690    V1=N[49]+N[50]+N[51]+N[52]+N[53]+N[54]+N[55]+N[56]+N[57]+N[58]+N[59]+N[60]
1700    Z1=E[25]+E[26]+E[27]+E[28]+E[29]+E[30]+E[31]+E[32]+E[33]+E[34]+E[35]+E[36]
1710    LET W1=(T1/Z1)*100
1720    LET O$="PENSACOLA, FLORIDA"
1730    LET X1=(U1/Z1)*100
1740    LET Y1=(V1/Z1)*100
1750    PRINT "------------------------------------------------------------------"
1760    PRINT
1770    PRINT "LOCATION       = ";O$
1780    PRINT "LATITUDE       =";A2;"DEGREES"
1790    PRINT "COLLECTOR TILT=";A1;"DEGREES"
1800    PRINT
1810    PRINT "THE TOTAL ENERGY LOAD (SPACE HEATING + DOMESTIC WATER"
1820    IMAGE "LOAD)= ",2D.2DE," JOULES/YEAR"
1830    PRINT  USING 1820;Z1
1840    PRINT
1850    PRINT "THE ANNUAL HEATING PERCENTAGE SUPPLIED BY SOLAR ENERGY FOR A"
1860    PRINT "COLLECTOR HAVING AN AREA:"
1870    IMAGE 24X,4D," SQUARE METERS IS ",3D," %"
1880    PRINT  USING 1870;M1,W1
1890    PRINT  USING 1870;N1,X1
1900    PRINT  USING 1870;O1,Y1
1910    PRINT
1920    PRINT "------------------------------------------------------------------"
1930    PRINT LIN(2)
1940    PRINT "Do you want an economic analysis to determine which of the three"
1950    PRINT "previously chosen collector sizes is the most economical to use"
1960    PRINT "(enter YES or NO)";
1970    INPUT P$
1980    IF P$="NO" THEN 4170
1990    REM ENTER H, Kt, Rb
2000    DATA 1.047E+07,.5,1.8,1.344E+07,.53,1.5,1.696E+07,.54,1.2
2010    DATA 2.131E+07,.59,.9,2.353E+07,.59,.75,2.378E+07,.58,.69
2020    DATA 2.248E+07,.56,.7,2.131E+07,.57,.8,1.8E+07,.55,1.18
2030    DATA 1.649E+07,.61,1.4,1.164E+07,.53,1.72,9.38E+06,.48,1.95
2040    REM ENTER THETAb, T/Tn B, ALPHA/ALPHAn THETAb, T/Tn 60
2050    DATA 36,.98,.99,.87,36,.98,.99,.87,39,.98,.99,.87
2060    DATA 44,.97,.97,.87,50,.96,.96,.87,53,.95,.96,.87
2070    DATA 51,.96,.96,.87,46,.97,.97,.87,41,.98,.98,.87
2080    DATA 37,.98,.99,.87,36,.98,.99,.87,36,.98,.99,.87
2090    REM ENTER DAYS PER MONTH, DEGREE DAYS
2100    DATA 31,237,28,179,31,117,30,21,31,0,30,0,31,0,31,0
2110    DATA 30,0,31,18,30,105,31,199
2120    REM ENTER SECONDS PER MONTH, TA
2130    DATA 2.68E+06,11,2.42E+06,12,2.68E+06,15,2.59E+06,20
2140    DATA 2.68E+06,23,2.59E+06,26,2.68E+06,27,2.68E+06,27
2150    DATA 2.59E+06,25,2.68E+06,21,2.59E+06,15,2.68E+06,12
2160    REM
2170    PRINT LIN(4)
2180    PRINT "           THIS NEXT PART OF THE PROGRAM WILL DETERMINE WHICH OF"
2190    PRINT "THE THREE COLLECTORS CHOSEN WILL BE THE MOST ECONOMICAL TO"
2200    PRINT "USE."
2210    PRINT LIN(2)
2220    PRINT "What is the annual mortgage interest rate (in decimal form)";
2230    INPUT A4
2240    PRINT
2250    PRINT "How many years is the mortgage for";
2260    INPUT B4
2270    PRINT
2280    PRINT "What is the down payment-as a fraction of the investment"
2290    PRINT "(in decimal form)";
```

```
2300  INPUT C4
2310  PRINT
2320  PRINT "What are the collector area dependent costs ($/square meter)";
2330  INPUT D4
2340  PRINT
2350  PRINT "What are the area independent costs";
2360  INPUT E4
2370  PRINT
2380  PRINT "What is the present cost of solar backup system fuel"
2390  PRINT "(in $/gigaJoule)";
2400  INPUT F4
2410  PRINT
2420  PRINT "What is the present cost of conventional system fuel"
2430  PRINT "(in $/gigaJoule)";
2440  INPUT G4
2450  PRINT
2460  PRINT "What is the efficiency of the solar backup furnace"
2470  PRINT "(in decimal form)";
2480  INPUT H4
2490  PRINT
2500  PRINT "What is the efficiency of the conventional system furnace"
2510  PRINT "(in decimal form)";
2520  INPUT I4
2530  PRINT
2540  PRINT "what is the property tax rate-as a fraction of investment"
2550  PRINT "(in decimal form)";
2560  INPUT J4
2570  PRINT
2580  PRINT "What is the effective income tax bracket---(State + Federal -"
2590  PRINT "State x Federal) (in decimal form)";
2600  INPUT K4
2610  PRINT
2620  PRINT "What are the extra insurance and maintenance costs-as a"
2630  PRINT "fraction of investment (in decimal form)";
2640  INPUT L4
2650  PRINT
2660  PRINT "What is the general inflation rate per year (in decimal form)";
2670  INPUT M4
2680  PRINT
2690  PRINT "What is the fuel inflation rate per year (in decimal form)";
2700  INPUT N4
2710  PRINT
2720  PRINT "What is the discount rate-after tax return on best "
2730  PRINT "alternative investment (in decimal form)";
2740  INPUT O4
2750  PRINT
2760  PRINT "What is the term of economic analysis (in years)";
2770  INPUT P4
2780  PRINT
2790  LET Q4=(Z1*G4/I4)/10^9
2800  PRINT "What is the depreciation lifetime (enter 0 for residential"
2810  PRINT "building)";
2820  INPUT R4
2830  PRINT
2840  PRINT "What is the salvage value-as a fraction of investment"
2850  PRINT "(enter 0 for residential building)";
2860  INPUT S4
2870  PRINT
2880  LET T4=1/(O4-N4)*(1-((1+N4)/(1+O4))^P4)
2890  LET U4=1/(O4-M4)*(1-((1+M4)/(1+O4))^P4)
2900  IF B4 >= P4 THEN 2930
2910  LET V4=1/(O4-A4)*(1-((1+A4)/(1+O4))^B4)
2920  GOTO 2940
2930  LET V4=1/(O4-A4)*(1-((1+A4)/(1+O4))^P4)
2940  IF P4 >= R4 THEN 2970
2950  LET W4=1/(O4-O)*(1-((1+O)/(1+O4))^P4)
2960  GOTO 2980
2970  LET W4=1/(O4-O)*(1-((1+O)/(1+O4))^R4)
2980  LET X4=1/(A4-O)*(1-((1+O)/(1+A4))^B4)
2990  IF B4 >= P4 THEN 3020
3000  LET Y4=1/(O4-O)*(1-((1+O)/(1+O4))^B4)
3010  GOTO 3030
3020  LET Y4=1/(O4-O)*(1-((1+O)/(1+O4))^P4)
3030  LET Z4=Y4/X4
3040  LET A5=Z4+V4*(A4-(1/X4))
3050  LET B5=C4+(1-C4)*(Z4-(A5*K4))
3060  LET C5=L4*U4
3070  LET D5=J4*U4*(1-K4)
```

```
3080    LET E5=S4/((1+O4)^P4)
3090    PRINT "Is this analysis for a residential or a commercial"
3100    PRINT "building (enter a 0 for residential and 1 for commercial)";
3110    INPUT S6
3120    IF S6=0 THEN 3150
3130    LET F5=K4*W4*(1-S4)/(R4)
3140    GOTO 3160
3150    LET F5=0
3160    LET G5=B5+C5+D5-E5
3170    LET H5=B5+C5*(1-K4)+D5-E5-F5
3180    LET I5=0
3190    LET J5=0
3200    LET K5=(D4*I5)+E4
3210    LET L5=Z1*(1-J5)*F4/H4/10^9
3220    LET M5=(Q4-L5)*(T4)
3230    LET N5=G5*K5
3240    LET O5=H5*K5
3250    LET P5=M5-N5
3260    LET Q5=M5*(1-K4)-O5
3270    LET R5=M1
3280    LET S5=W1/100
3290    LET T5=D4*(R5)+E4
3300    LET U5=Z1*(1-S5)*F4/H4/10^9
3310    LET V5=(Q4-U5)*T4
3320    LET W5=G5*T5
3330    LET X5=H5*T5
3340    LET Y5=V5-W5
3350    LET Z5=V5*(1-K4)-(X5)
3360    LET A6=N1
3370    LET B6=X1/100
3380    LET C6=D4*A6+E4
3390    LET D6=Z1*(1-B6)*F4/H4/10^9
3400    LET E6=(Q4-D6)*T4
3410    LET F6=G5*C6
3420    LET G6=H5*C6
3430    LET H6=E6-F6
3440    LET I6=E6*(1-K4)-G6
3450    LET J6=O1
3460    LET K6=Y1/100
3470    LET L6=D4*J6+E4
3480    LET M6=Z1*(1-K6)*F4/H4/10^9
3490    LET N6=(Q4-M6)*T4
3500    LET O6=G5*L6
3510    LET P6=H5*L6
3520    LET Q6=N6-O6
3530    LET R6=N6*(1-K4)-P6
3540    PRINT LIN(3)
3550    IF S6=0 THEN 3880
3560    PRINT "----------------------------------------------------------------------"
3570    PRINT
3580    IMAGE "COLLECTOR          ",6D.2D,6X,6D.2D,6X,6D.2D,6X,6D.2D
3590    PRINT   USING 3580;I5,R5,A6,J6
3600    PRINT "AREA"
3610    PRINT
3620    IMAGE "FRACTION           ",6D.2D,6X,6D.2D,6X,6D.2D,6X,6D.2D
3630    PRINT   USING 3620;J5,S5,B6,K6
3640    PRINT "BY SOLAR"
3650    PRINT
3660    IMAGE "INVESTMENT         ",6D.2D,6X,6D.2D,6X,6D.2D,6X,6D.2D
3670    PRINT   USING 3660;K5,T5,C6,L6
3680    PRINT "IN SOLAR"
3690    PRINT
3700    IMAGE "1ST YEAR           ",6D.2D,6X,6D.2D,6X,6D.2D,6X,6D.2D
3710    PRINT   USING 3700;L5,U5,D6,M6
3720    PRINT "FUEL EXPENSE"
3730    PRINT
3740    IMAGE "FUEL               ",6D.2D,6X,6D.2D,6X,6D.2D,6X,6D.2D
3750    PRINT   USING 3740;M5,V5,E6,N6
3760    PRINT "SAVINGS"
3770    PRINT
3780    IMAGE "EXPENSES           ",6D.2D,6X,6D.2D,6X,6D.2D,6X,6D.2D
3790    PRINT   USING 3780;O5,X5,G6,P6
3800    PRINT "(COMMERCIAL)"
3810    PRINT
3820    IMAGE "SAVINGS            ",6D.2D,6X,6D.2D,6X,6D.2D,6X,6D.2D
3830    PRINT   USING 3820;Q5,Z5,I6,R6
3840    PRINT "(COMMERCIAL)"
3850    PRINT
```

```
3860    PRINT "--------------------------------------------------------------
3870    GOTO 4120
3880    PRINT "--------------------------------------------------------------
3890    PRINT
3900    PRINT   USING 3580;I5,R5,A6,J6
3910    PRINT "AREA"
3920    PRINT
3930    PRINT   USING 3620;J5,S5,B6,K6
3940    PRINT "BY SOLAR"
3950    PRINT
3960    PRINT   USING 3660;K5,T5,C6,L6
3970    PRINT "IN SOLAR"
3980    PRINT
3990    PRINT   USING 3700;L5,U5,D6,M6
4000    PRINT "FUEL EXPENSE"
4010    PRINT
4020    PRINT   USING 3740;M5,V5,E6,N6
4030    PRINT "SAVINGS"
4040    PRINT
4050    PRINT   USING 3780;N5,W5,F6,O6
4060    PRINT "(RESIDENTIAL)"
4070    PRINT
4080    PRINT   USING 3820;P5,Y5,H6,Q6
4090    PRINT "(RESIDENTIAL)"
4100    PRINT
4110    PRINT "--------------------------------------------------------------
4120    PRINT LIN(2)
4130    PRINT "         BY NOW COMPARING THE SAVINGS ASSOCIATED WITH EACH"
4140    PRINT "COLLECTOR SIZE, WE CAN EASILY DETERMINE THE BEST COLLECTOR"
4150    PRINT "TO USE. WE CHOOSE THE COLLECTOR SIZE WHICH YIELDS THE "
4160    PRINT "GREATEST SAVINGS."
4170    END


RUN


        THIS PROGRAM WILL DETERMINE THE ANNUAL FRACTION OF
THERMAL ENERGY SUPPLIED TO A BUILDING BY SOLAR COLLECTORS
AND WILL THEN ECONOMICALLY DETERMINE, USING LIFE CYCLE
COSTING, THE MOST ECONOMICAL SIZE SOLAR COLLECTOR TO USE.
(THIS PROGRAM WILL WORK FOR BOTH LIQUID AND AIR TYPE SOLAR
COLLECTORS AND ALSO FOR BOTH RESIDENTIAL AND COMMERCIAL
BUILDINGS.)

        THE PROGRAM IS SET UP FOR THE PENSACOLA, FLORIDA AREA
(LATITUDE 30 DEGREES). FOR USE IN ANOTHER LOCATION BESIDES
PENSACOLA, NEW DATA WILL NEED TO BE ENTERED- STARTING IN LINE
2000. DATA FOR OTHER AREAS MAY BE FOUND IN THE BOOK WRITTEN BY
BECKMAN, KLEIN, AND DUFFIE ENTITLED; SOLAR HEATING DESIGN. (THIS
DATA MAY ALSO BE FOUND IN THIS PROGRAMMER'S THESIS.)



What is the latitude (in degrees)?30

What is the ground reflectance (between 0.2 and 0.7)?0.2

What is the design space heating load (in Watts)?24000

What is the design temperature difference (in degrees C.)?22

What is the water heating load (in liters/day)?400

What is the design hot water temperature (in degrees C.)?60

What is the average temperature of the water mains (in degrees C.)?11

What is the value for FRUL(F'R/FR)?3.64

What is the value for FR(Talpha)n(F'R/FR)?0.66

What is the storage size correction factor (X/Xo)?1.0

What is the load heat exchanger correction factor
(Y/Yo) [=1.0 for air systems]?1.0
```

What is the collector air flow rate correction factor
(X/Xo) [=1.0 for liquid systems]?1.0

What are the three collector areas to compute (in square meters)?25,50,100

Is this an air type or a liquid type solar collector (enter 1 or 2)?2

LOCATION   = PENSACOLA, FLORIDA
LATITUDE   = 30   DEGREES
COLLECTOR TILT= 45   DEGREES

THE TOTAL ENERGY LOAD (SPACE HEATING + DOMESTIC WATER
LOAD)= 11.25E+10 JOULES/YEAR

THE ANNUAL HEATING PERCENTAGE SUPPLIED BY SOLAR ENERGY FOR A
COLLECTOR HAVING AN AREA:
              25 SQUARE METERS IS   45 %
              50 SQUARE METERS IS   64 %
             100 SQUARE METERS IS   86 %

Do you want an economic analysis to determine which of the three
previously chosen collector sizes is the most economical to use
(enter YES or NO)?YES

          THIS NEXT PART OF THE PROGRAM WILL DETERMINE WHICH OF
THE THREE COLLECTORS CHOSEN WILL BE THE MOST ECONOMICAL TO
USE.

What is the annual mortgage interest rate (in decimal form)?0.09

How many years is the mortgage for?20

What is the down payment-as a fraction of the investment
(in decimal form)?0.1

What are the collector area dependent costs ($/square meter)?200

What are the area independent costs?1000

What is the present cost of solar backup system fuel
(in $/gigaJoule)?11.11

What is the present cost of conventional system fuel
(in $/gigaJoule)?11.11

What is the efficiency of the solar backup furnace
(in decimal form)?1.0

What is the efficiency of the conventional system furnace
(in decimal form)?1.0

what is the property tax rate-as a fraction of investment
(in decimal form)?0.0133

What is the effective income tax bracket---(State + Federal -
State x Federal) (in decimal form)?0.46

What are the extra insurance and maintenance costs-as a
fraction of investment (in decimal form)?0.01

What is the general inflation rate per year (in decimal form)?0.06

What is the fuel inflation rate per year (in decimal form)?0.10

What is the discount rate-after tax return on best
alternative investment (in decimal form)?0.08

What is the term of economic analysis (in years)?20

What is the depreciation lifetime (enter 0 for residential
building)?20

What is the salvage value-as a fraction of investment
(enter 0 for residential building)?0.10

Is this analysis for a residential or a commercial
building (enter a 0 for residential and 1 for commercial)?1

COLLECTOR AREA	0.00	25.00	50.00	100.00
FRACTION BY SOLAR	0.00	0.45	0.64	0.86
INVESTMENT IN SOLAR	1000.00	6000.00	11000.00	21000.00
1ST YEAR FUEL EXPENSE	1250.34	693.24	451.66	171.78
FUEL SAVINGS	0.00	12350.16	17705.53	23910.03
EXPENSES (COMMERCIAL)	758.05	4548.30	8338.55	15919.05
SAVINGS (COMMERCIAL)	-758.05	2120.79	1222.44	-3007.62

        BY NOW COMPARING THE SAVINGS ASSOCIATED WITH EACH
COLLECTOR SIZE, WE CAN EASILY DETERMINE THE BEST COLLECTOR
TO USE. WE CHOOSE THE COLLECTOR SIZE WHICH YIELDS THE
GREATEST SAVINGS.

DONE

RUN

        THIS PROGRAM WILL DETERMINE THE ANNUAL FRACTION OF
THERMAL ENERGY SUPPLIED TO A BUILDING BY SOLAR COLLECTORS
AND WILL THEN ECONOMICALLY DETERMINE, USING LIFE CYCLE
COSTING, THE MOST ECONOMICAL SIZE SOLAR COLLECTOR TO USE.
(THIS PROGRAM WILL WORK FOR BOTH LIQUID AND AIR TYPE SOLAR
COLLECTORS AND ALSO FOR BOTH RESIDENTIAL AND COMMERCIAL
BUILDINGS.)

        THE PROGRAM IS SET UP FOR THE PENSACOLA, FLORIDA AREA
(LATITUDE 30 DEGREES). FOR USE IN ANOTHER LOCATION BESIDES
PENSACOLA, NEW DATA WILL NEED TO BE ENTERED- STARTING IN LINE
2000. DATA FOR OTHER AREAS MAY BE FOUND IN THE BOOK WRITTEN BY
BECKMAN, KLEIN, AND DUFFIE ENTITLED; SOLAR HEATING DESIGN. (THIS
DATA MAY ALSO BE FOUND IN THIS PROGRAMMER'S THESIS.)

What is the latitude (in degrees)?30

What is the ground reflectance (between 0.2 and 0.7)?0.2

What is the design space heating load (in Watts)?24000

What is the design temperature difference (in degrees C.)?22

What is the water heating load (in liters/day)?400

What is the design hot water temperature (in degrees C.)?60

What is the average temperature of the water mains (in degrees C.)?11

What is the value for FRUL(F'R/FR)?3.64

What is the value for FR(Talpha)n(F'R/FR)?0.66

What is the storage size correction factor (X/Xo)?1.0

What is the load heat exchanger correction factor
(Y/Yo) [=1.0 for air systems]?1.0

What is the collector air flow rate correction factor
(X/Xo) [=1.0 for liquid systems]?1.0

What are the three collector areas to compute (in square meters)?25,50,100

Is this an air type or a liquid type solar collector (enter 1 or 2)?2

LOCATION   = PENSACOLA, FLORIDA
LATITUDE   = 30   DEGREES
COLLECTOR TILT= 45   DEGREES

THE TOTAL ENERGY LOAD (SPACE HEATING + DOMESTIC WATER
LOAD)= 11.25E+10 JOULES/YEAR

THE ANNUAL HEATING PERCENTAGE SUPPLIED BY SOLAR ENERGY FOR A
COLLECTOR HAVING AN AREA:
                          25 SQUARE METERS IS   45 %
                          50 SQUARE METERS IS   64 %
                         100 SQUARE METERS IS   86 %

Do you want an economic analysis to determine which of the three
previously chosen collector sizes is the most economical to use
(enter YES or NO)?YES

          THIS NEXT PART OF THE PROGRAM WILL DETERMINE WHICH OF
THE THREE COLLECTORS CHOSEN WILL BE THE MOST ECONOMICAL TO
USE.

What is the annual mortgage interest rate (in decimal form)?0.09

How many years is the mortgage for?20

What is the down payment-as a fraction of the investment
(in decimal form)?0.1

What are the collector area dependent costs ($/square meter)?200

What are the area independent costs?1000

What is the present cost of solar backup system fuel
(in $/gigaJoule)?11.11

What is the present cost of conventional system fuel
(in $/gigaJoule)?11.11

What is the efficiency of the solar backup furnace
(in decimal form)?1.0

What is the efficiency of the conventional system furnace
(in decimal form)?1.0

what is the property tax rate-as a fraction of investment
(in decimal form)?0.0133

What is the effective income tax bracket---(State + Federal -
State x Federal) (in decimal form)?0.46

What are the extra insurance and maintenance costs-as a
fraction of investment (in decimal form)?0.01

What is the general inflation rate per year (in decimal form)?0.06

What is the fuel inflation rate per year (in decimal form)?0.10

What is the discount rate-after tax return on best
alternative investment (in decimal form)?0.08

What is the term of economic analysis (in years)?20

What is the depreciation lifetime (enter 0 for residential
building)?0

What is the salvage value-as a fraction of investment
(enter 0 for residential building)?0

Is this analysis for a residential or a commercial
building (enter a 0 for residential and 1 for commercial)?0

COLLECTOR AREA	0.00	25.00	50.00	100.00
FRACTION BY SOLAR	0.00	0.45	0.64	0.86
INVESTMENT IN SOLAR	1000.00	6000.00	11000.00	21000.00
1ST YEAR FUEL EXPENSE	1250.34	693.24	451.66	171.78
FUEL SAVINGS	0.00	12350.16	17705.53	23910.03
EXPENSES (RESIDENTIAL)	1054.48	6326.88	11599.29	22144.10
SAVINGS (RESIDENTIAL)	-1054.48	6023.27	6106.25	1765.94

BY NOW COMPARING THE SAVINGS ASSOCIATED WITH EACH
COLLECTOR SIZE, WE CAN EASILY DETERMINE THE BEST COLLECTOR
TO USE. WE CHOOSE THE COLLECTOR SIZE WHICH YIELDS THE
GREATEST SAVINGS.

DONE

The following computer program written in BASIC is used to perform a life-cycle costing analysis for determining the cost benefits of energy-saving components used in buildings. For further discussion, see

Life-Cycle Costing: A Guide for Selecting Energy Conservation Projects for Public Buildings (1978) by the U. S. Department of Commerce, available from the U. S. Government Printing Office, Washington, D.C.

```
LCCEO

5 DIM E(8,6),E$(6),Q$(5),I(3,8),R(4,8),M(2,8)
1Ø DIM U$(6)
12 REM SINGLE PRESENT VALUE DISCOUNT FORMULA
15 DEF FNP(X,Z)=1/(1+X)↑Z
17 REM UNIFORM PRESENT VALUE DISCOUNT FORMULA INCLUDING ENERGY
   ESCALATION RATE
2Ø DEF FNU(X,Y,Z)=(1+Y)/(X-Y)*(1-((1+Y)/(1+X))↑Z)
25 PRINT "INPUT NAME OF AGENCY"
3Ø INPUT Q$(1)
35 PRINT "INPUT PROJECT NAME"
4Ø INPUT Q$(2)
45 PRINT "INPUT LOCATION OF AGENCY"
5Ø INPUT Q$(3)
55 PRINT "TYPE 1 IF PROJECT IS FOR AN EXISTING BUILDING, 2 IF"
6Ø PRINT "IT IS FOR A NEW BUILDING"
65 INPUT Q
7Ø PRINT "INPUT GROSS FLOOR AREA AFFECTED (IN SQUARE FEET)"
75 INPUT Q$(4)
8Ø PRINT "INPUT EXPECTED LIFE OF SYSTEM"
85 INPUT N1
9Ø PRINT "INPUT EXPECTED LIFE OF BUILDING"
95 INPUT Q$(5)
1ØØ PRINT "INPUT STUDY PERIOD"
1Ø5 INPUT N2
11Ø IF Q<>1 THEN 135
115 LET B$="SAVINGS"
12Ø LET C$="SAVED"
125 LET D$="EXISTING"
13Ø GO TO 15Ø
135 LET B$="COSTS"
14Ø LET C$="CONSUMED"
145 LET D$="NEW"
15Ø PRINT "INPUT DISCOUNT RATE"
155 INPUT D
16Ø LET V=Ø
165 LET R1=Ø
17Ø LET M1=Ø
175 PRINT
18Ø PRINT "ENERGY "B$
185 PRINT
19Ø PRINT "INPUT NUMBER OF ENERGY SOURCES USED"
195 INPUT N4
2ØØ MAT E=ZER(8,N4)
2Ø5 FOR J=1 TO N4
21Ø PRINT "INPUT ENERGY TYPE "J" AND UNIT (E.G. ELECTRICITY
    (HEATING),KWH)"
215 PRINT "NOTE: SEPARATE TYPE AND UNIT BY A COMMA"
22Ø INPUT E$(J),U$(J)
225 PRINT "INPUT ANNUAL AMOUNT OF "E$(J)" "C$" BY THIS SYSTEM"
23Ø INPUT E(1,J)
235 PRINT "INPUT LOCAL PRICE/"U$(J)
24Ø INPUT E(2,J)
245 PRINT "INPUT LONG TERM ENERGY ESCALATION RATE"
25Ø INPUT E(3,J)
255 PRINT "TYPE 1 IF YOU HAVE A SEPARATE SHORT TERM ENERGY"
26Ø PRINT "ESCALATION RATE, Ø IF NOT"
265 INPUT S
27Ø IF S=Ø THEN 295
275 PRINT "INPUT SHORT TERM ENERGY ESCALATION RATE"
28Ø INPUT E(4,J)
285 PRINT "INPUT EXPECTED NUMBER OF YEARS THAT THIS RATE CAN BE
    USED"
29Ø INPUT E(5,J)
```

```
295 PRINT
300 NEXT J
305 REM COMPUTE PRESENT VALUE ENERGY SAVINGS/COSTS
310 FOR J=1 TO N4
315 LET E(6,J)=E(1,J)*E(2,J)
320 IF S=0 THEN 415
325 IF D<>E(4,J) THEN 340
330 LET E(7,J)=E(6,J)*E(5,J)
335 GO TO 345
340 LET E(7,J)=E(6,J)*FNU(D,E(4,J),E(5,J))
345 LET E(8,J)=E(2,J)
350 FOR K=1 TO E(5,J)
355 LET E(8,J)=E(8,J)+E(8,J)*E(4,J)
360 NEXT K
365 IF D<>E(3,J) THEN 380
370 LET S(J)=E(1,J)*E(8,J)*(N2-E(5,J))
375 GO TO 390
380 LET S(J)=E(1,J)*E(8,J)*FNU(D,E(3,J),N2-E(5,J))
385 LET S(J)=S(J)*FNP(D,E(5,J))
390 LET S(J)=S(J)+E(7,J)
395 GO TO 425
400 IF D<>E(3,J) THEN 415
405 LET S(J)=E(6,J)*N2
410 GO TO 420
415 LET S(J)=E(6,J)*FNU(D,E(3,J),N2)
420 REM  S(J)=PRESENT VALUE OF ENERGY SAVINGS
425 LET V=V+S(J)
430 NEXT J
435 PRINT
440 PRINT "INVESTMENT COST"
445 PRINT
450 PRINT "INPUT INITIAL INVESTMENT COST (BASE YEAR)"
455 INPUT I
460 PRINT "INPUT NUMBER OF ADDITIONAL INVESTMENTS OR REPLACEMENTS"
465 PRINT "OVER THE STUDY PERIOD"
470 INPUT N5
475 FOR K=1 TO N5
480 PRINT "INPUT YEAR THAT COST OF ADDITIONAL INVESTMENT/REPLACEMENT
    # "K
485 PRINT "IS TO BE INCURRED"
490 INPUT I(1,K)
495 PRINT "INPUT COST IN TODAY'S DOLLARS"
500 INPUT I(2,K)
505 PRINT "INPUT SCRAP VALUE OF EQUIPMENT BEING REPLACED (IF NONE OR"
510 PRINT "NOT APPLICABLE, TYPE 0)"
515 INPUT I(3,K)
520 LET I=I+I(2,K)*FNP(D,I(1,K))-I(3,K)*FNP(D,I(1,K))
525 NEXT K
530 PRINT "INPUT RESIDUAL VALUE OF INVESTMENT AT END OF STUDY PERIOD"
535 INPUT I9
540 LET I9=I9*FNP(D,N2)
545 LET I=I-I9
550 REM I IS PRESENT VALUE OF INVESTMENT COST
555 PRINT
560 PRINT "NON-ENERGY COSTS"
565 PRINT
570 LET R$=D$
575 IF Q=1 THEN 585
580 GO TO 590
585 PRINT "CALCULATIONS FOR "R$" SYSTEM"
590 PRINT
595 PRINT "INPUT TOTAL YEARLY AMOUNT OF ANNUAL RECURRING COSTS
    IN TODAY'S $"
600 INPUT R
605 LET R=R*FNU(D,0,N2)
610 PRINT "INPUT NUMBER OF OTHER MAINTENANCE OR REPAIR COSTS
    THAT OCCUR"
615 PRINT "ON A PERIODIC BASIS"
620 INPUT N6
625 FOR J=1 TO N6
630 PRINT "INPUT AMOUNT OF RECURRING COST "J
635 INPUT R(1,J)
640 PRINT "INPUT FIRST YEAR THAT THIS COST IS INCURRED AND
    THE PERIOD"
645 PRINT "OVER WHICH IT WILL RECUR (EXAMPLE: FOR A RECURRING
    COST"
650 PRINT "BEGINNING IN YEAR 5 AND OCCURING EVERY 5 YEARS
    TYPE: 5,5)"
655 INPUT R(2,J),R(3,J)
660 LET R(4,J)=R(1,J)*FNP(D,R(2,J))
665 FOR K=1 TO INT((N2-R(2,J))/R(3,J))
670 IF R(2,J)+K*R(3,J)=N2 THEN 685
675 REM DON'T ADD RECURRING COSTS THAT OCCUR LAST YEAR OF
    STUDY PERIOD
```

```
680 LET R(4,J)=R(4,J)+R(1,J)*FNP(D,R(2,J)+K*R(3,J))
685 NEXT K
690 LET R1=R1+R(4,J)
695 NEXT J
700 LET R1=R1+R
705 REM NON-RECURRING COSTS
710 PRINT
715 PRINT "INPUT TOTAL NUMBER OF NON-RECURRING MAINTENANCE OR
    REPAIR"
720 PRINT "COSTS THAT YOU HAVE TO MAKE"
725 INPUT N7
730 IF N7=0 THEN 770
735 FOR J=1 TO N7
740 PRINT "INPUT COST IN TODAY'S DOLLARS AND YEAR OF OCCURRENCE
    FOR"
745 PRINT "NON-RECURRING COST "J" (EXAMPLE:  FOR A COST OF
    $500 IN"
750 PRINT "YEAR 15, TYPE: 500,15)"
755 INPUT M(1,J),M(2,J)
760 LET M1=M(1,J)*FNP(D,M(2,J))
765 NEXT J
770 IF R$="NEW" THEN 805
775 LET R$="NEW"
780 LET P=R1+M1
785 LET R1=0
790 LET M1=0
795 PRINT
800 GO TO 585
805 IF Q=2 THEN 825
810 LET W=R1+M1
815 LET X=P-W
820 GO TO 840
825 LET X=R1+M1
830 PRINT
835 PRINT
840 PRINT "                   PROJECT SUMMARY REPORT"
845 PRINT
850 PRINT
855 PRINT "    NAME OF AGENCY                    "Q$(1)
860 PRINT
865 PRINT "    PROJECT                          "Q$(2)
870 PRINT
875 PRINT "    LOCATION OF AGENCY               "Q$(3)
880 PRINT
885 PRINT "    GROSS FLOOR AREA AFFECTED        "Q$(4)" SQUARE FEET"
890 PRINT
895 PRINT "    EXPECTED LIFE OF SYSTEM          "N1" YEARS"
900 PRINT
905 PRINT "    EXPECTED LIFE OF BUILDING        "Q$(5)" YEARS"
910 PRINT
915 PRINT "    STUDY PERIOD                     "N2" YEARS"
920 PRINT
925 PRINT "    DISCOUNT RATE                    "D
930 PRINT
935 PRINT "              *         *         *"
940 PRINT
945 PRINT "                              (IN THOUSANDS $)"
950 PRINT
955 PRINT "    TOTAL PRESENT VALUE OF ENERGY "B$
960 PRINT
965 FOR K=1 TO N4
970 LET S(K)=S(K)/1000
975 PRINT "              "E$(K)"      "S(K)
980 NEXT K
985 PRINT
990 LET V=V/1000
995 PRINT "              TOTAL                    "V
1000 PRINT
1005 LET I=I/1000
1010 PRINT "        TOTAL PRESENT VALUE OF INVESTMENT COST "I
1015 PRINT
1020 IF Q=2 THEN 1150
1025 LET P=P/1000
1030 PRINT "        PRESENT VALUE OF NON-ENERGY COSTS FOR  "
1035 PRINT "        EXISTING SYSTEM                        "P
1040 PRINT
1045 LET W=W/1000
1050 PRINT "        PRESENT VALUE OF NON-ENERY COSTS FOR   "
1055 PRINT "        PROPOSED ALTERNATIVE                   "W
1060 PRINT
1065 LET X=X/1000
1070 PRINT "        PRESENT VALUE OF CHANGE IN NON-ENERGY  "
1075 PRINT "        COSTS                                  "X
1080 PRINT
```

```
1085 LET Z1=V+X
1090 PRINT "          TOTAL PRESENT VALUE SAVINGS            "Z1
1095 PRINT
1100 PRINT
1105 PRINT "                       LIFE-CYCLE COST MEASURES"
1110 PRINT
1115 LET Z2=Z1/I
1120 PRINT "          SAVINGS TO INVESTMENT RATIO            "Z2
1125 PRINT
1130 LET Z3=Z1-I
1135 PRINT "          NET SAVINGS IN PRESENT VALUE           "
1140 PRINT "          DOLLARS (IN THOUSANDS $)               "Z3
1145 GO TO 1190
1150 LET X=X/1000
1155 PRINT "          PRESENT VALUE OF NON-ENERGY COSTS      "X
1160 PRINT
1165 PRINT
1170 PRINT
1175 LET Z8=V+X+I
1180 PRINT "          TOTAL LIFE-CYCLE COSTS IN PRESENT      "
1185 PRINT "          VALUE DOLLARS (IN THOUSANDS $)         "Z8
1190 END
```

The following example output is from the previous

life-cycle costing computer listing.

```
INPUT NAME OF AGENCY
 ? NATIONAL ADMINISTRATION
INPUT PROJECT NAME
 ? ECS
INPUT LOCATION OF AGENCY
 ? WASHINGTON DC
TYPE 1 IF PROJECT IS FOR AN EXISTING BUILDING, 2 IF IT IS FOR A
NEW BUILDING
 ? 1
INPUT GROSS FLOOR AREA AFFECTED (IN SQUARE FEET)
 ? 2.3 MILLION
INPUT EXPECTED LIFE OF SYSTEM
 ? 40
INPUT EXPECTED LIFE OF BUILDING
 ? 30
INPUT STUDY PERIOD
 ? 25
INPUT DISCOUNT RATE
 ? .10

ENERGY SAVINGS

INPUT NUMBER OF ENERGY SOURCES USED
 ? 3
INPUT ENERGY TYPE 1 AND UNIT (E.G. ELECTRICITY (HEATING),KWH)
NOTE:   SEPARATE TYPE AND UNIT BY A COMMA
 ? ELECTRICITY (HEATING),KWH
INPUT ANNUAL AMOUNT OF ELECTRICITY (HEATING) SAVED BY THIS SYSTEM
 ? 4980000
INPUT LOCAL PRICE/KWH
 ? .033
INPUT LONG TERM ENERGY ESCALATION RATE
 ? .05
TYPE 1 IF YOU HAVE A SEPARATE SHORT TERM ENERGY ESCALATION
RATE, 0 IF NOT
 ? 1
INPUT SHORT TERM ENERGY ESCALATION RATE
 ? .10
INPUT EXPECTED NUMBER OF YEARS THAT THIS RATE CAN BE USED
 ? 2

INPUT ENERGY TYPE 2 AND UNIT (E.G. ELECTRICITY (HEATING),KWH)
NOTE:   SEPARATE TYPE AND UNIT BY A COMMA
 ? ELECTRICITY (COOLING),KWH
INPUT ANNUAL AMOUNT OF ELECTRICITY (COOLING) SAVED BY THIS SYSTEM
 ? 2930000
INPUT LOCAL PRICE/KWH
 ? .038
INPUT LONG TERM ENERGY ESCALATION RATE
 ? .05
TYPE 1 IF YOU HAVE A SEPARATE SHORT TERM ENERGY ESCALATION
RATE, 0 IF NOT
 ? 1
```

```
INPUT SHORT TERM ENERGY ESCALATION RATE
 ? .1Ø
INPUT EXPECTED NUMBER OF YEARS THAT THIS RATE CAN BE USED
 ? 2

INPUT ENERGY TYPE 3 AND UNIT (E.G. ELECTRICITY (HEATING),KWH)
NOTE:  SEPARATE TYPE AND UNIT BY A COMMA
 ? NATURAL GAS,THERM
INPUT ANNUAL AMOUNT OF NATURAL GAS SAVED BY THIS SYSTEM
 ? 468ØØØØ
INPUT LOCAL PRICE/THERM
 ? .225
INPUT LONG TERM ENERGY ESCALATION RATE
 ? .Ø7
TYPE 1 IF YOU HAVE A SEPARATE SHORT TERM ENERGY ESCALATION
RATE, Ø IF NOT
 ? 1
INPUT SHORT TERM ENERGY ESCALATION RATE
 ? .Ø6
INPUT EXPECTED NUMBER OF YEARS THAT THIS RATE CAN BE USED
 ? 2

INVESTMENT COST

INPUT INITIAL INVESTMENT COST (BASE YEAR)
 ? 725ØØØ
INPUT NUMBER OF ADDITIONAL INVESTMENTS OR REPLACEMENTS
OVER THE STUDY PERIOD
 ? 2
INPUT YEAR THAT COST OF ADDITIONAL INVESTMENT/REPLACEMENT # 1
IS TO BE INCURRED
 ? 1
INPUT COST IN TODAY'S DOLLARS
 ? 725ØØØ
INPUT SCRAP VALUE OF EQUIPMENT BEING REPLACED (IF NONE OR
NOT APPLICABLE, TYPE Ø)
 ? Ø
INPUT YEAR THAT COST OF ADDITIONAL INVESTMENT/REPLACEMENT # 2
IS TO BE INCURRED
 ? 1Ø
INPUT COST IN TODAY'S DOLLARS
 ? 5ØØØØ
INPUT SCRAP VALUE OF EQUIPMENT BEING REPLACED (IF NONE OR
NOT APPLICABLE, TYPE Ø)
 ? Ø
INPUT RESIDUAL VALUE OF INVESTMENT AT END OF STUDY PERIOD
 ? 2ØØØØØ

NON-ENERGY COSTS

CALCULATIONS FOR EXISTING SYSTEM

INPUT TOTAL YEARLY AMOUNT OF ANNUAL RECURRING COSTS IN TODAY'S $
 ?12ØØØØ
INPUT NUMBER OF OTHER MAINTENANCE OR REPAIR COSTS THAT OCCUR
ON A PERIODIC BASIS
 ? 1
INPUT AMOUNT OF RECURRING COST  1
 ?14ØØØØ
INPUT FIRST YEAR THAT THIS COST IS INCURRED AND THE PERIOD
OVER WHICH IT WILL RECUR (EXAMPLE: FOR A RECURRING COST
BEGINNING IN YEAR 5 AND OCCURRING EVERY 5 YEARS TYPE: 5,5)
 ? 5,5

INPUT TOTAL NUMBER OF NON-RECURRING MAINTENANCE OR REPAIR
COSTS THAT YOU HAVE TO MAKE
 ? Ø

CALCULATIONS FOR NEW SYSTEM

INPUT TOTAL YEARLY AMOUNT OF ANNUAL RECURRING COSTS IN TODAY'S $
 ? 18ØØØØ
INPUT NUMBER OF OTHER MAINTENANCE OR REPAIR COSTS THAT OCCUR
ON A PERIODIC BASIS
 ? 1
INPUT AMOUNT OF RECURRING COST 1
 ? 28ØØØØ
INPUT FIRST YEAR THAT THIS COST IS INCURRED AND THE PERIOD
OVER WHICH IT WILL RECUR (EXAMPLE: FOR A RECURRING COST
BEGINNING IN YEAR 5 AND OCCURRING EVERY 5 YEARS TYPE: 5,5)
 ? 5,5
INPUT TOTAL NUMBER OF NON-RECURRING MAINTENANCE OR REPAIR
COSTS THAT YOU HAVE TO MAKE
 ? Ø
```

PROJECT SUMMARY REPORT

NAME OF AGENCY	NATIONAL ADMINISTRATION
PROJECT	ECS
LOCATION OF AGENCY	WASHINGTON DC
GROSS FLOOR AREA AFFECTED	2.3 MILLION SQUARE FEET
EXPECTED LIFE OF SYSTEM	4Ø YEARS
EXPECTED LIFE OF BUILDING	3Ø YEARS
STUDY PERIOD	25 YEARS
DISCOUNT RATE	.1

```
            *            *            *
```

(IN THOUSANDS $)

TOTAL PRESENT VALUE OF ENERGY SAVINGS

```
        ELECTRICITY (HEATING)    2596.
        ELECTRICITY (COOLING)    1758.79
        NATURAL GAS      18404.3

        TOTAL                            22759.1
```

TOTAL PRESENT VALUE OF INVESTMENT COST   1384.91

PRESENT VALUE OF NON-ENERGY COSTS FOR
EXISTING SYSTEM   1284.47

PRESENT VALUE OF NON-ENERGY COSTS FOR
PROPOSED ALTERNATIVE   2Ø24.33

PRESENT VALUE OF CHANGE IN NON-ENERGY
COSTS   -739.852

TOTAL PRESENT VALUE SAVINGS   22019.3

LIFE-CYCLE COST MEASURES

SAVINGS TO INVESTMENT RATIO   15.8994

NET SAVINGS IN PRESENT VALUE
DOLLARS (IN THOUSANDS $)   20634.4

RUNNING TIME: 6.6 SECS I/O TIME: 37.3 SECS

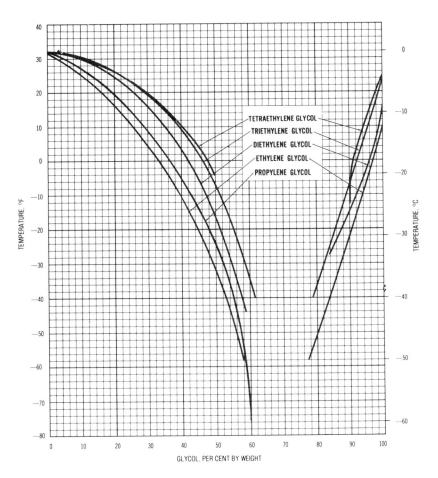

Figure 1. Freezing Points of Aqueous Glycol Solutions.

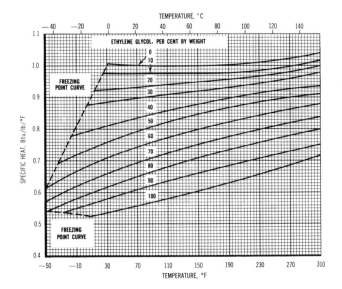

Figure 2. Specific Heats of Aqueous Ethylene Glycol Solutions.

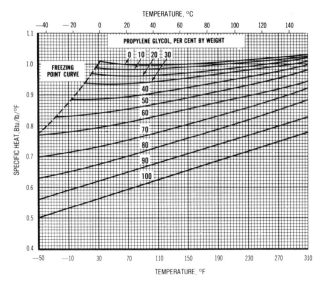

Figure 3. Specific Heats of Aqueous Propylene Glycol Solutions.

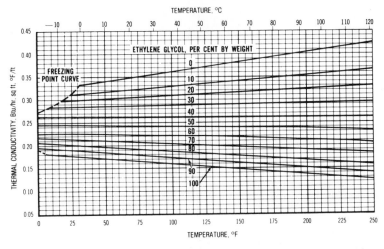

Figure 4. Thermal Conductivities of Aqueous Ethylene Glycol Solutions.

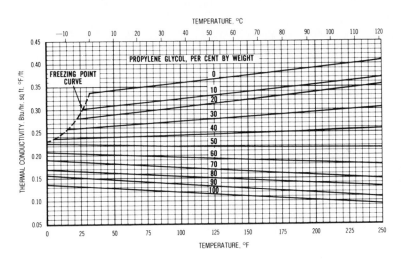

Figure 5. Thermal Conductivities of Aqueous Propylene Glycol Solutions.

REFERENCE

Union Carbide Corporation—Chemicals and Plastics Division, New York, N.Y., *Glycols*; 1978; pp. 17, 39, 41–43.

APPENDIX IX
CONSTRUCTION DETAILS FOR A WOODEN ROCK BIN

The following illustrations are presented as a general guide for the construction of a wooden rock bin. This design is intended for above-ground installation in the basement of a building. Since the filled rock bin is very heavy, the basement floor slab upon which the bin rests must be designed to handle the large weight.

In constructing the rock bin, Cole[2] recommends the following:

1. All reinforcing ribs shall be dense, Number 1 Douglas Fir or Southern Yellow Pine.
2. Exterior grade plywood with bonding glue capable of continuously withstanding a temperature of 140°F for a 20-year period shall be used.
3. All vertical and all horizontal structural support members shall be nailed and glued to the exterior grade plywood. All bonding glues shall meet the temperature specification in Item 2.
4. Timber Engineering Company (TECO)[7] connectors or equivalent shall be used to join all vertical ribs to the bottom connecting members.
5. All timber in contact with concrete shall be coated with asphalt paint.
6. To provide a finished appearance, the plywood storage container's outer cover shall be attached to the reinforcing ribs with screws. This exterior surface should be painted.
7. All joints, all timber–concrete contacts, and all cracks shall be caulked with a silicone caulk. A 3/8-inch caulk bead is recommended.

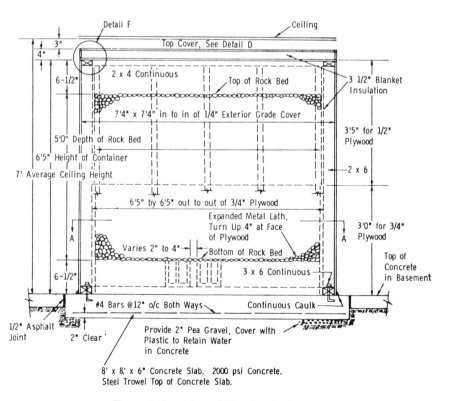

Figure 1. Elevation of Wooden Rock Bin.

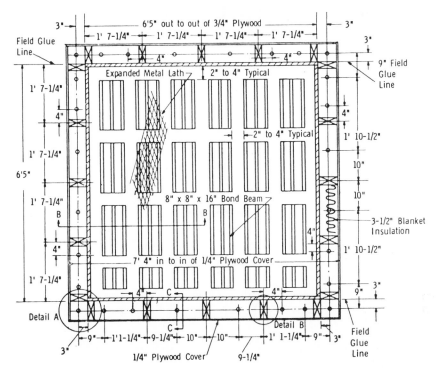

Figure 2. Section A-A of Wooden Rock Bin.

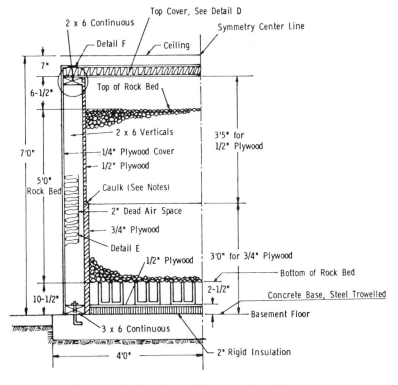

Figure 3. Section B-B of Wooden Rock Bin.

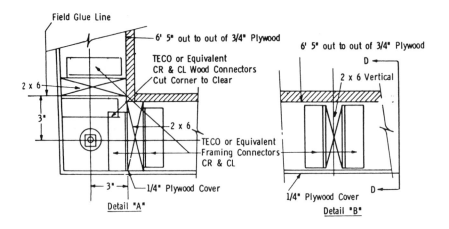

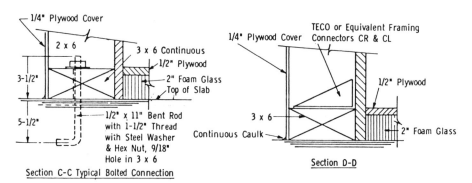

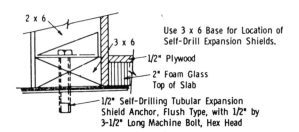

Figure 4. Corner and Base Details for Wooden Rock Bin.

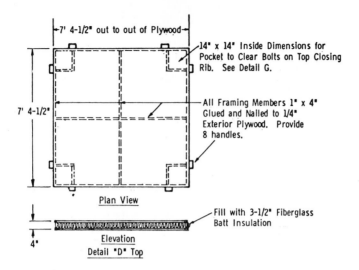

←7' 4-1/2" out to out of Plywood→

14" x 14" Inside Dimensions for
Pocket to Clear Bolts on Top Closing
Rib. See Detail G.

7' 4-1/2"

All Framing Members 1" x 4"
Glued and Nailed to 1/4"
Exterior Plywood. Provide
8 handles.

Plan View

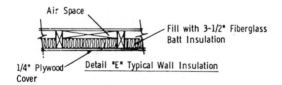

Fill with 3-1/2" Fiberglass
Batt Insulation

Elevation

4"

Detail "D" Top

Air Space

Fill with 3-1/2" Fiberglass
Batt Insulation

1/4" Plywood
Cover

Detail "E" Typical Wall Insulation

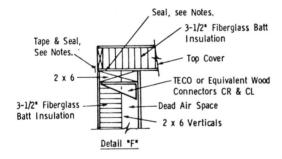

Seal, see Notes.

3-1/2" Fiberglass Batt
Insulation

Tape & Seal,
See Notes.

Top Cover

2 x 6

TECO or Equivalent Wood
Connectors CR & CL

3-1/2" Fiberglass
Batt Insulation

Dead Air Space

2 x 6 Verticals

Detail "F"

Figure 5. Top Details of Wooden Rock Bin.

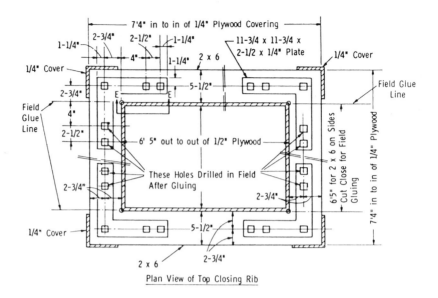

Plan View of Top Closing Rib

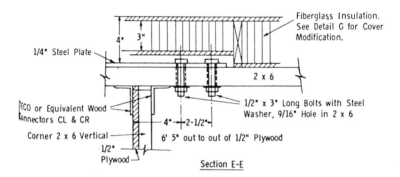

Section E-E

Figure 6. Top Closing Rib of Wooden Rock Bin.

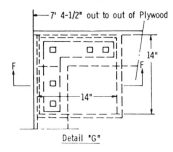

Detail "G"

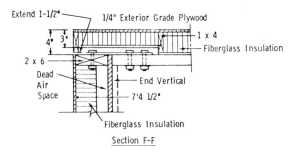

Section F-F

Figure 7. Closure at End of Cover over Bolted Joint.

REFERENCES AND SUGGESTED READING

1. Baumeister, Theodore, Avallone, Eugene A., and Baumeister, Theodore III, *Mark's Standard Handbook For Mechanical Engineers,* New York: McGraw-Hill Book Company, 1978.
2. Cole, Roger L., Nield, Kenneth J., Rohde, Raymond R., and Wolosewicz, Ronald, M., *Design and Installation Manual for Thermal Energy Storage,* Argonne, Illinois: Argonne National Laboratory, February, 1979, pp. F1 to F10.
3. Merritt, Frederick S., ed., *Standard Handbook for Civil Engineers,* New York: McGraw-Hill Book Company, 1976.
4. Parker, Harry, *Simplified Design of Structural Timber,* New York: John Wiley & Sons, Inc., 1963.
5. Parker, Harry, *Simplified Engineering for Architects and Builders,* New York: John Wiley & Sons, Inc., 1975.
6. Southern Pine Inspection Bureau, *Grading Rules,* Southern Pine Inspection Bureau, Pensacola, Florida, 1977.
7. Timber Engineering Company (TECO), 5530 Wisconsin Avenue, Washington, D.C. 20015.

APPENDIX X
CONCRETE STORAGE TANKS

The following illustrations show sample drawings of concrete storage tanks.

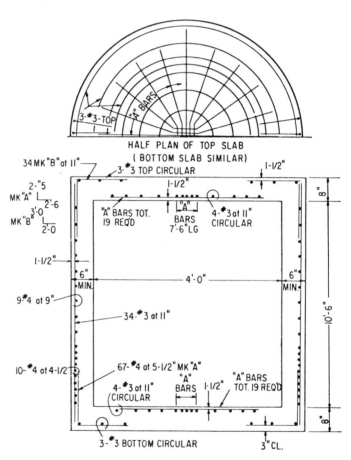

VERTICAL SECTION THRU ₵ (5000 GAL.) CIRCULAR CONCRETE TANK

Figure 1. Sample Drawing of a Circular Concrete Tank.

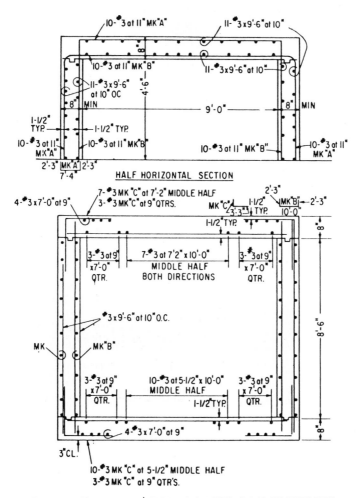

VERTICAL SECTION THRU ₵ (5000 GAL.) RECTANGULAR CONCRETE TANK

Figure 2. Sample Drawing of a Rectangular Concrete Tank.

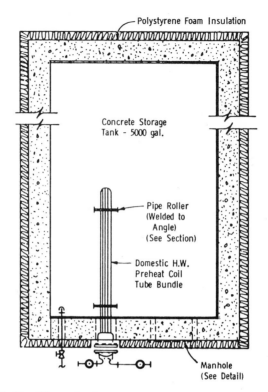

Figure 3. Plan View of a Rectangular Concrete Tank (not to scale).

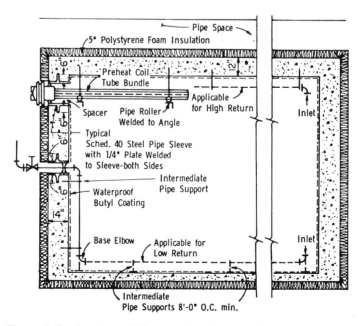

Figure 4. Section through Rectangular Concrete Tank (not to scale).

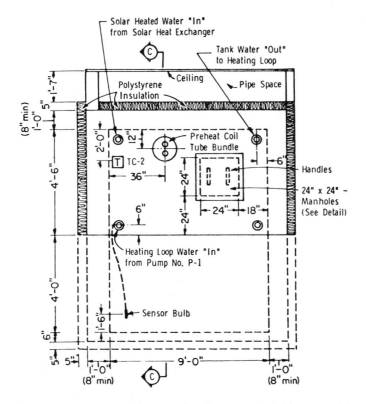

Figure 5. Front View of Rectangular Concrete Tank (not to scale).

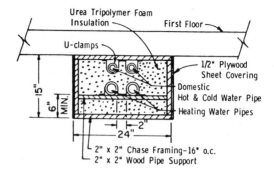

DETAIL OF PIPE CONDUIT

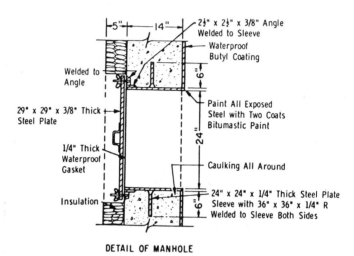

DETAIL OF MANHOLE

Figure 6. Details of Rectangular Concrete Tank.

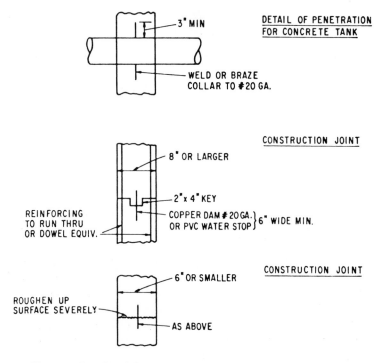

Figure 7. Details of Penetrations and Joints for Concrete Tanks.

REFERENCES AND SUGGESTED READING

1. American Concrete Institute, Detroit, Michigan, *Building Code Requirements for Reinforced Concrete (ACI 318-77)*, 1977.
2. Cole, Roger, L., Nield, Kenneth, J., Rohde, Raymond R., and Wolosewicz, Ronald M., *Design and Installation Manual for Thermal Energy Storage*, Argonne, Illinois: Argonne National Laboratory, February, 1979, pp. H1 to H43.
3. Everard, Noel J., and Tanner, John L. III, *Reinforced Concrete Design*, (Schaum's Outline Series) New York: McGraw-Hill Book Company, 1966.
4. Hurd, M. K., *Formwork for Concrete*, American Concrete Institute, Detroit, Michigan, 1973.
5. Parker, Harry, *Simplified Design of Reinforced Concrete*, New York: John Wiley & Sons., Inc., 1976.
6. Parker, Harry, *Simplified Engineering for Architects and Builders*, New York: John Wiley & Sons, Inc., 1975.
7. Portland Cement Association, Skokie, Illinois, *Design and Control of Concrete Mixtures*, 1968.
8. The Subcommittee on Placing Reinforcing Bars, *Placing Reinforcing Bars*, Concrete Reinforcing Steel Institute, Chicago, Illinois, 1976.

APPENDIX XI
SOLAR FUSION PROCESSES

The two fusion producing processes working on the sun are the proton–proton cycle and the carbon cycle.

The proton–proton cycle is shown in the following steps:

$$^1_1H + {}^1_1H \rightarrow {}^2_1D + e^+ + \nu + \text{energy (1.44 MeV)}$$ STEP 1

$$^2_1D + {}^1_1H \rightarrow {}^3_2He + \gamma + \text{energy (5.49 MeV)}$$ STEP 2

$$^3_2He + {}^3_2He \rightarrow {}^4_2He + 2{}^1_1H + \text{energy (12.85 MeV)}$$ STEP 3

The superscripts refer to the mass number which is the total of protons and neutrons in the nucleus, while the subscripts represent only the number of protons (i.e., atomic number).

The first step of the proton–proton cycle involves combining each proton from the two hydrogen atoms to form a deuterium nucleus (called a deuteron), a positron (e^+) and a neutrino (ν). In this process energy is also produced equaling 1.44 million electron volts (MeV).

The positron (e^+) will then interact with an electron (e^-) and turn into gamma rays. The neutrino (ν), being a neutral particle, will leave the sun without interacting with other particles. The energy produced in this reaction will generate heat. In *Step 2* the deuteron $({}^2_1D)$ left from the first step will combine with a free hydrogen proton to form a He-3 nucleus. In the process both a gamma ray (γ) is produced and energy of 5.49 MeV is liberated. In order for the third step of the fusion reaction to take place, Steps 1 and 2 must take place again. This is reasonable when one looks at Step 3 and sees that two He-3 nuclei $({}^3_2He)$ are needed for Step 3 to take place. In Step 3 the two helium-3 nuclei combine to form a helium-4 nucleus $({}^4_2\text{helium})$ and two protons $(2{}^1_1H)$, with 12.85 MeV of energy being released.

When the equations for the three steps are added, remembering that there are a total of five equations to be added because Steps 1 and 2 occur twice for every time Step 3 occurs, one obtains an overall fusion reaction equation of

$$4{}^1_1H + 2e^- \rightarrow {}^4_2He + 2\nu + \text{energy (26.73 MeV)}$$

The carbon cycle, first proposed simultaneously and independently by Hans Bethe and Carl Von Weizsacker, relies on a carbon atom acting as the catalyst for the fusion reaction.

The carbon cycle has six basic steps:

$$^{12}_6C + {}^1_1H \rightarrow {}^{13}_7N + \gamma = \text{energy (1.95 MeV)}$$ STEP 1

$$^{13}_7N \rightarrow {}^{13}_6C + e^+ + \nu + \text{energy (2.22 MeV)}$$ STEP 2

$$^{13}_6C + {}^1_1H \rightarrow {}^{14}_7N + \gamma + \text{energy (7.54 MeV)}$$ STEP 3

$$^{14}_7N + {}^1_1H \rightarrow {}^{15}_8O + \gamma + \text{energy (7.35 MeV)}$$ STEP 4

$$^{15}_8O \rightarrow {}^{15}_7N + e^+ + \nu + \text{energy (2.71 MeV)}$$ STEP 5

$$^{15}_7N + {}^1_1H \rightarrow {}^{12}_6C + {}^4_2He + \gamma + \text{energy (4.96 MeV)}$$ STEP 6

The symbols C and N represent the elements carbon and nitrogen, respectively, while the symbol $^{15}_8O$ represents a positive beta emitter.

When the six steps for the carbon cycle are added, one obtains:

$$4^1_1H + 2e^- \rightarrow {}^4_2He + 2\nu + \text{energy (26.71 MeV)}$$

The carbon cycle thus produces the same end results as the proton–proton cycle—namely four hydrogen atoms are converted into one helium atom along with the release of 26.71 MeV of energy (4.28×0.10^{-5} ergs, 4.06×0.10^{-15} Btu, or 1.02×10^{-12} calories).

Energy must be released in the fusion process of converting four hydrogen atoms into helium, in accordance with Einstein's famous equation $E = MC^2$, to satisfy the law of conservation of matter and energy. In the conversion of four (hydrogen) protons to helium, some matter is converted into energy.

Mass of four protons $= (4)(1.673 \times 0.10^{-24}g) - 6.692 \times 10^{-24}$ grams(g)

Mass of 1 helium-4 atom $= 6.646 \times 10^{-24}g$

$(6.692 \times 10^{-24}g) - (6.646 \times 10^{-24}g) = $ Loss of 4.600×10^{-26} grams

Since Energy (in ergs) equals Mass (in grams times) times C^2 (the speed of light in centimeters per second), then:

$$E = (4.600 \times 10^{-26}g)(3.00 \times 10^{10}\text{cm/sec})^2 = 4.14 \times 10^{-5}\text{ergs}$$

This energy figure of 4.14×10^{-5} ergs is essentially the same as the amount of energy previously computed both for the proton–proton and the carbon cycle equations (4.28×10^{-5} ergs). For all intents and purposes of this discussion, the two energy yield figures can be considered the same (i.e., approximately 4×10^{-5} ergs).*

Since the energy produced in the conversion of four hydrogen protons to a single helium-4 atom is known, the amount of fusion energy contained in a single gram of hydrogen atom can be determined.

$$\frac{4 \text{ protons}}{6.692 \times 10^{-24}g} = \frac{x \text{ protons}}{1 \text{ g}}$$

This ratio shows that 5.977×10^{23} hydrogen atoms (protons) are in one gram.

By knowing there is a 4 to 1 relationship between the number of hydrogen atoms needed to produce one helium-4 atom, it can then be shown that 5.977×10^{23} hydrogen atoms (protons) are required to produce 1.494×10^{23} atoms of helium-4.

$(5.977 \times 10^{23}$ protons$)(1.673 \times 10^{-24}g) = 1.000$ g

$(1.494 \times 10^{23}$ helium-4 atoms$)(6.646 \times 10^{-24}g) = 0.9929$ g

1.000 g $- 0.9929$ g $= 0.0071$ grams of matter converted to energy

Since $E = MC^2$, then

$$E = (0.0071 \text{ g})(3.000 \times 10^{10} \text{ cm/sec})^2 = 6.390 \times 10^{18} \text{ ergs}$$

Therefore, when one gram of protons (hydrogen atoms) are converted in the fusion process to helium-4, 6.390×10^{18} ergs, 6.060×10^8 Btu's, 1.527×10^{11} calories, or 177.5 megawatt-hours of energy are produced. Assuming a home used 1500 kilowatt-hours of electricity per month, the household could run almost ten years on one gram of the fusionable material! Remember that one gram is only 0.0022 pounds or 35/100th of an ounce.

One final fusion calculation that should be made is the determination of how much energy the sun produces in one second. One source predicts that each year the sun converts 2.0×10^{19} kilograms of hydrogen into helium (see Ref. 4). This means that 6.3×10^{14} grams of hydrogen are converted into helium each second. In the previous calculations it was shown that 6.390×10^{18} ergs of energy are produced from one gram of hydrogen atoms, that there are 5.977×10^{23} hydrogen atoms (protons) in one gram, and that 1.494×10^{23} atoms of helium-4 can be produced from the gram of hydrogen.

$$\left(6.390 \times 10^{18} \frac{\text{ergs}}{\text{gram}}\right) \left(6.3 \times 10^{14} \frac{\text{grams}}{\text{sec}}\right) = 4.0 \times 10^{33} \frac{\text{ergs}}{\text{sec}}$$

Therefore, the sun produces energy at the rate of 4.0×10^{33} ergs/sec, 5.2×10^{23} horsepower, 9.3×10^{25} cal/sec or 3.8×10^{23} Btu/sec.

REFERENCES

1. Baker, Robert H., and Frederick, Laurence W., *Astronomy*, New York: Van Nostrand Reinhold Co., 1971.
2. Gamon, George, *A Star Called the Sun*, New York: Viking Press, Inc., 1964.
3. Gibson, E. G., *The Quiet Sun*, Washington, D.C.: U. S. Government Printing Office, 1972.
4. Glasstone, Samuel, *Sourcebook on Atomic Energy*, New York: Van Nostrand Reinhold, 1967.
5. Henderson, S. T., *Daylight and Its Spectrum*, New York: American Elsevier Publishing Company, Inc., 1970.
6. Hoyle, Fred, *Astronomy and Cosmology—A Modern Course*, W. H. Freeman and Company, 1975.
7. Inglis, S., *Planets, Stars, and Galaxies*, New York: John Wiley and Sons, Inc., 1972.
8. Masterton, W. L., and Slowinski, E. J., *Chemical Principles*, Philadelphia: W. B. Saunders Company, 1973.
9. Menzel, Donald H., *Our Sun*, Cambridge, Massachusetts: Harvard University Press, 1959.
10. Motz, Lloyd, and Duveen, Aneta, *Essentials of Astronomy*, New York: Columbia University Press, 1977.
11. Weast, R., *Handbook of Chemistry and Physics*, Cleveland, Ohio: CRC Press, 1975.
12. Zirin, Harold, *The Solar Atmosphere*, Waltham, Massachusetts: Blaisdell Publishing Company, 1966.

*The slight discrepancy in the two energy yield figures result from the fact that in the two fusion cycles previously discussed, the actual total yields for each fusion cycle will be slightly less than the yields stated (i.e., 26.7 MeV) due to the effects of the neutrinos produced in the fusion reaction which act to lower the actual energy yields (See Ref. 3).

APPENDIX XII
STATE SOLAR LEGISLATION*

State	Law	Description	Contact
I. TAX INCENTIVES			
Alaska	CH. 94 Laws of 1977	Alaska allows a 10% residential fuel conservation credit of up to $200 per individual or married couple for money spent on the following: 1) insulation; 2) insulating windows; 3) labor related to items 1 and/or 2; 4) alternate energy systems which are not dependent on fossil fuel, including solar, wind, tidal, and geothermal. Expires 12/31/82.	State Dept. of Revenue Income Tax Division Pouch SA State Office Building Juneau, AK 99811 (907) 465-2326
Arizona	CH. 93 Laws of 1975, as amended by CH. 129 Laws of 1976, as amended by CH. 112 Laws of 1978	When solar energy devices are added to business or investment property, the taxpayer may elect to depreciate, to amortize over 36 months, or to claim an income tax credit. This credit, during 1979, may equal 30% of the cost of the solar energy system. It is reduced 5% each year until the law expires, 12/31/84. The maximum credit is $1000. Credit is available to individual taxpayers at the same rate.	State Dept. of Revenue Box 29002 Phoenix, AZ 85038
	CH. 165 Laws of 1974	Solar energy devices are exempt from property taxes through 12/31/84.	
	CH. 42 Laws of 1977	Solar energy devices are exempt from Transaction Privilege and Use Taxes.	
	CH. 112 Laws of 1978	A 25% credit is allowed for residential insulation and ventilation devices, such as insulating doors and windows. Maximum credit is $100. Credit expires 12/31/84.	
Arkansas	Act 535 Laws of 1977	Individuals may deduct the entire cost of solar heating and cooling equipment from taxable income. The cost of other energy-conserving devices, such as insulation, storm windows, and motor-driven ventilation, may also be deducted.	State Dept. of Revenue Income Tax Section 7th & Wolfe Streets Little Rock, AR 72201 (501) 371-2193
California	CH. 168 Laws of 1976 CH. 1082 Laws of 1977 CH. 1154 Laws of 1978	California provides personal income tax credit of 55% of the cost of a solar energy system, up to a maximum of $3000. If a system is installed in other than a single-family dwelling and the cost exceeds $6000, the credit equals 25% of the cost, or $3000, whichever is greater. In both single-family dwellings and other buildings, the cost of energy-conserving devices installed in conjunction with the solar energy system may also be included in the total cost used to calculate the tax credit. If a federal credit is claimed, the state credit is reduced by the amount of the federal credit. The same provisions apply to corporate taxpayers. The system must meet the criteria of the California Energy Commission. Eligible expenses for credit include attorney's fees, compensation, and recording fees associated with obtaining a solar easement. Credit expires 12/31/80.	Franchise Tax Board Attn: Correspondence Sacramento, CA 95807 (916) 355-0370

*U. S. Department of Housing and Urban Development, *State Solar Legislation,* Rockville, Maryland: National Solar Heating and Cooling Information Center, January, 1979.

State	Law	Description	Contact
I. TAX INCENTIVES			
Colorado	CH. 344 Laws of 1975	Solar heating and cooling devices are assessed for property tax at 5% of their value.	Local Assessor or Board of Assessors
	CH. 512 Laws of 1977	Individual taxpayers may deduct the cost of a solar energy device from taxable income. Corporate taxpayers may use the deduction in lieu of depreciation. Individuals and corporations may also deduct the cost of wind and geothermal energy systems.	State Dept. of Revenue Income Tax Division 1375 Sherman Street Denver, CO 80261 (303) 839-3781
Connecticut	PA 77-457	Solar collectors are exempt from sales tax through 10/1/82.	State Tax Department Audit Division 92 Farmington Avenue Hartford, CT 06115 (203) 566-2501
	PA 76-409 and 77-490	Municipalities are authorized to exempt solar heating and cooling or solar electrical systems from property tax for 15 years. Windmills and waterwheels are exempt. The amount of exemption equals the difference between an assessment including the solar energy system and an assessment including a conventional energy system. Installation must take place before 10/1/91. The taxpayer must apply to the local Board of Assessors. The system must meet the standards of the Commission of Planning and Energy Policy.	Local Assessor or Board of Assessors
Delaware	CH. 512 Laws of 1978	This law provides an income tax credit of $200 for solar energy devices designed to produce domestic hot water. Systems must meet HUD Intermediate Minimum Property Standards Supplement for Solar Heating and Domestic Hot Water Systems. Systems must be warranteed according to criteria set out in the law.	Division of Revenue State Office Building 820 French Street Wilmington, DE 19801 (302) 571-3360
Georgia	Act 1030 Laws of 1976, as amended by Act 1309 Laws of 1978	Real estate owners may claim a refund of sales tax paid for the purchase of solar equipment. Expires 7/1/86.	State Dept. of Revenue Sales Tax Division 309 Trinity-Washington B Atlanta, GA 30334 (404) 656-4065
	Georgia Constitution, Article VII, Section 1, Paragraph IV.	Any county or municipality may exempt solar heating and cooling equipment and machinery used to manufacture solar equipment from property taxes. Expires 7/1/86.	Local City Council or County Board of Supervisors
Hawaii	Act 189 Laws of 1976	A 10% income tax credit is provided to individuals and corporations who purchase solar energy devices that are placed in service by 12/31/81. The law also provides property tax exemptions for solar energy systems through 12/31/81. This exemption also applies to any non-nuclear and non-fossil fuel system and to any improvement that increases the efficiency of systems which use fossil fuel.	State Tax Department P. O. Box 259 Honolulu, HI 96809 (808) 548-3270
Idaho	CH. 212 Laws of 1976	This law allows an income tax deduction for a solar heating/cooling or solar electrical system installed in the taxpayer's residence. The deduction equals 40% of the cost in the first year and 20% of the cost in	State Tax Commission 5257 Fairview Boise, ID 83722 (208) 384-3290

State	Law	Description	Contact

I. TAX INCENTIVES

State	Law	Description	Contact
		each of the next 3 years; the maximum deduction in any year is $5000. This deduction also applies to systems fueled by wind, geo-thermal energy, wood, or wood products. Built-in fireplaces qualify if they have control doors, regulated draft, and heat exchangers that deliver heated air to substan-tial portions of the residence.	
Illinois	PA 79-943 (1975), as amended by PA 80-430 (1977)	A property owner who installs a solar or wind energy system may claim an alterante valuation for property taxes. The property is assessed twice: with the solar or wind energy system and also as though it were equipped with a conventional system. The lesser of the two assessments is used to compute the tax due. Owners must file a claim with the local Board of Assessors. Systems must meet the standards of the Division of Energy of the Department of Business and Economic Development.	Local Assessor or Board of Assessors
Indiana	P.L. 15 (1974) as amended by P.L. 68 (1977)	The law permits the property owner who installs a solar heating and cooling system to have property assessment reduced by the difference between the assessment of the property with the system and the assessment of the property without the system. The owner must apply to the county auditor.	Local Assessor or Board of Assessors
Iowa	Section 441.21, code of 1979	Installation of solar energy system will not increase the assessed, actual, or taxable values of property for 1979-1985.	Local Assessor or Board of Assessors
Kansas	CH. 434 Laws of 1976, as amended by CH. 346 Laws of 1977	The individual taxpayer is allowed an income tax credit of 25% of the cost of a resi-dential solar energy system to a maximum of $1000. A solar energy installation on a busi-ness or investment property receives a credit equal to 25% of the system cost, $3000, or that year's tax bill, whichever is the least amount. The cost of an installation on a business or investment property can be amor-tized over 60 months. Wind energy systems are also covered by this law. Credit expires 7/1/83.	State Dept. of Revenue P. O. Box 692 Topeka, KS 66601 (913) 296-3909
	CH. 345 Laws of 1977, as amended by CH. 419 Laws of 1978	If a solar energy system supplies 70% of the energy for heating and cooling, the property owner may be reimbursed for 35% of his property tax for up to 5 consecu-tive years. Applies through 1985. Claims must be filed with the Department of Revenue.	
Louisiana	Act 591, 1978	Solar energy equipment installed in owner-occupied residential buildings or in swimming pools are exempt from property tax.	Local Parish Tax Assessor
Maine	CH. 542 Laws of 1977	Solar space or water heating systems are exempt from property tax for 5 years after installa-tion. Eligible taxpayers must apply to the local Board of Assessors. Purchasers of solar energy systems may also receive a sales tax rebate from the Office of Energy Resources.	Local Assessor or Board of Assessors (For Property Tax) Office of Energy Resources 55 Capitol Street Augusta, ME 04330 (207) 289-2196 (For Sales Tax)

State	Law	Description	Contact
I. TAX INCENTIVES			
Maryland	CH. 509 Laws of 1975, as amended by CH. 509 Laws of 1978	A solar energy unit will be assessed at no more than a conventional system needed to serve the building	Local Assessor or Board of Assessors
	CH. 740 Laws of 1976	Baltimore City and any other city or county may offer property tax credits for the use of solar energy systems in any type of building. Credit may be applied over a 3-year period.	Local City or County Dept. of Revenue
Massachusetts	CH. 734 Laws of 1975, as amended by CH. 388 Laws of 1978	Solar energy systems are exempt from property tax for 20 years from the date of installation.	Local Assessor or Board of Assessors
	CH. 989, 1977	Sales of equipment for residential solar energy systems, wind power systems, or heat pumps are exempt from sales tax.	State Dept. of Corporations & Taxation 100 Cambridge Street Boston, MA 02204 (617) 727-4601
	CH. 487, 1977	Corporations may deduct the cost of a solar or wind energy system from income. The system will also be exempt from tangible property tax.	State Dept. of Corporations & Taxation 100 Cambridge Street Boston, MA 02204 (617) 727-4201
Michigan	PA 132, 1976	Proceeds from sales of solar, wind, or water energy conversion devices used for heating, cooling, or electrical generation in new or existing residential or commercial buildings are excluded from business activities tax. Expires 1/1/85.	State Dept. of Tresury State Tax Commission State Capitol Building Lansing, MI 48922 (517) 373-2910
	PA 133, 1976	Tangible property used for solar, wind, or water energy devices is excluded from excise tax if it is used to heat, cool, or electrify a new or existing commercial or residential building. Expires 1/1/85.	State Dept. of Tresury State Tax Commission State Capitol Building Lansing, MI 48922 (517) 373-2910
	PA 135, 1976	This law exempts solar, wind, or water energy conversion devices from real and personal property tax. An application must be filed with local tax assessor, who will submit it to the state tax commission for certification. Authority to exempt expires 7/1/85, but exemptions made by that time stay in force.	Local Government Services Treasury Building Lansing, MI 48922 (517) 373-3232
	PA 605, 1979	Income tax credit may be claimed for a residential solar, wind, or water energy device that is used for heating, cooling, or electricity. This includes devices designed to use the difference between water temperatures in a body of water. Energy conservation measures installed in connection with such devices are also eligible; these include insulation and water-flow reduction devices. Swimming pool heaters are eligible only if 25% or more of their heating capacity is used for residential purposes. The credit may be carried from year to year until it is expended. The law instructs the Department of Commerce to establish system eligibility standards within 180 days of the law's passage. To be eligible, expenditures must be made by 12/31/83. The rate of credit changes annually. For	State Dept. of Treasury State Tax Commission State Capitol Building Lansing, MI 48922 (517) 373-2910

State	Law	Description	Contact

I. TAX INCENTIVES

		1979, the rate for single-family dwellings is 25% of the first $2000 spent, plus 15% of the next $8000 spent. In 1979, the rate for other buildings is 25% of the first $2000, plus 15% of next $13,000.	
Minnesota	CH. 786 Laws of 1978	The market value of solar, wind, or agriculturally derived methane gas systems used for heating, cooling, or electricity in a building or structure is excluded from property tax. The installations must be done prior to 1/1/84.	Local Assessor or Board of Assessors
Montana	CH. 548 Laws of 1975 as amended by CH. 574 Laws of 1977	Energy systems using non-fossil fuel energy (such as solar, wind, decomposition of organic wastes) installed in a taxpayer's dwelling before 12/31/82 are eligble for a tax credit of 10% for the first $1000 and 5% of the next $3000. If a federal tax credit is also claimed, the state credit is reduced to 5% of the first $1000 and 2-1/2% of the next $3000.	State Dept. of Revenue Income Tax Section Mitchell Building Helena, MT 59601 (406) 449-2837
	CH 576 Laws of 1977	This law provides individual or corporate deductions for energy conservation improvements, including storm windows and insulation. It applies to all types of buildings at the following rates: Residential buildings: 100% of 1st $1000 50% of 2nd $1000 20% of 3rd $1000 10% of 4th $1000 Non-residential buildings: 100% of 1st $2000 50% of 2nd $2000 20% of 3rd $2000 10% of 4th $2000	State Dept. of Revenue Income Tax Section Mitchell Building Helena, MT 59601 (406) 449-2837
Nevada	CH. 345 Laws of 1977	This law establishes a property tax allowance on solar, wind geothermal, water-powered, or solid waste energy systems in residential buildings. The property tax allowance equals the difference in tax on the property with the energy system and the tax on the property without the energy system. The allowance may not exceed the tax accrued or $2000, whichever is less. Claims are to be filed with the county assessor.	Local County Assessor
New Hampshire	CH 391 Laws of 1975, as amended by CH 5202 Laws of 1977	Cities and towns are enabled to grant a property tax exemption to property owners with solar heating, cooling, or hot water systems and will decide the amount of the exemption and the manner of determination. An application for the exemption must be filed with the local assessor.	Local Assessor or Board of Assesors
New Jersey	CH. 256 Laws of 1977	Solar heating and cooling systems, including sea thermal gradients and wind-powered systems, are exempt from property tax. Systems must be certified under the State Uniform Construction Act on forms designated by the Division of Taxation. Systems must meet standards established by the State Energy Office. Expires 12/31/82.	Local Assessor or Board of Assessors

State	Law	Description	Contact
I. TAX INCENTIVES			
	CH. 465 Laws of 1977	Solar energy devices designed to provide heating, cooling, electrical or mechanical power are exempt from sales tax. These systems must meet the standards established by the state Department of Energy.	State Div. of Taxation Tax Counselors P. O. Box 999 Trenton, NJ 08646 (609) 292-6400
New Mexico	CH. 12 Laws of 1975, as amended by CH. 170 Laws of 1978	This law provides for an income tax credit of 25% of the cost of a solar energy system or a minimum of $1000. It is available for solar energy systems which heat and/or cool the taxpayer's residence and for swimming pool heating systems. The credit is not available if federal credit is claimed. The criteria of the Solar Heating and Cooling Act of 1974 (42 USC 5506) must be met. Credit in excess of the taxes due will be refunded.	State Dept. of Taxation & Revenue Income Tax Division P. O. Box 630 Santa Fe, NM 87503 (505) 827-3221
	CH. 114 Laws of 1977	Individuals may claim an income tax credit for a solar energy system used in an irrigation pumping system. The system design must be approved by the energy resources board prior to installation, and it must result in a 75% reduction in the use of fossil fuel. This law is not applicable if federal credit were claimed or if credit were claimed for this equipment under other provisions of the state law. Credit in excess of the taxes due will be refunded.	
New York	CH. 322 Laws of 1977	This law provides for property tax reductions for owners of solar or wind energy systems. The reduction of assessment is equal to the difference between assessment of the property with the energy system and assessment of the property without the system. The system must conform to guidelines of the state energy office and must be installed before 7/1/88. The exemption is good for 15 years after it is granted.	Local Assessor or Board of Assessors
North Carolina	CH. 792 Laws of 1977	This law provides for a corporate and individual income tax credit of 25% of the cost of a solar heating, cooling, or hot water system. There is a maximum credit of $1000 per unit or building. Although this credit may be taken only once, the amount of credit may be spread over 3 years. The system may be in any type of building, and it must meet the criteria of the Solar Heating and Cooling Demonstration Act of 1974.	State Dept. of Revenue Income Tax Division P. O. Box 25000 Raleigh, NC 27640 (919) 733-3991
	CH. 965 Laws of 1977	Buildings with solar heating or cooling systems shall be assessed as though they had a conventional system. Expires 12/1/85.	Local Assessor or Board of Assessors
North Dakota	CH. 537 Laws of 1977	This law provides for an income tax credit for solar or wind energy devices. The credit is 5% per year for 2 years. The system must provide heating, cooling, mechanical, or electrical power.	State Tax Commission Income Tax Division Capitol Building Bismarck, ND 58505 (707) 224-3450

State	Law	Description	Contact
		I. TAX INCENTIVES	
	CH. 508 Laws of 1975	Solar heating or cooling systems in any building are exempt from property tax for 5 years after installation.	Local Assessor or Board of Assessors
Oklahoma	CH. 209 Laws of 1977	Individuals may claim an income tax credit for solar energy devices used to heat, cool, or furnish electrical or mechanical power at principal residence. The credit is equal to 25% of the cost of the system or a maximum of $2000. Although this credit may be taken only once, the amount of credit may be spread over 3 years. Expires on 1/1/88.	State Tax Commission Income Tax Division 2501 Lincoln Blvd. Oklahoma City, OK 73194 (405) 521-3125
Oregon	CH. 196 Laws of 1977	This law provides for income tax credit for the installation of solar, wind, or geothermal energy systems in a dwelling for purposes of heating, cooling, hot water, or electrical power. The credit equals 25% of the cost of the system. The maximum credit is $100. Although this credit may be taken only once, the amount of credit may be spread over 3 years. The systems must meet the criteria of the Department of Energy. The credit expires on 1/1/85.	State Dept. of Revenue State Office Building Salem, OR 97310 (503) 378-3366
		This law also provides for a property tax exemption for property equipped with solar energy systems. The exemption equals the value of the property with the solar energy system minus the value of the property without the system. This credit expires 1/1/98.	Local Assessor or Board of Assessors
Rhode Island	CH. 202 Laws of 1977	Solar heating or cooling systems in residential or non-residential buildings shall be assessed at no more than the value of a conventional system necessary to serve the building. Law expires 4/1/97.	Local Assessor or Board of Assessors
South Dakota	CH. 74 Laws of 1978	This law provides property tax assessment credit for renewable resource energy systems (sun, wind, goethermal and biomass). For residential property, the amount of the credit equals the assessed value of the property with the system, minus the assessed value of the property without the system, but not less than the actual installation cost of the system. The credit for systems in commercial buildings is equal to 50% of the cost of installation. For residential buildings, full credit is given for 5 years. For the next 3 years, the credit is 75%, 50%, and 25% of the full credit. For commercial buildings full credit for 3 years, and for the next 3 years credit is 75%, 50% and 25% of the full credit. Taxpayers must apply to the county auditor.	Local County Assessor
Tennessee	CH. 837 Laws of 1978	Solar or wind energy systems for heating, cooling, or electrical power shall be exempt from property taxation. Law expires 1/1/88.	Local Assessor or Board of Assessors

State	Law	Description	Contact

I. TAX INCENTIVES

State	Law	Description	Contact
Texas	Article VII, Sec. 2(a) of Texas Constitution 1978	The legislature is allowed to exempt solar- or wind-powered energy devices from property tax.	
	CH. 719 Laws of 1975	Solar energy systems used for heating, cooling, or electrical power are exempt from sales tax. Corporations may deduct from taxable capital the amortized cost of a solar energy device over a period of 60 months or more.	Comptroller of Public Accounts Capitol Station Drawer SS Austin, TX 78775 (512) 475-2206
	CH. 584 Laws of 1977	This law provides a franchise tax exemption for corporations exclusively engaged in manufacturing, selling, or installing solar energy devices for heating, cooling, or electrical power.	Comptroller of Public Accounts Capitol Station Drawer SS Austin, TX 78775 (512) 475-2206
Vermont	Act 226 Laws of 1976	Towns may enact a property tax exemption for alternate energy systems. Systems exempted are grist mills, windmills, solar energy systems, and devices to convert organic matter to methane. All components are exempt, including land on which the facility is situated, up to one-half acre.	Local Assessor or Board of Assessors
	Act 210 Laws of 1978	Wood-fired central heating and solar or wind systems for heating, cooling, or electrical power are eligible for income tax credit if they are installed in the taxpayer's dwelling before 7/1/83. The credit is equal to 25% of the cost of the system or $1,000, whichever is less. Businesses may deduct 25% of the cost of the system or $3,000, whichever is less.	State Tax Dept. Income Tax Division State Street Montpelier, VT 05602 (802) 828-2517
Virginia	CH. 561 Laws of 1977	Any county, city, or town may exempt solar energy equipment used for heating, cooling, or other applications from property tax. The State Board of Housing must certify the system. The exemption is good for not less than 5 years.	Local Tax Governing Body
Washington	CH. 364 Laws of 1977	Solar water and space heating or solar power systems are exempt from property taxation. Claims must be filed with the county assessor. The exemption is valid for 7 years. Claims must be filed by 12/31/81.	Local Assessor or Board of Assessors
Wisconsin	CH. 313 Laws of 1977	Alternative energy systems installed by businesses may be used as a tax deduction in the year paid for, may be depreciated, or may be amortized over 5 years. This law covers solar, waste conversion, and wind systems but does not include solid fuel consuming devices used for residential purposes. The system must be certified by the Department of Industry, Labor and Human Relations. Individuals may claim income tax credit on systems costing at least $500. The credit may not exceed $10,000. The following tax credit applies to systems installed on buildings that existed before 4/20/77.	State Dept. of Revenue Income Tax Division 3648 University Avenue Madison, WI 53705 (608) 266-2772

If the system were installed in:	The tax credit equals this % of the system cost:
1979, 1980	24%
1981, 1982	18%
1983, 1984	12%

State	Law	Description	Contact

I. TAX INCENTIVES

For building constructed after 4/20/77 the following credit rate applies:

If the system were installed in: The tax credit equals this % of the system cost:

1979, 1980	16%
1981, 1982	12%
1983, 1984	8%

Any tax credit in excess of taxes due may be refunded. Expenses must be incurred before 12/31/84.

II. GRANTS AND LOANS

State	Law	Description	Contact
California	CH. 1 and CH. 7 Laws of 1978	This law creates the Solar Energy Demonstration Loan Program that provides $2,000 interest-free loans for solar space heating and domestic hot water systems in areas where a state of emergency has been declared.	Dept. of Housing and Community Development Division of Research and Policy Development 921 10th Street Fifth Floor Sacramento, CA 95814 (916) 445-4728
	CH. 1243 Laws of 1978	The maximum loan available to veterans for home mortgages is increased if the home is equipped with a solar energy system.	Dept. of Veterans' Affairs 1227 "O" Street P. O. Box 1559 Sacramento, CA 95807 (916) 445-2347 or any local district office of the Dept. of Veterans' Affairs
Illinois	PA 80-430, 1977	This law establishes a $5 million solar energy demonstration program.	Divison of Energy Ill. Dept. of Business & Economic Development 222 South College Springfiled, IL 62706 (217) 782-7500
Iowa	CH. 1086 Laws of 1978	This law establishes a loan and grant fund for property improvements and mortgages for low-income families. Solar energy systems qualify as improvements.	William H. McNarey, Executive Director Iowa Housing Finance Authority 218 Liberty Bldg. Des Moines, IA 50319 (515) 281-4058
Maine	CH. 685 Laws of 1978	Maine provides $16,000 for a solar hot water demonstration program; grants are $400 each.	Office of Energy Resources 55 Capitol Street Augusta, ME 04330 (207) 289-2196
Massachusetts	CH. 28 Laws of 1977 CH. 260 Laws of 1977 CH. 73 Laws of 1978	Banks and credit unions are authorized to make loans with extended maturation periods and increased maximum amounts for home improvements, including solar energy systems. Banks may lend up to $15,000, and credit unions may lend up to $12,000.	Local Bank or Credit Union
Montana	CH. 548 Laws of 1975	This law allows a public utility to lend money for the installation of solar energy systems in dwellings at an interest rate not to exceed 7% per year. The utility will then receive a license tax credit for the difference between 7% and the prevailing rate of interest.	Public Service Commission 1227 11th Avenue Helena, MT 59601 (406) 449-3008
Oregon	CH. 732 Laws of 1977	This law creates a loan fund for alternate energy projects and authorizes the Director of the Department of Energy to sell bonds to finance the loan fund.	Oregon Dept. of Energy Room 111 Labor & Industries Bldg. Salem, OR 97310 (503) 378-4128

State	Law	Description	Contact

II. GRANTS AND LOANS

State	Law	Description	Contact
	CH. 315 Laws of 1977	To finance a domestic solar energy system, veterans can obtain a loan in excess of the maximum allowed under the War Veterans Fund. The system must provide at least 10% of the home's energy requirements and must meet performance criteria established by the State Department of Energy.	State Dept. of Veterans' Affairs 3000 Market Street Plaza Suite 522 Salem, OR 97310 (503) 378-6438
Tennessee	CH. 884 Laws of 1978	Provides loans to low- and moderate-income persons to make energy conserving improvements, including the installation of solar hot water systems.	Tennessee Housing Development Authority Hamilton Bank Building Nashville, TN 32219 (615) 741-3023

III. LAND USE PROVISIONS

State	Law	Description	Contact
California	CH. 1366 Laws of 1978	Anyone who owns, occupies, or controls real estate is prohibited from allowing a tree or shrub to cast a shadow on a solar collector between 9:30 a.m. and 2:30 p.m. Trees casting a shadow before the installation of a collector are excluded.	
	CH. 1154 Laws of 1978	This law declares that any restriction on real property purporting to prohibit the installation and use of a solar energy system is void and unenforceable. It recognizes solar easements and prescribes their contents. City and county governments may not prohibit or restrict solar energy systems except to ensure the public health. The law requires that the subdivision maps be designed to accomodate passive solar energy systems to the maximum extent possible. It permits city and county governments to require the dedication of solar easements before approving the map.	Local Planning/ Zoning Body
Colorado	CH. 326 Laws of 1975	Solar easements are recognized and their contents are prescribed.	Local Clerk or Recorder
Connecticut	PA 314 Laws of 1978	The zoning commission of each city, town, or borough is authorized to regulate development to encourage energy efficiency and the use of renewable forms of energy, including solar.	Local Planning/ Zoning Body
Florida	CH. 309 Laws of 1978	Solar easements are regognized and subject to the same requirements as other easements; the contents are prescribed.	Local Clerk or Recorder
Georgia	Act 1446 Laws of 1978	Solar easements are recognized and subject to the same requirements as other easements; the contents are prescribed.	Local Clerk or Recorder
Idaho	CH. 294 Laws of 1978	Solar easements are recognized and are made subject to the same requirements as other easements; the contents are prescribed.	Local Clerk or Recorder
Kansas	CH. 227 Laws of 1977	Solar easements are recognized and are subject to the same requirements as other easements; the contents are prescribed.	Local Clerk or Recorder
Maryland	CH. 934 Laws of 1977	Solar easements are recognized as a lawful restriction on land.	Local Clerk or Recorder

State	Law	Description	Contact
III. LAND USE PROVISIONS			
Minnesota	CH. 786 Laws of 1978	Zoning ordinances may provide for the protection of solar access for solar energy systems. Solar easements are recognized and the contents are prescribed; they are enforceable in civil actions. Depreciation resulting from easements (but not any appreciation) shall be included in revaluation for property tax.	Local Planning/ Zoning Body (For Zoning) Local Clerk or Recorder (For Easements)
New Jersey	CH. 152 Laws of 1978	Solar easements are recognized and subject to the same requirements as other easements; the contents are prescribed.	Local Clerk or Recorder
New Mexico	CH. 169 Laws of 1977	The right to use solar energy is a property right of landowners; disputes regarding access will be settled by rule of prior appropriation.	
North Dakota	CH. 425 Laws of 1977	Solar easements are recognized and subject to the same requirements as other easements; the contents are prescribed.	Local Clerk or Recorder
Virginia	CH. 323 Laws of 1978	This law subjects solar easements to the same legal requirements as other easements and mandates contents of the agreement.	Local Clerk or Recorder
IV. STANDARDS AND REGULATION OF CONSTRUCTION			
California	CH. 670 Laws of 1976	Any city or county may require that new buildings subject to the State Housing Law be constructed in a manner that permits the installation of solar heating.	Local Building Inspector
	CH. 1081 Laws of 1977	The State Energy Resources, Conservation and Development Commission is required to adopt regulations and standards governing solar energy equipment.	Energy Resources Conservation and Development Commission 111 Howe Avenue Sacramento, CA 95825 (916) 322-3690
Connecticut	PA 76-409, 1976	The Commissioner of Planning and Energy Policy is required to establish standards for solar energy systems.	Connecticut Energy Division - OPM 20 Grand Street Hartford, CT 06115 (203) 566-5765
Florida	CH. 76-246 Laws of 1976	The Florida Solar Energy Center is required to establish standards for solar energy systems.	Florida Solar Energy Center 300 State Road 401 Cape Canaveral, FL 32920 (305) 783-0300
	CH. 74-361 Laws of 1974	The law stipulates that no single-family dwelling shall be constructed unless it is designed to facilitate future installation of a solar hot water system.	Local Building Inspector
	CH. 78-309 Laws of 1978	All solar energy systems manufactured or sold in Florida must meet the standards established by the Florida Solar Energy Center.	Bureau of Codes and Standards 2571 Executive Center Circle Tallahassee, FL 32301 (904) 488-3581
Illinois	PA 80-430, 1977	The Division of the Department of Business and Economic Development is required to establish guidelines and regulations for solar energy systems.	Illinois Institute of Natural Resources 325 West Adams Room 300 Springfield, IL 62706 (217) 785-2800

State	Law	Description	Contact
IV. STANDARDS AND REGULATION OF CONSTRUCTION			
Louisiana	Act 542, 1978	The law requires the Department of Natural Resources and Development to adopt regulations and standards governing solar energy systems.	Keith Overdyke Dept. of Natural Resources and Development P. O. Box 44396 Baton Rouge, LA 70804 (504) 342-4500
Michigan	PA 605, 1979	The Department of Commerce is required to formulate standards for solar energy systems.	Enerty Extension Service Michigan Energy Admin. Dept. of Commerce P. O. Box 30228 Lansing, MI 48909 (517) 373-6430
Minnesota	CH. 333 Laws of 1976	The Building Code Division of the Department of Administration is required to promulgate performance standards for solar energy systems (SBC 6101-6108).	Minnesota Dept. of Administration Building Code Division 408 Metro Square Bldg. St. Paul, MN 55101 (612) 296-4639
Nevada	Ch. 513 Laws of 1977	The law stipulates that the Public Works Board shall establish standards for new buildings. The standards shall include provisions for the design and construction of solar energy systems.	State Public Works Bd. 505 East King St. Room 400 Carson City, NV 89710 (702) 885-4870
New Jersey	CH. 256 Laws of 1977	The Division of Energy Planning and Conservation of the State Energy Office is required to adopt standards for solar energy systems.	Dept. of Energy 101 Commerce Street Newark, NJ 07102 (201)648-3290
New Mexico	CH. 347 Laws of 1977	This law directs the New Mexico Solar Energy Research and Development Institute to develop performance standards for solar energy equipment.	New Mexico Solar Energy Research & Development Institute Box 3 SOL New Mexico State Univ. Las Cruces, NM 88003 (505) 646-1846
New York	CH. 322 Laws of 1977	This law requires that the Commissioner of the State Energy Office promulgate guidelines and definitions for solar energy systems.	New York State Energy Office Agency Bldg. 2 Empire State Plaza Albany, NY 12223 (518) 474-8181
Oregon	CH. 196 Laws of 1977	The Department of Energy is required to adopt rules prescribing performance criteria for solar energy systems.	Oregon Dept. of Energy Room 111 Labor & Industries Bldg. Salem, OR 97310 (503) 378-4128

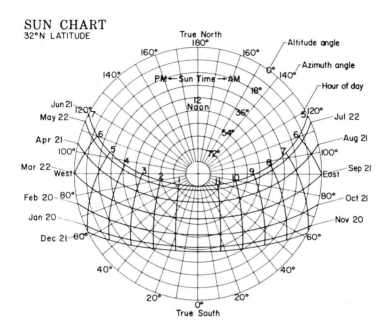

SUN CHART
32°N LATITUDE

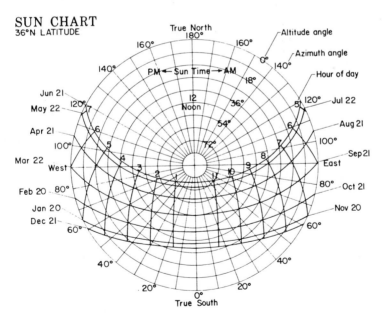

SUN CHART
36°N LATITUDE

*Los Alamos Scientific Laboratory, *Pacific Regional Solar Heating Handbook,* Washington, D.C.: U. S. Government Printing Office, November, 1976, pp. 100–102.

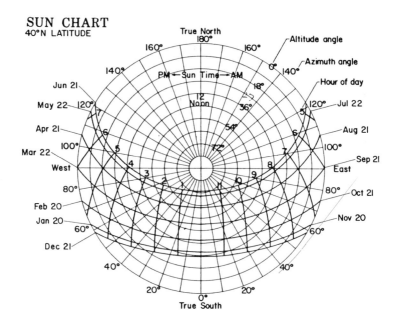

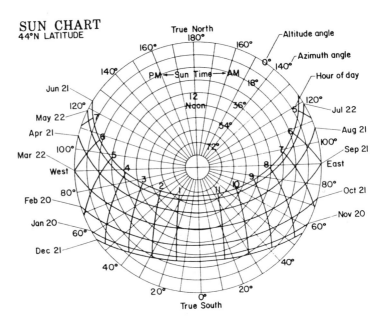

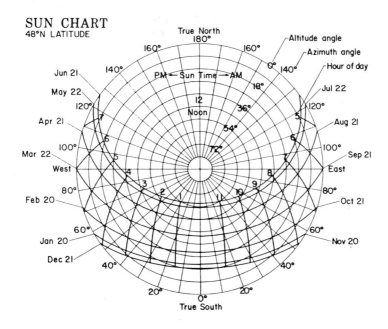

SUN CHART
48°N LATITUDE

True North
180°

160°   160°

Altitude angle
Azimuth angle
Hour of day

Jun 21   140°

May 22   120°

PM ← Sun Time → AM

0° 140°

18°

Jul 22

12
Noon

36°

120°

Apr 21

7

6

54°

5

Aug 21

Mar 22   100°

72°

100°

Sep 21

West

5

4

8

East

3

80°

Feb 20

2

9

80°

Oct 21

1

10

60°

Jan 20

60°

Nov 20

Dec 21   40°

40°

20°   20°

0°
True South

APPENDIX XIV
WORKSHEETS

WORKSHEET 1.

A1. Location = _____   D1. Ground Reflectance = (p) = _____
B1. Latitude = _____   E1. Solar Reflectance Component = $p(1 - \cos s)/2$ = _____
C1. Collector Tilt = (s) = _____   F1. $(1 + \cos s)/2$ = _____

G1. Month	H1. $\bar{K}_T$ From Appendix II	I1. $\bar{H}_d/\bar{H}$ From Eq. 6.24	J1. $\dfrac{1-\bar{H}_d/\bar{H}}{1-I1}$	K1. $\bar{R}_b$ From Tables 6.5 to 6.7 or Fig. 6.7	L1. Beam Component J1*K1	M1. Diffuse Component F1*I1	N1. Reflective Component E1.	O1. $\bar{R}$ L1+M1+N1	P1. $\bar{H}$ MJ/day-m^2 From Appendix II	Q1. $\bar{H}_T$ O1*P1*10^6 J/day-m^2
JAN										
FEB										
MAR										
APR										
MAY										
JUN										
JUL										
AUG										
SEP										
OCT										
NOV										
DEC										

WORKSHEET 2.

A2. Location = _____   D2. Azimuth (γ) = _____

B2. Latitude = _____   E2. bo = _____

C2. Tilt = _____

F2. Month	G2. δ	H2. θ	I2. $K\tau\alpha$ or $(\overline{\tau\alpha})/(\tau\alpha)_n$
Jan			
Feb			
Mar			
Apr			
May			
Jun			
Jul			
Aug			
Sep			
Oct			
Nov			
Dec			

WORKSHEET 3. Summary Building Load.

A3. Month	B3. Heating Degree Days (DD)	C3. Monthly Space Heating Loads (Q_s) (J/Mo.) (from Eq. 6.34)	D3. Domestic Water Heating Load (Q_w) (J/mo.) (from Eq. 6.30 and Table 6.17)	E3. Total Building Load (Q_t) (J/mo.) (C3.+D3.)
JAN				
FEB				
MAR				
APR				
MAY				
JUN				
JUL				
AUG				
SEP				
OCT				
NOV				
DEC				
TOTALS				

WORKSHEET 4. x/A and y/A Calculation.

A4. $F_R(\tau\alpha)_n (F_R'/F_R)$ = _____   B4. $F_R U_L (F_R'/F_R)$ = _____

C4. Month	D4. N	E4. ΔT	F4. $\tau\alpha/(\tau\alpha)_n$	G4. (H_T) (J/day-m^2)	H4. Q_T (J/mo)	I4. (100-$\bar{T}a$)	J4. (x/A) B4*I4*E4/H4	K4. (y/A) A4*F4*G4*D4/H4
JAN								
FEB								
MAR								
APR								
MAY								
JUN								
JUL								
AUG								
SEP								
OCT								
NOV								
DEC								

WORKSHEET 5. Determination of Annual Solar Load Fraction (F).

A5. Collector Area (a) = _____
B5. Storage Size Correction Factor (x/xo) = _____
C5. Collector Air Flow Rate Correction Factor (x/xo) = ____ (if liquid system, x/xo = 1.0)
D5. Load Heat Exchanger Correction Factor (y/yo) = ____ (if air system, y/yo = 1.0)

E5. Month	F5. Corrected x/A (J4)(B5)(C5)	G5. Corrected y/A (K4)(D5)	H5. x	I5. y	J5. F	K5. Monthly Solar Load (J5)(E3)
JAN						
FEB						
MAR						
APR						
MAY						
JUN						
JUL						
AUG						
SEP						
OCT						
NOV						
DEC						

Annual Fraction of Total Building Thermal   L5 TOTAL = _____
Load Supplied by Solar Energy (L5 Total/E3 Total) = _____

WORKSHEET 6.*

H.	Annual mortgage interest rate	____ %/100
I.	Term of mortgage	____ Yrs.
J.	Down payment (as fraction of investment)	____ %/100
K.	Collector area dependent costs	____ $/m²
L.	Area independent costs	____ $
M.	Present cost of solar backup system fuel	____ $/GJ
N.	Present cost of conventional system fuel	____ $/GJ
O.	Efficiency of solar backup furnace	____ %/100
P.	Efficiency of conventional system furnace	____ %/100
Q.	Property tax rate (as fraction of investment)	____ %/100
R.	Effective income tax bracket (state+federal- state x federal)	____ %/100
S.	Estra ins. & maint. costs (as fraction of investment)	____ %/100
T.	General inflation rate per year	____ %/100
U.	Fuel inflation rate per year	____ %/100
V.	Discount rate (after tax return on best alternative investment)	____ %/100
W.	Term of economic analysis	____ Yrs.
X.	First year non-solar fuel expense (total, E3.)(N.)/(P.)÷10⁹	____ $
Y.ᵃ	Depreciation lifetime	____ Yrs.
Z.	Salvage value (as fraction of investment)	____ %/100
AA.	Table 7.8 with Yr=(W.), Column=(U.) and Row=(V.)	____
BB.	" (W.) " (T.) " (V.)	____
CC.ᵃ,ᵇ	" MIN(I.,W.) " (H.) " (V.)	____
DD.	" MIN(W.,Y.) " (Zero) " (V.)	____
EE.	" (I.) " (Zero) " (H.)	____
FF.	" MIN(I.,W.) " (Zero) " (V.)	____
GG.	(FF.)/(EE.), Loan Payment	____
HH.	(GG.)+(CC.)[(H.)-1/(EE.)], Loan Interest	____
II.	(J.)+(1-J)[(GG.)-(HH.)(R.)], Capital cost	____
JJ.	(S.)(BB.), I&M cost	____
KK.	(Q.)(BB)(1-R.), Property tax	____
LL.	(Z.)/(1+V.)(W.), Salvage value	____
MM.	(R.)(DD.)(1-Z)/(Y.), Depreciation	____
NN.	Other costs	____
OO.	(II.)+(JJ.)+(KK.)-(LL.)+(NN.), Residential costs	____
PP.	(II.)+(JJ.)(1-F)+(KK.)-(LL.)-(MM.)+(NN.)(1-R.), Commercial Costs	____

    a. Commercial only.

    b. Straight line only. Use Tables 7.12 and 7.13 for other depreciation methods.

*Adapted from Beckman, William A., Klein, Sanford A., and Duffie, John A., *Solar Heating Design by the F-Chart Method,* New York: John Wiley & Sons, 1977. Used with permission.

WORKSHEET 7.*

R1. Collector Area (Worksheet 5)				
R2. Fraction By Solar (Worksheet 5)				
R3. Investment in Solar (K.)(R1.)+(L.)				
R4. 1st Year Fuel Expense (Total,E3)(1-R2.)(M.)/(O.)÷10^9				
R5. Fuel Savings (X.-R4.)(AA.)				
R6. Expenses (Residential) (OO.)(R3.)				
R7. Expenses (Commercial) (PP.)(R3.)				
R8. Savings (Residential) (R5.)-(R6.)				
R9. Savings (Commercial) (R5.)(1-R.)-(R7.)				

*Adapted from Beckman, William A., Klein, Sanford A., and Duffie, John A., *Solar Heating Design by the F-Chart Method,* New York: John Wiley & Sons, 1977. Used with permission.

WORKSHEET 8.*

	1	2	3	4
R10. Year (n) (first year=1)				
R11.[b] Current Mortgage [(R11.)-(R14.)+(R18.)]				
R12. Fuel Savings (X.-R4.)$(1+U.)^{n-1}$				
R13. Down Payment (1st year only) (R3.)(J.)				
R14. Mortgage Payment (1-J.)(R3.)/(EE.)				
R15. Extra Insur. & Maint. (S.)(R3.)$(1+T.)^{n-1}$				
R16. Extra Property Tax (R3.)(Q.)$(1+T.)^{n-1}$				
R17. Sum (R13.+R14.+R15.+R16.)				
R18. Interest on Mortgage (R11.)(H.)				
R19. Tax Savings (R.)(R16.+R18.)				
R20.[c] Depreciation (st.line) (R3.)(1-Z.)/(Y.)				
R21.[c] Business Tax Savings (R.)(R20.+R15.-R12.)				
R22. Salvage Value (R2.)(Z.)(Last year only)				
R23. Solar Savings (R12.-R17.+R19.+R21.+R22.)				
R24.[d] Discounted Savings (R23.)/$(1+V.)^n$				

b. For the first year use [(R3.)(1-J.)]; for subsequent years use equation with previous years values.

c. Income producing property only.

d. The down payment should not be discounted.

*Adapted from Beckman, William A., Klein Sanford A., and Duffie, John A., *Solar Heating Design by the F-Chart Method,* New York: John Wiley & Sons, 1977. Used with permission.

INDEX